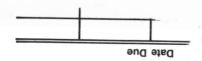

Optical Properties of Ions in Solids

NATO ADVANCED STUDY INSTITUTES SERIES

A series of edited volumes comprising multifaceted studies of contemporary scientific issues by some of the best scientific minds in the world, assembled in cooperation with NATO Scientific Affairs Division.

Series B: Physics

Volume 1 – Superconducting Machines and Devices
edited by S. Foner and B. B. Schwartz

Volume 2 – Elementary Excitations in Solids, Molecules, and Atoms (Parts A and B)
edited by J Devreese, A. B. Kunz, and T. C. Collins

Volume 3 – Photon Correlation and Light Beating Spectroscopy
edited by H. Z. Cummins and E. R. Pike

Volume 4 – Particle Interactions at Very High Energies (Parts A and B)
edited by David Speiser, Francis Halzen, and Jacques Weyers

Volume 5 – Renormalization and Invariance in Quantum Field Theory
edited by Eduardo R. Caianiello

Volume 6 – Interaction between Ions and Molecules
edited by Pierre Ausloos

Volume 7 – Low-Dimensional Cooperative Phenomena: *The Possibility of High-Temperature Superconductivity*
edited by H. J. Keller

Volume 8 – Optical Properties of Ions in Solids
edited by Baldassare Di Bartolo

Volume 9 – Electronic Structure of Polymers and Molecular Crystals
edited by Jean-Marie André and Janos Ladik

Volume 10 – Progress in Electro-Optics: *Reviews of Recent Developments*
edited by Ezio Camatini

The series is published by an international board of publishers in conjunction with NATO Scientific Affairs Division

A	**Life Sciences**	Plenum Publishing Corporation
B	**Physics**	New York and London
C	**Mathematical and Physical Sciences**	D. Reidel Publishing Company Dordrecht and Boston
D	**Behavioral and Social Sciences**	Sijthoff International Publishing Company Leiden
E	**Applied Sciences**	Noordhoff International Publishing Leiden

Optical Properties of Ions in Solids

Edited by

Baldassare Di Bartolo

Department of Physics
Boston College
Chestnut Hill, Massachusetts

Assistant Editor

Dennis Pacheco

Department of Physics
Boston College
Chestnut Hill, Massachusetts

PLENUM PRESS • NEW YORK AND LONDON

Published in cooperation with NATO Scientific Affairs Division

Library of Congress Cataloging in Publication Data

NATO Advanced Study Institute on Optical Properties of Ions in Solids, Erice, Italy, 1974.
Optical properties of ions in solids.

(NATO advanced study institutes series: Series B, physics; v. 8)
Includes bibliographical references and index.
1. Solids–Optical properties–Congresses. 2. Ions–Congresses. 3. Luminescence–Congresses. I. Di Bartolo, Baldassare. II. Title. III. Series.
QC176.8.06N23 1974 541'.372 75-11910
ISBN 0-306-35708-9

Lectures presented at the 1974 NATO Advanced Study Institute on Optical Properties of Ions in Solids, held at Erice, Italy, June 6-21, 1974

A Division of Plenum Publishing Corporation
227 West 17th Street, New York, N.Y. 10011

United Kingdom edition published by Plenum Press, London
A Division of Plenum Publishing Company, Ltd.
Davis House (4th Floor), 8 Scrubs Lane, Harlesden, London, NW10 6SE, England

Printed in the United States of America

Preface

These proceedings report the lectures and seminars presented at the NATO Advanced Study Institute on "Optical Properties of Ions in Solids," held at Erice, Italy, June 6-21, 1974. The Institute was the first activity of the International School of Atomic and Molecular Spectroscopy of the "Ettore Majorana" Centre for Scientific Culture.

The Institute consisted of a series of lectures on optical properties of ions in solids that, starting at a fundamental level, finally reached the current level of research. The sequence of lectures and the organization of the material taught were in keeping with a didactical presentation. In essence the Institute had the two-fold purpose of organizing what was known on the subject, and updating the knowledge in the field. Fifteen series of lectures for a total of 44 hours were given. Five one-hour seminars and five twenty-minute seminars were presented.

A total of 57 participants came from 40 laboratories in the following countries: Belgium, Canada, France, Germany, Ireland, Israel, Italy, Netherlands, Poland, Romania, Switzerland, the United Kingdom, and the United States. The secretaries of the Institute were: D. Pacheco for the scientific aspects and A. La Francesca for the administrative aspects of the meeting.

These proceedings report the lectures, the one-hour seminars (abstracts only) and the twenty-minute seminars (titles only). The proceedings report also the contributions sent by Prof. K. Rebane and Dr. L. A. Rebane who, unfortunately, were not able to come.

I would like to acknowledge the sponsorship of the Institute by the North Atlantic Treaty Organization, the National Science Foundation, the Italian Ministry of Public Education, the Italian Ministry of Scientific and Technological Research, the Regional Sicilian Government and the Department of Physics of Boston College.

I would like to thank for their help Prof. A. Zichichi, Director of the "Ettore Majorana" Centre for Scientific Culture, the members of the organizing committee (Professors D. Curie, R. Orbach and F. Williams), Professor R. L. Carovillano, Dr. A. Gabriele, Ms. M. Zaini, Ms. P. Savalli, Prof. V. Adragna, Dr. G. Denaro, Dr. C. La Rosa, and Dr. T. L. Porter of the National Science Foundation.

It was a pleasure to direct this Institute and to edit these lectures.

December 1974,
Chestnut Hill, Massachusetts

B. Di Bartolo
Editor, and Director
of the Institute

Contents

HISTORICAL SURVEY OF STUDIES OF THE OPTICAL PROPERTIES OF IONS IN SOLIDS

Ferd Williams

Department of Physics, University of Delaware

Newark, Delware 19711

ABSTRACT

The history of scholarly work on ions in solids, with particular emphasis on optical properties, is reviewed from the very earliest work until the present. Early empirical studies are first discussed; then the basic quantum mechanical interpretations, including the adiabatic and Hartree-Fock approximations, are presented; crystal field theory and ligand field theory are discussed and applied to transition metal and rare earth ions; resonance energy transfer between ions is noted; and finally ion pairs and multiphoton transitions are reviewed. The principal emphasis is on radiative transitions between electronic states, both optical absorption and luminescent emission, with attention both to the evolution of concepts and to recent experimental methods.

I. INTRODUCTION

In the introductory chapter, we shall be primarily concerned with the historical development of concepts which are important in the interpretation of the optical properties of ions in solids. Metals will not be included, nor will the infrared absorption of non-metals involving transitions between vibrational levels within a given electronic state. We are primarily concerned with radiative transitions between electronic states of ions as perturbed by their being in the condensed phase. This is essentially the spectroscopy of ions in solids. The effects of phonons on the electronic transitions are included.

We begin by describing some of the early observations and interpretations of optical absorption and of luminescence. We consider the problem of separating the many-body system of electrons and nuclei by means of the adiabatic approximation. This provides the basis for the configuration coordinate model. The many-electron problem is then reduced to the one-electron problem by the Hartree-Fock approximation. The electronic states of ions in crystals are then described in terms of crystal field theory and also are considered more generally by ligand field theory. Examples are cited for very weak, weak and strong interactions with the crystal or the ligand field. We shall also briefly discuss the Jahn-Teller effect.

In addition to the tight-binding electronic states of ions in ionic crystals, we shall also consider effective mass electronic states of quasi-ions of charged dopants in semiconductors. Associated defects or impurities in both ionic crystals and in semiconductors will be discussed in addition to the analyses on point defects or impurities. Charge transfer transitions, as well as intra-ion transitions, are included. Finally, we shall review research on multiphoton transitions.

II. EARLY INVESTIGATIONS OF OPTICAL TRANSITIONS

The early efforts to understand optical absorption and luminescence developed quite separately. These were, of course, without the benefit of quantum theory and, therefore, were entirely empirical. The simplest characteristics were not understandable during the earliest work.

In 1652 Zecchi (1) advanced the understanding of phosphorescence by noting that the color of the phosphorescent light was the same after illumination by light of different colors. Exactly 200 years later, Stokes (2) clarified the nature of fluorescence by showing that the scattered and incident light differed in refrangibility (color). He concluded that the fluorescent light was usually less refrangible (longer wavelength) than was the exciting light. This is Stokes' law of luminescence. In 1867, E. Becquerel (3) investigated uranyl salts and distinguished two types of afterglow; he attributed these to monomolecular and to bimolecular decay mechanisms.

As regards optical absorption and dispersion, Lorentz (4) showed in 1880 that the dispersion of light in insulators near an absorption line could be explained by a simple electronic model. He assumed that non-metallic solid bodies contain electrons which are bound to equilibrium positions in accordance with Hooke's law. The optical absorption, therefore, is describable in terms of a system of oscillators and the interaction of radiation with matter is expressable in terms of a frequency dependent complex dielectric constant. In the simplest analysis the effects of the local field are neglected in deriving the optical properties of the system of oscillators. Smakula (5) in 1930 applied the theory to defects in alkali halides and obtained his now well-known equation relating the concentration of defects or impurity ions to their oscillator strength and their experimental absorption spectrum.

The earliest systematic study of the effects of ligands on the optical absorption arising from transition metal ions was reported by Werner (6) in 1892. He was investigating the properties of metal-ammonia complexes in aqueous solutions, and related their colors to the coordination numbers of the metal ions and structures of the complexes. This provided the basis for the stereochemistry of inorganic complexes and was also applied to crystalline complex salts and their hydrates.

III. QUANTUM MECHANICAL CONSIDERATIONS

Matter in the solid state is a many-body system consisting of interacting electrons and nuclei. Optical absorption and luminescence involve transitions between electronic states. Therefore, we are primarily interested in the electronic part, ψ^{el}, of the complete wave function ψ. Born and Oppenheimer (7) in 1927 provided the basis for separating ψ^{el} from the vibrational part, ψ^{vib}. The complete Hamiltonian for a system of k nuclei of mass M_k and of j electrons of mass m_j is:

$$H = \sum_k [-\frac{\hbar^2}{2M_k}\Delta_R + V(R_k)] - \sum_j \frac{\hbar^2}{2m_j}\Delta_r + \sum_j\sum_k V(r_j,R_k) + \sum_{i<j}\frac{e^2}{|r_i-r_j|}, \quad (1)$$

where $V(R_k)$ is the potential operator for the k nucleus. $V(r_j,R_k)$ is the potential operator for the interaction of the j electron with the k nucleus and $e^2/|r_i-r_j|$ are the electron-electron interactions. The complete wave function ψ is assumed to be of a form:

$$\psi(r_j,R_k) = \psi_R^{el}(r_j)\psi^{vib}(R_k) \quad (2)$$

where the subscript R on ψ^{el} specifies a parametric dependence of the electronic wave function on the nuclear coordinates. The many-electron Schrödinger equation is, therefore,

$$\left[-\sum_j \frac{\hbar^2}{2m_j}\Delta_r + \sum_{jk}\sum V_R(r_j) + \sum_{i<j}\frac{e^2}{|r_i - r_j|}\right]\psi_R^{el}(r_j) = E_R^{el}\psi_R^{el}(r_j) \tag{3}$$

where E_R^{el} is the eigenvalue of $\psi_R^{el}(r_j)$. The coupling term between electronic and vibrational parts of the wave function is neglected in obtaining eq. (3) and is of course the electron-phonon interaction.

The basic idea of the adiabatic approximation is that there exist stationary electronic states described by $\psi_R^{el}(r)$. This many-electron wave function is smoothly or adiabatically perturbed as a consequence of displacements of the nuclei. The validity of this approximation depends on the period for orbital electronic motion being short compared to the period of lattice vibrations. This is the case for tight-binding states of ions in solids. The adiabatic approximation provides the basis for approximating the electronic states of ions in solids as stationary states perturbed by the field from the other ions in the solid and also for the configuration coordinate model extensively used in interpreting optical absorption and luminescence. The first application of the configuration coordinate model to ionic crystals was by Von Hippel (8), who calculated in 1936 the transition energy for charge transfer between anion and cation for alkali halide crystals. Seitz (9) in 1938 first used the configuration coordinate model in interpreting optical properties of impurity ions in ionic crystals. Schön (10) in 1948 noted the contribution of the quantum mechanical zero point energy to the widths of low temperature absorption and emission bands of impurity ions. In 1951, Williams (11) approximated the configuration coordinate with a single real coordinate of the impurity ion and its nearest neighbors, and computed by semiempirical methods the interactions of the excited and, separately, the unexcited ion with the crystal as a function of the configuration coordinate, thus providing the basis for computation of approximate absorption and emission spectra. Shortly thereafter, Williams and Hebb (12) included the effects of the zero point energy in the theoretical calculation of spectra. The theoretical spectra are of the form:

$$P(h\nu) = \left[\int \psi_{R_o,f}^{el*}(e\underline{r})\psi_{R_o,i}^{el}d\underline{r}\right]^2 (K/2\pi kT)^{1/2} \times \{\exp[-K(\Delta R)^2/2k\theta\coth(\theta/T)]\}dR/d(h\nu) \tag{4}$$

where K is the force constant for the displacement ΔR of R from the equilibrium position R_o and $k\theta$ is the zero point energy of the initial state. In 1952 Lax (13) developed a general analysis which provided justification for a single configuration coordinate for interpreting the optical properties of impurity ions in crystals. Since then Kirstofel (14) and Potekhina (15) have reported calculations of spectra by more rigorous quantum mechanical methods. Keil (16) and Kelley (17) have reported theoretical spectra including structure from zero phonon and phonon-assisted transitions.

For both the calculations just described and the crystal and ligand field theory analyses to be described in the next section ψ^{el} must first be determined for the free ion. This many-electron function is separable into one-electron functions by means of the Hartree-Fock (18) approximation. In accordance with the Pauli principle and with the antisymmetry consistent with Fermi's statistics, ψ^{el} is the following for an ion with N electrons:

$$\psi^{el}(r) = (N!)^{-\frac{1}{2}} \begin{vmatrix} \phi_1(r_1) & \phi_1(r_2) & \cdots & \phi_1(r_N) \\ \phi_2(r_1) & & & \cdot \\ \cdot & & & \cdot \\ \cdot & & & \cdot \\ \cdot & & & \cdot \\ \phi_N(r_1) & & & \phi_N(r_N) \end{vmatrix} \qquad (5)$$

The $\phi_j(r_i)$ are one-electron wave function in which the coordinate r_i includes the spin as well as the space coordinates. By substitution of eq. (5) into eq. (4) and application of the variational principle the Hartree-Fock integral equations are obtained, which are then solved self-consistently, that is, the ϕ_j's used in the interaction terms of each integral equation must be solutions of the other members of the set of N equations. These are the best one-electron functions available in most cases for an ion with N electrons. For some magnetic ions unrestricted Hartree-Fock functions, which allow for different spatial dependencies of electrons differing only in spin, are available.

IV. CRYSTAL AND LIGAND FIELD THEORY

The basic concepts of crystal field theory were developed by J. Becquerel (19) and by Bethe (20) in 1929. The central ion was considered to be perturbed by the electrical field of the crystal, particularly by the nearest neighbors or ligands. The analysis is especially applicable to orbitally degenerate ions , that is, ions with incompletely filled inner shells. The crystal field may partly or completely remove the orbital degeneracy. In other words, the electronic states which in the free ion are invariant under infinitesimal rotations become in the crystal invariants only under finite rotations. The electronic part of the Hamiltonian for an ion in the crystal field, including spin-orbit interaction, is the following:

$$H_{el} = -\frac{\hbar^2}{2m}\sum_j \Delta_r - \sum_j \frac{Ze^2}{r_j} + \sum_{i<j} \frac{e^2}{|r_i - r_j|} + \sum_j \xi_j(r) l_j \cdot s_j + V_c(r) \quad (6)$$

The ions in crystals can be subdivided on the basis of the magnitude of $V_c(r)$ compared to the other terms in eq. (6). The crystal field interaction is considered very weak for $V_c(r)$ smaller than the spin-orbit term in eq. (6). Examples of very weak crystal field interaction are the trivalent rare earth ions for which the complete multiplet structure of the free ion is preserved, with some further splitting in the crystal field. For $V_c(r)$ greater than the spin-orbit interaction but smaller than the term in eq. (6) for electron-electron repulsion, the interaction is considered weak. Examples of weak interaction are the first transition metal sequence. For $V_c(r)$ greater than electron-electron repulsion, the crystal field interaction is strong. Examples of strong interaction are ions in the second and third transition metal sequence and also ions with incompletely filled outer shells, for example those with the $6s^2$ configuration.

Calculations based on crystal and ligand field theory have developed separately for the weak field transition metal ions on one hand and for the very weak field rare earth ions on the other. Van Vleck (21) in 1939, and Polder (22) in 1942, obtained surprisingly good results for the crystal field parameter using a point charge model for transition metal ions. In 1948 Hellwege (23) systematized the group theoretical aspects of crystal field theory. Kleiner (24) in 1952, took account of the overlap of the central ion and the ligands and obtained poorer agreement with the experimental values of the crystal field parameter; in fact, the calculated splitting was of the wrong sign. Tanabe and Sugano (25) in 1954, removed this discrepancy by making the orbitals of the transition metal and the ligands orthogonal and found a large resonance energy between these orbitals. This calculation is essentially a molecular orbital analysis and generalizes crystal field theory to ligand field theory. In 1955, Orgel (26) explained the absorption band widths for transition metal ions on the basis of the dependence of the crystal field splitting parameter on the nuclear coordinates. This effect is, of course, related to the dependence of transition energy on nuclear coordinate in the configurational coordinate model.

The very weak field case is evident in the extensive experimental studies of Dieke and associates (27) on rare earth ions in solids. Judd et al.(28) developed the formal methods for calculating not only the splitting of the free ion states by the crystal field, but also methods for calculating the probabilities of the transitions in the optical spectra.

In addition to the effects of the crystal field on the electronic states of ions, there are two other aspects to the degeneracies of these states. Kramers (29) showed in 1930 that the spin degeneracy is not removed by an electrostatic field. States for systems with odd numbers of electrons must, therefore, have even degeneracy in an electrostatic field. On the other hand, Jahn and Teller (30) demonstrated in 1937 that for non-linear molecules with orbital degeneracy, the molecule spontaneously distorts to a lower symmetry and a lower degeneracy. If the magnitude of the energy change accompanying the distortion is large compared to the zero point energy, then we have the static Jahn-Teller effect, whereas if the zero point energy is the larger, then we have the dynamic Jahn-Teller effect.

As noted by Jorgensen (31), the pioneer work on the electrostatic model occurred during the period 1929-51. By 1956 molecular orbital effects involving the ligands had been shown to be important. The concept that the effective field in an ion in a solid is different from the field in an isolated ion has survived. The group theoretical classifications remain valid. However, the quantitative calculations of the magnitude of crystal or ligand field splitting have become complex and somewhat controversial.

V. ENERGY TRANSFER

In the case of solids containing more than one type of impurity ion , the optical properties can be affected by energy transfer processes. One important effect involves optical absorption in an ion of one type, energy transfer to an ion of another type, which then decays by luminescent emission. The dominant energy transfer mechanism was explained quantum mechanically by Förster (32) in 1946 in the dipole approximation for energy transfer between organic molecules. In 1953, Dexter (33) generalized the theory to higher order interactions including exchange effects and applied the theory to energy transfer between ions in inorganic solids. In this resonance energy transfer mechanism the sensitizer ion in its excited electronic state, equilibrates with the lattice modes before transfer occurs to the fluorescer. In the dipole approximation the interaction

Hamiltonian for resonance energy transfer is the same as for Van der Waals' (34) interaction and is separable into factors depending on the sensitizer and fluorescer. The transfer probability can be put in the form:

$$P_{SF} = \frac{3}{4\pi} \frac{h^4 c^4}{\kappa^2 R^6} A_S \sigma_F \int_0^\infty \frac{P_S'(E) P_F(E)}{E^4} dE \qquad (7)$$

where κ is the optical dielectric constant, R is the sensitizer-fluorescer distance, A_S is the radiative transition probability of the sensitizer, σ_F is the absorption cross-section of the fluorescer, and P_S' and P_F are, respectively, the normalized emission spectrum of the sensitizer and absorption spectrum of the fluorescer. If selection rules eliminate the dipole-dipole term, then higher terms such as dipole-quadrupole, quadrupole-quadrupole or exchange may dominate. The lifetime of the excited sensitizer is reduced by resonance transfer. In the aforementioned cases the transfer probability depends on the overlap of the emission spectrum of the sensitizer and the absorption spectrum of the fluorescer, transfer occurring with energy conservation. Following a suggestion of Imbusch based on measurements on ruby, Orbach (35) showed in 1966 the effectiveness of phonon assistance in quadrupole-quadrupole energy transfer.

VI. ION PAIRS

Additional effects on the optical properties of ions in solids occur for systems in which ions of the same or of different type are sufficiently close to each other so that their respective electronic states are perturbed. The simplest case involves impurity ions of the same type and the same valence as the ions they replace in the crystal. These will be distributed to a good approximation at random. At higher concentrations, pairs will be present. For example, Cr^{+3}-Cr^{+3} pairs in ruby. Varsanyi and Dieke (36) in 1961 reported cooperative optical absorption by pairs in which two impurity ions are excited simultaneously by a single photon. Partlow and Moos (37) in 1967 reported two-photon decay from metastable interacting ions. Nakazawa and Shionoya (38) in 1970 reported two rare earth ions cooperating in producing a single photon with twice the excitation energy of each ion.

Another class of ion pairs involves impurity ions of a different valence state than that of the ions they replace in the solid and, consequently, there occurs charge compensation by ions or defects of opposite net charge. Because of the Coulomb interaction between these impurities, there may be a substantially greater concentration of nearest-neighbor pairs than corresponds to a random distribution. An interesting practical example of this type of ion pair exists in calcium halophosphate doped with antimony. The Sb^{3+} substitutes at type II Ca^{2+} sites, and is charge-compensated and paired with O^{2-} at a nearest neighbor X sites. The electronic states of this pair have been investigated by Soules et al.(39) using molecular orbital theory.

In addition to the tight-binding ion pairs, we also note the existence of effective mass quasi-ions and their pairs in semiconductors. The electronic states of these quasi-ions and of their pairs are describable in terms of the continuum states of the semiconductor perturbed by the Coulomb field of the charged dopants. In other words, the electronic particle is bound by the Coulomb field and moves with an effective mass characteristic of the continuum states. The importance of pairs of oppositely charged impurities, referred to as donor-acceptor pairs in semiconductors, to luminescent phenomena was originally reported by Prener and Williams (40) in 1956 relevant to zinc sulfide phosphors. A spectroscopy of donor-acceptor pairs has developed from the identification by Hopfield et al.(41) of line spectra from distant pairs in gallium phosphide.

VII. MULTIPHOTON TRANSITIONS

The availability of lasers has made possible spectroscopy of multiphoton transitions. This spectroscopy provides additional information on the optical properties of ions in solids over that available from single-photon spectroscopy. For example, two-photon absorption is an allowed dipole transition between states of the same parity, whereas single photon absorption is an allowed dipole transition between states of opposite parity. The theory of two-photon transition was discussed by Goeppert-Mayer (42) in 1931. However, experimental work on this type of spectroscopy was deferred until the invention of the laser in 1961. The two-photon absorption coefficient was estimated by Kleinman (43) in 1962 for impurity lines in solids, and Gold and Hernandez (44) calculated the two-photon absorption for ions strongly coupled to phonons. Besides providing additional transitions, multiphoton spectroscopy is valuable in investigating the bulk optical properties of ions in solids because of the low absorption coefficient for these transitions.

Sequential multiphoton transitions have also been investigated. Auzel (45) in 1966 reported sequential multiple excitation involving materials containing two different rare earth ions. He thus observed anti-Stokes luminescence. The incident radiation is absorbed in the sensitizer, and the fluorescer is multiply excited by several sequential energy transfers. Luminescent emission occurs in a single-photon decay.

A related effect involves single photon excitation and cascade luminescent decay via intermediate states. This is a quantum splitting phenomena. Sommerdijk et al. (46) and Piper et al. (47) recently reported cascade fluorescent decay in fluorides doped with trivalent praseodymium, the latter investigators reporting quantum efficiencies in excess of unity.

VIII. CONCLUDING REMARKS

In this chapter we have been primarily concerned with the evolution in concepts used in interpreting the optical properties of ions in solids. This review is in no sense complete, but is intended only to cover some of the key ideas. Many important contributions have not been mentioned, particularly in the area of crystal and ligand field theory. The literature is voluminous, and the many fine textbooks such as those by Orgel (48), Griffith (49), Ballhausen (50) and Jorgensen (31) are recommended for more complete presentations. For more complete reports on the luminescence of ions in solids the textbook by Curie (51) and proceedings of conferences edited by Crosswhite and Moos (52) and by Williams (53, 54) are suggested.

In addition, some recent experimental methods for studying the optical properties of ions in solids should be emphasized. Laser spectroscopy can be applied more generally than the multiphoton spectroscopy just described. Excited state spectroscopy can provide the kind of detailed information on excited electronic states of ions in solids which has previously been available only for the ground states. X-ray photoelectronic spectroscopy can be used to directly determine the binding energy of electrons of ions in solids.

REFERENCES

1. See Encyclopedia Britannica 9 (Encyl. Brit., Inc., Chicago 1951), p. 423-427.

2. G. Q. Stokes, Phil. Trans. Roy. Soc. London A142 II, 463 (1852).

3. E. Becquerel, "La Lumière, ses causes et ses effects"

(Gauthier-Villars, Paris 1867).

4. H. A. Lorentz, "Theory of Electrons" (Teubner, Leipzig, 1906).

5. A. Smakula, Zeits. f. Physik 59, 609 (1930).

6. See F. J. Moore, "A History of Chemistry" (McGraw-Hill, New York 1939), pp. 365-370.

7. M. Born and J. R. Oppenheimer, Ann. d. Phys. 84, 457 (1927).

8. A. von Hippel, Zeits. f. Physik 101, 680 (1936).

9. F. Seitz, J. Chem. Phys. 6, 150 (1938).

10. M. Schön, Zeits. f. Physik 119, 463 (1942).

11. F. Williams, J. Chem. Phys. 19, 457 (1951).

12. F. Williams and M. Hebb, Phys. Rev. 84, 1181 (1951).

13. M. Lax, J. Chem. Phys. 20, 1752 (1952).

14. N. N. Kristofel, Opt. Spectros. USSR (Engl. trans.) 7, 45 (1959).

15. N. D. Potekhina, Opt. Spectros. USSR (Engl. trans.) 8, 437 (1960).

16. T. H. Keil, Phys. Rev. 140, A601 (1965).

17. C. S. Kelley, Phys. Rev. B6, 4763 (1972).

18. D. R. Hartree, Proc. Cambridge Phil. Soc. 24, 89 (1928); V. Fock, Zeits. f. Physik 61, 126 (1930).

19. J. Becquerel, Zeits. f. Physik 58, 205 (1929).

20. H. Bethe, Ann. Physik 3 (5), 135 (1929).

21. J. H. Van Vleck, J. Chem. Phys. 7, 72 (1939).

22. D. Polder, Physica 9, 709 (1942).

23. K. H. Hellwege, Ann. Physik 4, 95, 127, 136, 143, 150, 357

(1948).

24. W. H. Kleiner, J. Chem. Phys. 20, 1784 (1952).

25. Y. Tanabe and S. Sugano, J. Phys. Soc. Japan 9, 753, 766 (1954); 11, 864 (1956).

26. L. Orgel, J. Chem. Phys. 23, 1824 (1955).

27. G. H. Dieke, Spectra and Energy Levels of Rare Earth Ions in Crystals (Wiley, New York 1968).

28. B. R. Judd, Proc. Roy. Soc. A227, 552 (1955); J. P. Elliot, B. R. Judd and W. A. Runciman, Proc. Roy. Soc. A240, 509 (1957); J. Chem. Phys. 44, 839 (1966).

29. H. A. Kramers, Proc. Acad. Sci. Amsterdam 33, 753 (1930).

30. H. A. Jahn and E. Teller, Proc. Roy. Sco. Lond. A161, 220 (1937).

31. C. K. Jorgensen, "Modern Aspects of Ligand Field Theory" (North Holland, Amsterdam 1971).

32. Th. Förster, Naturwiss.33, 166 (1946).

33. D. L. Dexter, J. Chem. Phys. 21, 836 (1953).

34. See H. Margenau, Rev. Mod. Phys. 11, 1 (1939).

35. R. Orbach "Optical Properties of Ions in Crystals" (Ed. by Crosswhite and Moos, Interscience, N. Y. 1967),p. 445-456.

36. F. Varsanyi and G. H. Dieke, Phys. Rev. Lett. 7, 442 (1961).

37. W. D. Partlow and H. W. Moos, Phys. Rev. 157, 252 (1967).

38. E. Nakazawa and S. Shionoya, Phys. Rev. Lett. 25, 1710 (1970).

39. T. F. Soules, T. S. Davis and E. R. Kreidler, J. Chem. Phys. 55, 1056 (1971).

40. J. Prener and F. Williams, J. Electrochem. Soc. 103, 342 (1956); J. Phys. et Rad. 17, 667 (1956).

41. J. J. Hopfield, D. G. Thomas and M. Gershenzon, Phys. Rev. Lett. 10, 162 (1963).

42. M. Goeppert-Mayer, Ann. Physik 9, 273 (1931).

43. D. A. Kleinman, Phys. Rev. 125, 87 (1962).

44. A. Gold and J. P. Hernandez, Phys. Rev. 139A, 2002 (1965).

45. F. Auzel, Compt. rend. Acad. Sci. Paris 262B, 1016 (1966); 263B, 819 (1966).

46. J. L. Sommerdijk, A. Bril and A. W. de Jager, J. Luminescence 8, 341 (1974).

47. W. W. Piper, J. A. DeLuca and F. S. Ham, J. Luminescence 8, 344 (1974).

48. L. E. Orgel, "Transition-Metal Chemistry" (Methuen, London, 1960).

49. J. S. Griffith, "Theory of Transition-Metal Ions" (University Press, Cambridge 1961).

50. C. J. Ballhausen, "Introduction to Ligand Field Theory" (McGraw-Hill, N. Y. 1962).

51. D. Curie, "Luminescence Crystalline" (Dunod, Paris, 1960); "Luminescence in Crystals" (Translated by Garlick) (Wiley, N. Y. 1963).

52. "Optical Properties of Ions in Crystals," Ed. by H. Crosswhite and H. Moos (Interscience, N. Y. 1967).

53. "Luminescence," Ed. by F. Williams (North Holland, Amsterdam,

1970).

54. "Luminescence of Crystals, Molecules, and Solutions," Ed. by F. Williams (Plenum Press, N. Y. 1973).

MAGNETIC IONS IN SOLIDS

B. Di Bartolo

Department of Physics, Boston College

Chestnut Hill, Massachusetts 02167, U.S.A.

ABSTRACT

This series of four lectures present in a comprehensive way the energy levels of ions in solids. The first lecture deals with basic concepts of group theory and with the interrelations between group theory and quantum mechanics. The second lecture starts with the consideration of the groups of interest for free atoms and atoms in a homogeneous magnetic field; the concept of a complete set of commuting operators is stressed. The energy levels of atoms are then considered with a perturbational approach making use of both symmetry and quantum mechanical concepts. The third lecture deals with magnetic ions in solids; the concept of crystalline field is introduced. By proper use of symmetry concepts and of the complete set of commuting operators, the various schemes (weak field, intermediate field, strong field) corresponding to the strength of the crystalline interaction are treated. The last lecture deals with the effect of covalent bonding on the energy levels of magnetic ions in crystalline solids; this more comprehensive treatment of ions in crystals illustrates the occurrence of states generally high in energy which produce very strong absorption transitions and that are associated with the phenomenon of charge transfer.

I. INTRODUCTION

The most relevant consideration is studying the energy levels of ions in solids is that these ions are in environments of definite symmetry. Quantitative evaluations are necessary in order to find the energies of the ions' quantum states. These calculations may be based on the use of one or more adjustable parameters whose values may be found by fitting the "theoretical" energy level scheme with the one observed experimentally. In a more sophisticated approach one may attempt a solution of the problem by first principle calculations. In any case, however, the symmetry aspect of the situation remains as the underlying reality. For this reason in the present article I have considered symmetry as the guiding principle and organized accordingly the subject matter. I have not discarded altogether quantitative information, but rather used it solely to produce a step by step perturbational approach. At each step I have relied on the connections between group theory and quantum mechanics.

For a more detailed knowledge the reader is referred to the books cited in the bibliography at the end of this article (1-8); no attempt was made to make this bibliography complete.

II. SYMMETRY CONCEPTS

II.A. Properties of a Group

An ensemble of elements forms a <u>group</u> if the elements have the following properties:

1) The product of two elements is an element of the ensemble;

2) the element E = identity is an element of the ensemble;

3) the product of elements is associative, i.e. (AB)C = A(BC) and

4) for each element A, there is an element B (reciprocal of A) such that AB = BA = E.

If an ensemble of elements forms a group, the number of elements in the ensemble is called <u>order of the group</u>.

In general $AB \neq BA$; if AB = BA, the group is called <u>Abelian</u>.

II.B. Example

Consider the equilateral triangle in Figure 1 and the following geometrical operations that can be performed on it.

E = identity
C_3 = 120° clockwise rotation about the axis Z
C_3^2 = 240° clockwise rotation about the axis Z
σ_1 = reflection through the plane containing the axes Z and 11
σ_2 = reflection through the plane containing the axes Z and 22
σ_3 = reflection through the plane containing the axes Z and 33

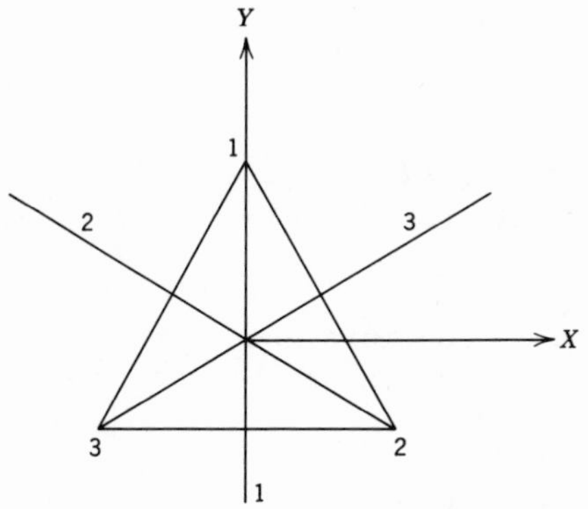

Figure 1. Equilateral Triangle: A Figure of Symmetry C_{3v}. (The axis Z is ⊥ to the X and Y axes.)

These operations leave the triangle invariant, i.e. if one of these operations is performed on the triangle, we would not be able to tell afterwards.

The operations above form a group that is called C_{3v}; they may be arranged in a multiplication table as follows.

	E	C_3	C_3^2	σ_1	σ_2	σ_3
E	E	C_3	C_3^2	σ_1	σ_2	σ_3
C_3	C_3	C_3^2	E	σ_3	σ_1	σ_2
C_3^2	C_3^2	E	C_3	σ_2	σ_3	σ_1
σ_1	σ_1	σ_2	σ_3	E	C_3	C_3^2
σ_2	σ_2	σ_3	σ_1	C_3^2	E	C_3
σ_3	σ_3	σ_1	σ_2	C_3	C_3^2	E

It is easy to verify that if the elements of an ensemble can be put in a multiplication table similar to the one above (in which no element repeats itself in a row or in a column), the ensemble forms a group.

II.C. Classes

Given an element A of a group, an element B of the same group such that

$$B = X^{-1}AX, \tag{1}$$

X being also an element of the group, is called <u>conjugate</u> of A with respect to X. A collection of elements of a group conjugate to each other is called a <u>class</u>.

A group can be decomposed into nonoverlapping classes. The group C_{3v} of the previous section can be decomposed into three classes:

$$C_{3v} = \mathcal{C}_1 + \mathcal{C}_2 + \mathcal{C}_3, \tag{2}$$

where

$$\mathcal{C}_1 = E$$

$$\mathcal{C}_2 = C_3, C_3^2$$

$$\mathcal{C}_3 = \sigma_1, \sigma_2, \sigma_3.$$

II.D. Representations

A <u>matrix representation</u> of a group is a set of matrices that multiply according to the multiplication table of the group. We may list the following three representations for the group C_{3v}.

C_{3v}	E	C_3	C_3^2	σ_1	σ_2	σ_3
Γ_1	1	1	1	1	1	1
Γ_2	1	1	1	-1	-1	-1
Γ_3	$\begin{pmatrix} 1 & 0 \\ 0 & 1 \end{pmatrix}$	$\begin{pmatrix} -\frac{1}{2} & \frac{\sqrt{3}}{2} \\ -\frac{\sqrt{3}}{2} & -\frac{1}{2} \end{pmatrix}$	$\begin{pmatrix} -\frac{1}{2} & -\frac{\sqrt{3}}{2} \\ \frac{\sqrt{3}}{2} & -\frac{1}{2} \end{pmatrix}$	$\begin{pmatrix} -1 & 0 \\ 0 & 1 \end{pmatrix}$	$\begin{pmatrix} \frac{1}{2} & -\frac{\sqrt{3}}{2} \\ -\frac{\sqrt{3}}{2} & -\frac{1}{2} \end{pmatrix}$	$\begin{pmatrix} \frac{1}{2} & \frac{\sqrt{3}}{2} \\ \frac{\sqrt{3}}{2} & -\frac{1}{2} \end{pmatrix}$

A representation Γ is reducible if a matrix β can be found such that, when used in a similarity transformation, produces the following effect on each matrix of Γ:

$$a' = \beta^{-1} a \beta = \begin{pmatrix} a_1' & & & 0 \\ & a_2' & & \\ & & a_3' & \\ 0 & & & a_4' \end{pmatrix}, \tag{3}$$

where a_1', a_2', etc. are square matrices. For a matrix b of Γ we would get a similar effect with the square matrices b_i' having the same dimension of a_i'.

A representation Γ is irreducible if no matrix β can be found that puts its matrices in the form (3) above. The three representations Γ_1, Γ_2 and Γ_3 of C_{3v} reported previously are irreducible.

Consider the coordinate transformation brought about by a clockwise rotation by an angle θ about the Z axis; the new (primed) coordinates and the old (unprimed) coordinates are related as follows:

$$\begin{aligned} x' &= x \cos\theta - y \sin\theta \\ y' &= x \sin\theta + y \cos\theta . \end{aligned} \tag{4}$$

For θ = 120° we obtain

$$\begin{aligned} x' &= -\frac{1}{2} x - \frac{\sqrt{3}}{2} y \\ y' &= \frac{\sqrt{3}}{2} x - \frac{1}{2} y . \end{aligned} \tag{5}$$

In matrix form

$$\begin{pmatrix} x' \\ y' \end{pmatrix} = (x \quad y) \begin{pmatrix} -\frac{1}{2} & \frac{\sqrt{3}}{2} \\ -\frac{\sqrt{3}}{2} & -\frac{1}{2} \end{pmatrix}. \tag{6}$$

We recognize here the square matrix in (6) as being the matrix of the representation Γ_3 of the group C_{3v} that corresponds to the operation C_3 (clockwise rotation by 120°).

We could repeat the above for each operation of C_{3v}; we would get back all the matrices of Γ_3. In a case like this it is said that (x, y) form a basis for the irreducible representation Γ_3 of C_{3v}; in particular it is said that x transforms according to the first column and that y transforms according to the second column of the matrices of Γ_3.

II.E. Characters

The characters of a representation are the traces of the matrices of the representation. In the case of the group C_{3v} the characters of the three representations previously reported are

C_{3v}	E	C_3	C_3^2	σ_1	σ_2	σ_3
Γ_1	1	1	1	1	1	1
Γ_2	1	1	1	-1	-1	-1
Γ_3	2	-1	-1	0	0	0

Let $\chi_\alpha(R)$ denote the character of the representation Γ_α in correspondence to the operator R. Irreducible representations have characters for which the following orthogonality relation is valid

$$\sum_R \chi^*_\alpha(R)\ \chi_\beta(R) = g\delta_{\alpha\beta}, \tag{7}$$

where g = order of the group. If $\chi(R)$ is the character of a reducible representation in correspondence to the operator R, we can write

$$\chi(R) = \sum_j c_j\ \chi_j(R) \tag{8}$$

where $\chi_j(R)$ = character of an irreducible representation. The orthogonality relation (7) allows us to find the coefficients in the expansion above:

$$c_j = \frac{1}{g} \sum_R \chi(R)\ \chi_j(R). \tag{9}$$

The following rules should also be kept in mind:

1) Two inequivalent irreducible representations have different character systems.

2) The number of irreducible representations is equal to the number of classes.

Finally

$$\int \psi_a \, f \, \psi_b \, Q\tau = 0$$

if the product representation $\Gamma_a \, \Gamma_f \, \Gamma_b$ does not contain the totally symmetrical representation Γ_1.

II.F. Group Theory and Quantum Mechanics

A quantum mechanical system with a Hamiltonian H has a wave function given by the Schroedinger equation

$$H\psi = E\psi. \tag{10}$$

All the operations R that leave the Hamiltonian invariant, i.e.

$$RHR^{-1} = H, \tag{11}$$

form the <u>group of the Schroedinger equation</u>. From (10)

$$RH\psi = ER\psi \tag{12}$$

and, because of (11)

$$H(R\psi) = E(R\psi). \tag{13}$$

$R\psi$ is a solution of the Schroedinger equation that belongs to the same eigenvalue E. If E is not degenerate

$$R\psi = c\psi \qquad (c = \text{const}). \tag{14}$$

If E is m-fold degenerate

$$R\psi_l = \sum_{i=1}^{m} a_{il} \, \psi_i . \tag{15}$$

Similarly, if S is another operation of the group

$$S\psi_i = \sum_{j=1}^{m} b_{ji} \, \psi_j . \tag{16}$$

Also

$$(SR)\psi_l = \sum_{j=1}^{m} c_{jl} \, \psi_j . \tag{17}$$

But also

$$SR\psi_l = S(R\psi_l) = \sum_{i=1}^{m} a_{il}\, S\psi_i =$$
$$\sum_{i=1}^{m} a_{il} \sum_{j=1}^{m} b_{ji}\, \psi_j = \sum_{j=1}^{m} \left[\sum_{i=1}^{m} b_{ji}\, a_i \right] \psi_j . \tag{18}$$

Comparing (17) with (18)

$$c_{jl} = \sum_{i=1}^{m} b_{ji}\, a_{il} \tag{19}$$

or

$$\underset{\sim}{C} = \underset{\sim}{B} \cdot \underset{\sim}{A} \tag{20}$$

where $\underset{\sim}{C}$, $\underset{\sim}{B}$ and $\underset{\sim}{A}$ are the matrices of the coefficients c_{jl}, b_{ji} and a_{il}, respectively.

The representation generated by applying the different operations to the degenerate wave functions is irreducible. In general we can state the following:

> "Wave functions that belong to the same energy eigenvalue form a basis for an irreducible representation of the group of operations that leave the Hamiltonian invariant. The dimension of this irreducible representation is equal to the degree of the degeneracy."

III. ENERGY LEVELS OF ATOMS

III.A. Some Groups of Interest

1. Full Rotational Group. It is the group of all the operations that leave the Hamiltonian of a field-free atom invariant. It consists of all the real coordinate transformations, proper and improper. Some of the characters of the irreducible representations of such a group are given in Table 1.

We note here that all the proper rotations through an angle θ belong to the same class; likewise all the improper rotations through an angle θ belong to the same class. We note also that the irreducible representations corresponding to half integer angular momenta are double valued (1).

2. The Group $C_{\infty h}$. It is the group of all the operations

TABLE 1. CHARACTERS OF THE FULL ROTATIONAL GROUP

J	E	$\bar{E}$	C_4	$\bar{C}_4$	C_3	I_3	I	$I\bar{E}$	IC_4	$I\bar{C}_4$	IC_3	$I\bar{C}_3$
0	1	1	1	1	1	1	1	1	1	1	1	1
	1	1	1	1	1	1	-1	-1	-1	-1	-1	-1
$\frac{1}{2}$	2	-2	$\sqrt{2}$	$-\sqrt{2}$	1	-1	2	-2	$\sqrt{2}$	$-\sqrt{2}$	1	-1
	2	-2	$\sqrt{2}$	$-\sqrt{2}$	1	-1	-2	2	$-\sqrt{2}$	$\sqrt{2}$	-1	1
1	3	3	1	1	0	0	3	3	1	1	0	0
	3	3	1	1	0	0	-3	-3	-1	-1	0	0
$\frac{3}{2}$	4	-4	0	0	-1	1	4	-4	0	0	-1	1
	4	-4	0	0	-1	1	-4	4	0	0	1	-1
2	5	5	-1	-1	-1	-1	5	5	-1	-1	-1	-1
	5	5	-1	-1	-1	-1	-5	-5	1	1	1	1

that leave the Hamiltonian of an atom in a homogeneous magnetic field invariant. Some of the characters of the irreducible representations of this group are reported in Table 2. As expected, all the irreducible representations are one-dimensional, i.e. all the degeneracies are lifted in a magnetic field.

TABLE 2. CHARACTERS OF THE GROUP $C_{\infty h}$

$C_{\infty h}$	E	$C(\phi)$	I	$IC(\phi)$
A_g	1	1	1	1
A_u	1	1	-1	-1
C_{ng}	1	$e^{in\phi}$	1	$e^{in\phi}$
C_{nu}	1	$e^{in\phi}$	-1	$-e^{in\phi}$
S_{ng}	1 -1	$e^{in\phi/2}$	1 -1	$e^{in\phi/2}$
S_{nu}	1 -1	$e^{in\phi/2}$	-1 1	$-e^{in\phi/2}$

III.B. Complete Set of Commuting Operators

Consider two quantum-mechanical operators Q and R and assume the following:

a) $[Q, R] = 0$ and (21)

b) one of them, say Q, has non-degenerate eigenvalues

$$Q\psi_i = q_i\psi_i. \tag{22}$$

Then

$$Q(R\psi_i) = R(Q\psi_i) = q_i(R\psi_i). \tag{23}$$

$R\psi_i$ is an eigenfunction of Q which differs from ψ_i only by a constant

$$R\psi_i = r_i\psi_i. \tag{24}$$

ψ_i is a simultaneous eigenfunction of R and Q.

Consider now the case in which

a) $[Q, R] = 0$, and

b) both Q and R have degenerate eigenvalues.

In this case there exists a complete set of functions which are simultaneous eigenfunctions of Q and R.

Consider a set of operators which have the following properties:

1) they commute with each other, and

2) only one simultaneous eigenstate corresponds to a given set of eigenvalues.

Such a set of operators is called a complete set of commuting operators. There is one good quantum number for each operator in the complete set; all the quantum numbers together individuate a simultaneous eigenstate.

Assume that $|k\rangle$ and $|s\rangle$ are two eigenstates of an operator, say L_z; assume also that

$$[L_z, H_1] = 0, \tag{25}$$

where H_1 = perturbing Hamiltonian. It is then

$$\begin{aligned} \langle k|[H_1, L_z]|s\rangle &= \langle k|H_1 L_z|s\rangle - \langle k|L_z H_1|s\rangle \\ &= (m_s - m_k)\,\hbar\,\langle k|H_1|s\rangle = 0 \end{aligned} \tag{26}$$

If $m_s \neq m_k$, it is always

$$\langle k|H_1|s\rangle = 0.$$

On the other hand $\langle k|H_1|s\rangle$ may be different from zero only if $m_s = m_k$. A general result can be derived from this simple argument:

"Matrix elements of a perturbing Hamiltonian H_1 taken between eigenstates of an operator Q that commutes with H_1 are zero unless the eigenstates refer to the same eigenvalue of Q."

II.C. Atomic States

1. The Hamiltonian. The Hamiltonian of an atom in a magnetic field H in the z direction is given by

$$H = H_o + H_{so} + H_z, \tag{27}$$

where

$$H_o = \sum_{i=1}^{n} \left[\frac{p_i^2}{2m} - \frac{Ze^2}{r_i} + \sum_{j<i} \frac{e^2}{r_{ij}} \right] \tag{28}$$

$$H_{so} = \sum_{i=1}^{n} \xi(\underline{r}_i)\, \underline{l}_i \cdot \underline{s}_i \tag{29}$$

$$H_z = \frac{eH}{2mc} (L_z + 2S_z) \tag{30}$$

where r_i = distance of the ith electron from the nucleus, r_{ij} = $|\underline{r}_i - \underline{r}_j|$ and n = number of electrons.

2. The Unperturbed Hamiltonian. The unperturbed Hamiltonian represents the situation in which each electron sees an "effective charge" Z_{pi}:

$$H_u = \sum_{i=1}^{n} \left(\frac{p_i^2}{2m} - \frac{Z_{pi}}{r_i} e^2 \right) = \sum_{i=1}^{n} h_i. \tag{31}$$

An eigenfunction of this Hamiltonian is a product of one electron function, each consisting of an orbital part and of a spin part.

The complete set of commuting operators is given by

$$h_1, l_1^2, l_{1z}, s_1^2, s_{1z}; h_2, l_2^2, l_{2z}, s_z^2; \ldots$$

$$[H_u, L_z, S_z, P]. \tag{32}$$

where P = parity operator. In the above set the operators in square brackets are constants of the motion, but are no strictly necessary to define the complete set; we shall use the convention of putting operators of this type in square brackets when reporting a complete set of commuting operators.

The resultant energy levels are in this case "electronic configurations" which have parity $(-1)^{\sum_i l_i}$.

The group of operators that leave H_u invariant is given by the product group

$$R_{el_1} \times R_{sp_1} \times R_{el_2} \times R_{sp_2} \times \ldots \times R_{el_n} \times R_{sp_n}, \tag{33}$$

where

R_{el_i} = full rotational group for orbital i

R_{sp_i} = full rotational group for spin i.

All the transformations on the coordinates and spins of the individual electrons are indepedent. Degenerate eigenfunctions transform according to the representation

$$d_{l_1} \times d_{s_1} \times d_{l_2} \times d_{s_2} \times \ldots \ldots d_{l_n} \times d_{s_n} \tag{34}$$

where d_{l_1} is an irreducible representation of R_{el_1}, d_{s_1} is an irreducible representation of R_{sp_1} etc.

We may note at this point that we have not taken into account the fact that the Hamiltonian is also invariant with respect to the n! permutation operations that can be performed on the n electrons. These operations also form a group which is called the symmetric group of order n. When this group is taken into account we can no longer say that electron 1 is in a certain orbit, electron 2 in another orbit and so on; this corresponds group-theoretically to the fact that the eigenfunctions must transform according to the representations of the symmetric group. Fortunately the Pauli principle comes to our help by imposing the condition that all eigenfunctions must transform according to the anti-symmetric representation of the symmetric group. This condition is easily enforced by building up determinantal eigenfunctions which are antisymmetrized products of one electron function. This condition reduces the number of possible electron configurations.

3. The Electron-Electron Interaction. At this stage we examine the effect of the electron-electron interaction by considering the Hamiltonian

$$H_o = \sum_{i=1}^{n} \left[\frac{p_i^2}{2m} - \frac{Ze^2}{r_i} + \sum_{j<i} \frac{e^2}{r_{ij}} \right]. \tag{35}$$

An eigenstate of the system is defined by the eigenvalues of the operators

$$H_o,\ L^2,\ \underline{L_z},\ S^2,\ \underline{S_z},\ [\underline{J_z} = L_z + S_z,\ \underline{P}]. \tag{36}$$

The operators which are underlined are also part of the previous complete set (32). It is

$$[H_o, H_u] \neq 0. \tag{37}$$

H_o may connect eigenstates of H_u with equal M_L, M_S, P; it may cause interaction between different electronic configurations.

The electron-electron interaction produces a correlation of the following kind: the rotational operations on all the orbits must now be referred to the same axes, i.e. the full rotational groups d_{l_1}, d_{l_2}, etc. are no longer independent. The direct product (34) is now reducible

$$d_{l_1} \times d_{l_2} \times \ldots \ldots = \sum_L D_L, \tag{38}$$

where

$$L = (l_1 + l_2 + \ldots l_n), (l_1 + l_2 + \ldots l_n - 1), \ldots \ldots$$

Similarly

$$d_{s_1} \times d_{s_2} \times \ldots \ldots = \sum_S D_S \tag{39}$$

where

$$S = \frac{n}{2}, \frac{n}{2} - 1, \ldots$$

Some values of L and S are excluded by the Pauli principle. The total symmetry group is

$$R_{el} \times R_{sp}, \tag{40}$$

where R_{el} = full rotational group for the orbital part and R_{sp} = field rotational group for the spin part. These two groups are not referred to the same axes.

An energy level of the system is at this stage called a spectral term and is designated by the symbol ^{2S+1}L.

4. The Spin-Orbit Interaction. The term of the Hamiltonian which corresponds to the spin-orbit interaction is given by

$$H_{so} = \sum_i \xi(\underline{r}_i) \, \underline{l}_i \cdot \underline{s}_i. \tag{41}$$

An eigenstate of the system is defined by the eigenvalues of the operators

$$H_{so},\ J^2,\ \underline{J_z},\ [\underline{P}]. \tag{42}$$

The operators which are underlined are also part of the previous complete set (36), since

$$[H_{so},\ H_o] \neq 0, \tag{43}$$

H_{so} may connect states of H_o with equal M_J and P; it may connect even states with different L and S.

If the spin-orbit interaction is much smaller than the electron-electron interaction, use is made of the Russell-Saunders approximation which consists in disregarding the (off-diagnol) matrix elements of the H_{so} Hamiltonian between states with the same J but different L and S; in doing so H_{so} takes for form

$$H_{so} = \lambda \underline{L} \cdot \underline{S} \tag{44}$$

and the previous set becomes

$$H_{so},\ \underline{L}^2,\ \underline{S}^2,\ J^2,\ \underline{J_z},\ [\underline{P}]. \tag{45}$$

The operators which are underlined are also part of the complete set (36). It is

$$[H_{so},\ H_o] \neq 0 \tag{46}$$

as previously. H_{so} may connect states of H_o with the same L, S, M_J quantum numbers (and parity); it splits terms giving rise to multiplet levels $^{2S+1}L_J$, it does not connect spectral terms with different L and/or different S.

The introduction of the spin-orbit interaction produces a correlation between rotations in coordinate space and in spin space; operations on orbits and spins must now be referred to the same axes. $D_L \times D_S$ is still a representation of the symmetry group but it is no longer irreducible; it reduces in the following way

$$D_L \times D_S = \sum_J D_J, \tag{47}$$

where

$$J = L + S,\ L + S - 1,\ .\ .\ .,\ |L - S|.$$

5. The Zeeman Interaction. The application of a homogenous and constant magnetic field results in the perturbing Hamiltonian

$$H_z = \mu \mathcal{H}(L_z + 2S_z). \tag{48}$$

An eigenstate of the system is defined by eigenvalues of the operators

$$H_z,\ \underline{J_z},\ \underline{L}^2,\ \underline{S}^2,\ [\underline{P}]. \tag{49}$$

Considering the Russell-Saunders approximation valid in this case, this set has in common with the previous set (45) the operators which are underlined. It is

$$[H_z, H_{so}] \neq 0\,. \tag{50}$$

Therefore H_z may connect states of H_{so} with the same L, S and M_J quantum numbers (and parity); it may also connect multiplet levels of different J.

If the applied $\mathcal{H}$ field is such that the resulting perturbation is much smaller than the spin-orbit perturbation, the interaction between multiplets of different J is neglected, i.e. J^2 is introduced in the set

$$H_z,\ \underline{J}^2,\ \underline{J_z},\ \underline{L}^2,\ \underline{S}^2,\ [\underline{P}]. \tag{51}$$

Considering still the Russell-Saunders approximation valid, this set has in common with the set (45) the operators which are underlined. It is

$$[H_z, H_{so}] = 0; \tag{32}$$

therefore H_z simply splits levels within the same J manifold.

The presence of the perturbation H_z changes the group to $C_{\infty h}$. The splitting of levels corresponds to the reduction of the representation D_J as follows

$$D_J = \sum_{M_J} C_{M_J}\,, \tag{53}$$

where $M_J = J, J - 1, \ldots -J$.

In Figure 2 the example of an sp electronic configuration is reported.

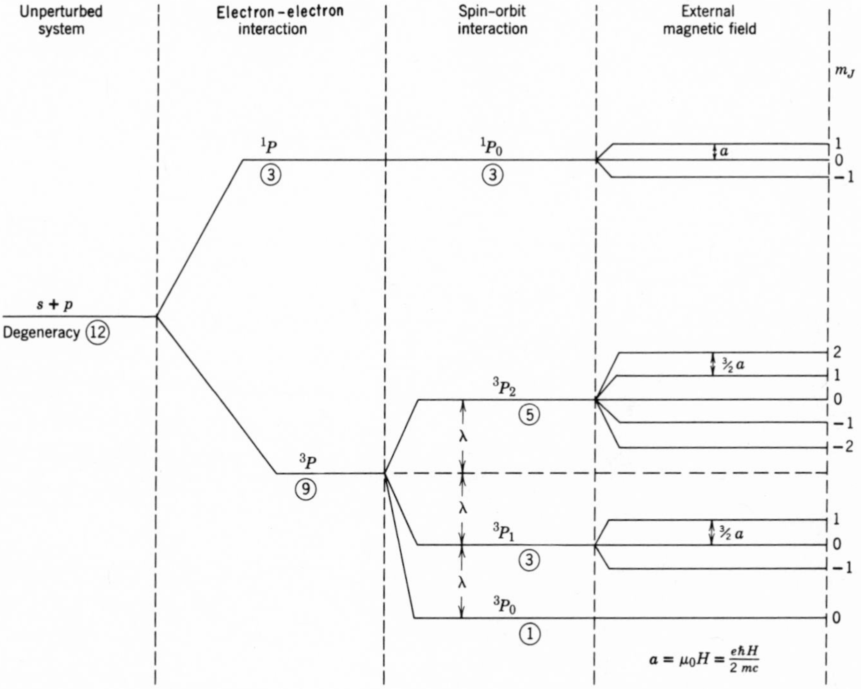

Figure 2. Splitting of the Energy Levels of an sp Electronic Configuration.

IV. MAGNETIC IONS IN CRYSTALS

IV.A. Magnetic Ions

Several groups of atoms in their ionic states have one or more of their shells partly filled with electrons: this fact reflects itself in some magnetic and optical properties of these ions when they are present as impurities in a crystal. The electrons of the unfilled shell provide a net magnetic moment and account for the magnetic properties of these ions; the same electrons set the energy levels of the ion. In general the absorption and emission impurity spectra correspond to transitions among these levels.

The magnetic ions can be divided in five categories:

1. <u>Transition Metal Ions of the First Series (Iron Group)</u>. The atoms of this series have electronic configurations of the type

$$1s^2\ 2s^2\ 2p^6\ 3s^2\ 3p^6\ 3d^n\ 4s^m = (\text{Ar core})^{18}\ 3d^n\ 4s^m$$

where m = 1, 2 and n = 1, 2, . . .10. Their atomic numbers range from Z = 21(Sc) to Z = 30(Zn). The divalent and trivalent ions of this series present configurations of the type $(\text{Ar core})^{18}\ 3d^n$ (n = 1, 2, . . .10) in which the unfilled 3d shell is the outermost shell. When in a crystal the orbital motion of the 3d electrons is strongly perturbed by the Coulomb field of the surrounding ions and energies of the order of 10,000 cm^{-1} (≡ 15,000 °K ≡ 1.25 eV ≡ 2 x 10^{-6} erg) are associated with the coupling of the orbitals with this field.

2. <u>Transition Metal Ions of the Second Series (Palladium Group</u>. The atoms of this series have electronic configurations of the type

$$(\text{Kr core})^{36}\ 4d^n 5s^m$$

where m = 1, 2 and n = 1, 2, . . .10. Their atomic numbers range from Z = 39(Y) to Z = 48(Cd). The divalent ions of this series present configurations of the type $(\text{Kr core})^{36}\ 4d^n$, where the unfilled 4d shell is the outermost shell.

3. <u>Transition Metal Ions of the Third Series (Platinum Group)</u>. The atoms of this series have electronic configurations of the type

$$(\text{Pd core})^{46}\ 4f^{14}\ 5s^2\ 5p^2\ 5d^n\ 6s^m$$

where m = 1, 2 and n = 2, 3, 4, 5, 6, 9, and 10. Their atomic numbers range from Z = 72(Hf) to Z = 80(Hg). The divalent ions present configurations of the type $(\text{Pd core})^{46} 4f^{14} 5s^2 5p^6 5d^n$, where the unfilled 5d shell is the outermost shell.

4. <u>Rare Earth Ions (Group of the Lanthanides)</u>. The atoms of this group have electronic configurations of the type

$$(\text{Pd core})^{46}\ 4f^n\ 5s^2\ 5p^6\ 5d^m\ 6s^2$$

where m = 1, 2 and n = 2, 3, . . .13. Their atomic numbers range from Z = 58(Ce) to Z = 70(Yb). The trivalent ions present configurations of the type $(\text{Pd core})^{46} 4f^n 5s^2 5p^6$ and the unfilled 4f shell is an inner shell screened by the 5s and 5p shells. (The 4f electrons have orbital radii of 0.6 to 0.35 from Ce to Lu; for the latter this is less than one half the ionic radius.)

When in a crystal the orbital motion of the 4f electrons are not much affected by the Coulomb field of the surrounding ions, as the interaction energy is in most cases much less than 1000 cm^{-1}; on the other hand, the spin-orbit energy is greater than for the 3d ions (600-3000 cm^{-1} from Ce to Yb).

5. Actinide Ions. The atoms of this group have electronic configurations of the type

$$(\text{Pt core})^{78}\ 5f^n\ 6d^m\ 7s^2$$

where m = 1, 2 and n = 0, 2, 3, . . . and their atomic numbers range from Z = 90(Th) up. The trivalent ions present configurations of the type $(\text{Pt core})^{78}5f^n7s^2$.

IV.B. The Crystalline Field

The electrons of a magnetic ion impurity occupy localized orbitals; each electron feels the influence of the electrons of the other ligand ions (a repulsion) and of the nuclei of the other ions (an attraction). Such influence is taken into account by assuming that the electrons of the magnetic impurity are subjected to the action of a crystalline field.

The crystalline field is considered to be external to the magnetic ion, i.e. the charges of the ligand ions do not penetrate into the region occupied by the magnetic ion; an electron of the magnetic ion feels the presence of a potential V that satisfies Laplace's equation

$$\nabla^2 V = 0. \tag{54}$$

It is then

$$V(r, \theta, \psi) = \sum_{l,m} A_{lm}\, r^l\, Y_l^m(\theta, \phi), \tag{55}$$

where Y_l^m are spherical harmonics. The crystalline field potential V has a definite symmetry that is set by the configuration of the ligand ions.

The Hamiltonian of a magnetic ion in a crystal is then given by

$$H = \sum_{i=1}^{n} \left[\frac{p_i^2}{2m} - \frac{Ze^2}{r_i} + \sum_{j<i} \frac{e^2}{r_{ij}} \right] + H_{so} + \sum_{i=1}^{n} eV(\underline{r}_i). \tag{56}$$

The crystalline field perturbation destroys the spherical symmetry of the environment.

It is useful to consider at this time the relative strengths of the crystalline field perturbation, the spin-orbit interaction and the electron-electron interaction. If the Hamiltonian of an ion in a solid were written as a sum of terms of decreasing "strength," the crystalline perturbation would appear in different positions, depending on its relative strength; the following table illustrate the positioning of the crystalline field perturbation for the three cases of weak, intermediate and strong crystalline field.(*)

	Term		Multiplet	
	$\sum_{i>j} \frac{e^2}{r_{ij}}$		H_{so}	
STRONG		INTERMEDIATE		WEAK
Pd group (4d) Pt group (5d)		Iron group (3d)		Rare earths (4f) Actinides (5f)

IV.C. The Weak Field Scheme

1. <u>The Hamiltonian of the Free Ion</u>. The Hamiltonian of the free ion is given in this case by

$$H_J = H_o + H_{so}. \tag{57}$$

An eigenstate of the system is defined by the eigenvalues of the operators

$$H_J,\ J^2,\ J_z,\ [\underline{P}]. \tag{58}$$

In the Russell-Saunders approximation also L and S are considered good quantum numbers.

(*) F. Williams in his article "Historical Survey of the Optical Properties of Ions in Solids," which appears in this book, calls the "intermediate" and "weak" field cases "weak" and "very weak," respectively.

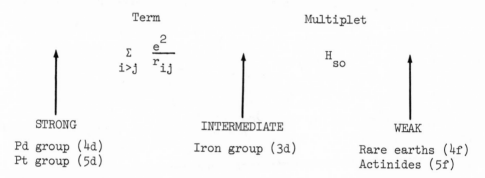

The symmetry group of the Hamiltonian H_J is

$$R_{el} \times R_{sp} \tag{59}$$

where R_{el} = full rotational group for the orbital part and R_{sp} = full rotational group for the spin part; here both R_{el} and R_{sp} are referred to the same axes. The eigenstates corresponding to a certain J value (multiplet states) transform according to the representation D_J. The degeneracy is 2J + 1.

2. The Crystalline Field Perturbation. The crystalline field perturbation is represented by

$$H_{cryst} = \sum_{i=1}^{n} eV(\underline{r}_i). \tag{60}$$

The introduction of the free ion in the crystal results in the reduction of the representation D_J into irreducible representations of the group of operators that leave H_{cryst} invariant; we call this group the crystal group or G_c. It is then

$$D_J = \sum_i \Gamma_i. \tag{61}$$

Degenerate eigenfunctions transform according to an irreducible representation Γ_i of G_c; the degree of degeneracy is equal to the dimension of Γ_i.

In the present case the quantum numbers J and M_J are replaced by a quantum number Γ that individuates the irreducible representation of G_c and a quantum number M_Γ that specifies an eigenfunction in the Γ manifold. An eigenstate of the system is defined by the eigenvalues of the operators

$$H_{cryst},\ \Gamma,\ \Gamma_z,\ (\underline{P}). \tag{62}$$

No significance has to be attached to the subscripts z in Γ_z, as it indicates only that this operator plays a role similar to that of J_z. The parity operator P, which is in any case redundant, may be included in (62) if the crystalline symmetry has a center of inversion. It is

$$[H_{cryst}, H_J] \neq 0; \tag{63}$$

therefore H_{cryst} may split multiplet (J) levels. It may also cause interaction between different J levels, different spectral terms and different electronic configurations; parity has to be conserved if H_{cryst} is invariant under inversion.

3. Examples. Consider an ion in a cubic crystal occupying a site of octahedral symmetry. The character table of the octahedral group follows

$G_c \equiv O$		E	$3C_2$	$6C_4$	$6C_2'$	$8C_3$
A_1	Γ_1	1	1	1	1	1
A_2	Γ_2	1	-1	-1	-1	-1
E	Γ_3	2	2	0	0	-1
T_1	Γ_4	3	-1	1	-1	0
T_2	Γ_5	3	-1	-1	1	0

The characters of the D_j representations can be found by using the formula

$$X_J(\omega) = \frac{\sin (J + 1/2)\, \omega}{\sin (1/2)\, \omega} \tag{64}$$

where ω = angle of rotation. We find

G_c	E	$3C_2$	$6C_4$	$6C_2'$	$8C_3$
D_0	1	1	1	1	1
D_1	3	-1	1	-1	0
D_2	5	1	-1	1	-1
D_3	7	-1	-1	-1	1
D_4	9	1	1	1	0
D_5	11	-1	1	-1	-1
D_6	13	1	-1	1	1

The various D_J representations are then reduced as follows

$D_0 = \Gamma_1,$

$D_1 = \Gamma_4,$

$D_2 = \Gamma_3 + \Gamma_5,$

$D_3 = \Gamma_2 + \Gamma_4 + \Gamma_5,$

$D_4 = \Gamma_1 + \Gamma_3 + \Gamma_4 + \Gamma_5,$

$D_5 = \Gamma_3 + 2\Gamma_4 + \Gamma_5,$

$D_6 = \Gamma_1 + \Gamma_2 + \Gamma_3 + \Gamma_4 + 2\Gamma_5.$

The energy level of an atom with J = 2 will be split by the crystalline field as in Figure 3.

If J is a half-integer, the representation D_J is double-valued and reduces in terms of double-valued representations of the symmetry group of the ion in the crystal. If this symmetry is octahedral, the group to consider is the octahedral double group O'

G_c = O'		E	$\bar{E}$	$3C_2$	$6C_4$	$\overline{6C_4}$	$6C_2'$	$8C_3$	$\overline{8C_3}$
A_1	Γ_1	1	1	1	1	1	1	1	1
A_2	Γ_2	1	1	1	-1	-1	-1	1	1
E	Γ_3	2	2	2	0	0	0	-1	-1
T_1	Γ_4	3	3	-1	1	1	-1	0	0
T_2	Γ_5	3	3	-1	-1	-1	1	0	0
$E_{1/2}$	Γ_6	2	-2	0	$\sqrt{2}$	$-\sqrt{2}$	0	1	-1
$E_{5/2}$	Γ_7	2	-2	0	$-\sqrt{2}$	$\sqrt{2}$	0	1	-1
G	Γ_8	4	-4	0	0	0	0	-1	1

The characters of the D_J representations of even dimensions in correspondence to the operations of the double group O' are given by

G_c	E	$\bar{E}$	$3C_2$	$6C_4$	$\overline{6C_4}$	$6C_2'$	$8C_3$	$\overline{8C_3}$
$D_{1/2}$	2	-2	0	$\sqrt{2}$	$-\sqrt{2}$	0	1	-1
$D_{3/2}$	4	-4	0	0	0	0	-1	1
$D_{5/2}$	6	-6	0	$-\sqrt{2}$	$\sqrt{2}$	0	0	0
$D_{7/2}$	8	-8	0	0	0	0	1	-1
$D_{9/2}$	10	-10	0	$\sqrt{2}$	$-\sqrt{2}$	0	-1	1
$D_{11/2}$	12	-12	0	0	0	0	0	0

The various D_J representations are then reduced as follows

$D_{3/2} = \Gamma_8$,

$D_{5/2} = \Gamma_7 + \Gamma_8$,

$D_{7/2} = \Gamma_6 + \Gamma_7 + \Gamma_8$,

$D_{9/2} = \Gamma_6 + 2\Gamma_8$,

$D_{11/2} = \Gamma_6 + \Gamma_7 + 2\Gamma_8$.

IV.D. The Intermediate Field Scheme

1. The Hamiltonian of the Free Ion. When neglecting the spin-orbit interaction the Hamiltonian of a free ion can be written as

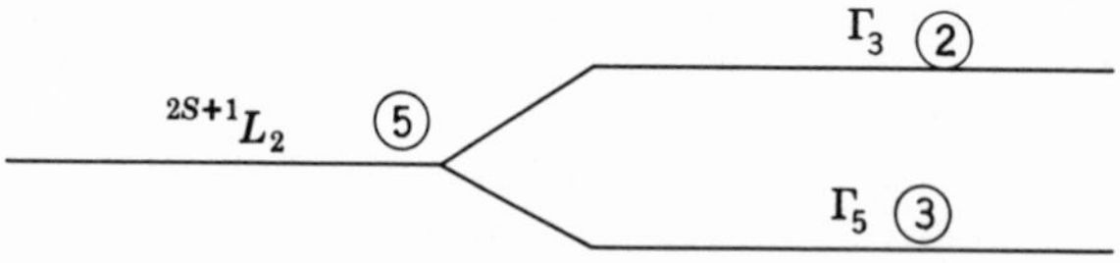

Figure 3. Atom with J = 2 in an Octahedral Field. The numbers in O give the degeneracies of the levels.

$$H_o = \sum_{i=1}^{n} \left[\frac{p_i^2}{2m} - \frac{Ze^2}{r_i} + \sum_{j<i} \frac{e^2}{r_{ij}} \right]. \quad (65)$$

An eigenstate of the system is defined by the eigenvalues of the operators

$$H_o,\ L^2,\ L_z,\ S^2,\ S_z,\ [P]. \quad (66)$$

The total symmetry group is

$$R_{el} \times R_{sp} \quad (67)$$

where R_{el} = full rotational group for the orbital part and R_{sp} = full rotational group for the spin part. Degenerate eigenfunctions transform as

$$D_L \times D_S. \quad (68)$$

2. The Crystalline Field Perturbation. The crystalline field perturbation is represented by

$$H_{cryst} = \sum_{i=1}^{n} eV(\underline{r}_i). \quad (69)$$

The introduction of the ion in the crystal does not affect the spin representation D_S; on the other hand, the representation D_L becomes reducible. D_L is reduced in terms of the irreducible representations of G_c:

$$D_L = \sum_i c_i \Gamma_i. \quad (70)$$

An eigenstate of the system is defined by the eigenvalues of the operators

$$H_{cryst},\ \underline{S}^2,\ \underline{S_z},\ \Gamma,\ \Gamma_z,\ (\underline{P}). \quad (71)$$

The subscript z in Γ_z indicates only that this operator plays a role similar to L_z, in that it identifies an eigenfunction in the Γ manifold. The operator P, which is in any case redundant, may be included in (71) if the crystalline symmetry has a center of inversion. It is

$$[H_{cryst}, H_o] \neq 0; \quad (72)$$

therefore H_{cryst}, besides splitting spectral terms, may cause interaction between different eigenstates of H_o

which have equal quantum numbers S and M_S; parity has to be conserved if H_{cryst} is invariant under inversion. The interacting spectral terms may belong to different electronic configurations.

The relevant group at this stage is

$$R_{sp} \times G_c. \tag{73}$$

Degenerate eigenfunctions transform according to a representation

$$D_S \times \Gamma_i. \tag{74}$$

An energy level is represented by a symbol

$$^{2S+1}\Gamma_i \tag{75}$$

and has a degeneracy

$$(2S + 1) \times (\text{dimension of } \Gamma_i). \tag{76}$$

3. The Spin-Orbit Interaction. The spin-orbit interaction is represented by

$$H_{so} = \sum_{i=1}^{n} \xi(\underline{r}_i)\, \underline{l}_i \cdot \underline{s}_i \tag{77}$$

and produces a correlation between rotations in coordinate space and rotations in spin space. Now operations of both types must be referred to the same axes; this results in the representation $D_S \times \Gamma_i$ of (74) becoming reducible and being reduced in terms of the irreducible representations of G_c:

$$D_s \times \Gamma_i = (\sum_k \Gamma_k) \times \Gamma_i = \sum_j \Gamma_j. \tag{78}$$

An eigenstate of the system is defined by the eigenvalues of the operators

$$H_{so}, \Gamma, \Gamma_z, (\underline{P}). \tag{79}$$

The subscript z in Γ_z indicates only that this operator plays a role similar to J_z, in that it identifies an eigenfunction in the Γ manifold. It makes sense to include the operator P (which commutes with H_{so}) in (79) if it is also part of the set (71). It is

$$[H_{so}, H_{cryst}] \neq 0; \tag{80}$$

therefore H_{so}, besides splitting the levels $^{2S+1}\Gamma_i$, may cause interaction between two different levels of this type, spectral term interaction and electronic configuration interaction.

The relevant group at this stage is G_c; degenerate eigenfunctions transform according to the irreducible representations of G_c.

4. Example. Consider an ion in a state ^4F (L = 3, S = 3/2) as for Co^{2+}. A crystalline field of octahedral symmetry will cause the splitting

$$D_3 = \Gamma_2 + \Gamma_4 + \Gamma_5.$$

The resulting energy levels will be

$$D_{3/2} \times \Gamma_2;\ D_{3/2} \times \Gamma_4;\ D_{3/2} \times \Gamma_5,$$

labeled, respectively

$$^4\Gamma_2,\ ^4\Gamma_4,\ ^4\Gamma_5$$

with total degeneracies 4 x 1, 4 x 3, 4 x 3. The introduction of the spin-orbit coupling causes the relabeling of $D_{3/2}$ as Γ_8 and

$$D_{3/2} \times \Gamma_2 = \Gamma_8 \times \Gamma_2 = \Gamma_8$$

$$D_{3/2} \times \Gamma_4 = \Gamma_8 \times \Gamma_4 = \Gamma_6 + \Gamma_7 + 2\Gamma_8$$

$$D_{3/2} \times \Gamma_5 = \Gamma_8 \times \Gamma_5 = \Gamma_6 + \Gamma_7 + 2\Gamma_8.$$

The energy levels for such a system are shown in Figure 4.

IV.E. The Strong Field Scheme

1. The Unperturbed Hamiltonian. The unperturbed Hamiltonian represents the situation in which each electron sees an effective charge Z_{pi}:

$$H_u = \sum_{i=1}^{n} \left(\frac{p_i^2}{2m} - \frac{Z_{pi}}{r_i} e^2\right) = \sum_{i=1}^{n} h_i. \tag{81}$$

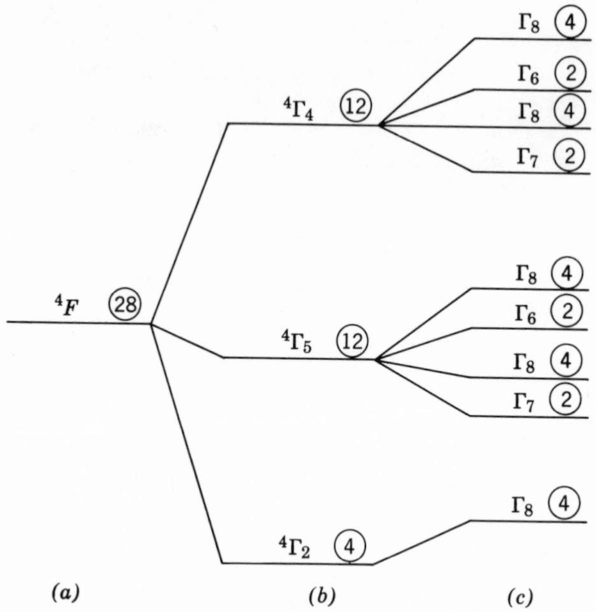

Figure 4. Ion in a ^4F State in an Octahedral Symmetry. (a) Free ion's spectral term; (b) ion in an intermediate field (LS coupling not included); (c) ion in an intermediate field (LS coupling included).

The complete set of commuting operators is given by

$$h_1, l_1^2, l_{1z}, s_1^2, s_{1z}; h_2, l_2^2, l_{2z}, s_2^2, s_{2z}; \ldots$$

$$[H_u, L_z, S_z, P]. \tag{82}$$

The relevant group as in (33) is

$$R_{el_1} \times R_{sp_1} \times R_{el_2} \times R_{sp_2} \times \ldots\ldots \tag{83}$$

Degenerate eigenfunctions correspond to the same electronic configuration and transform according to the product representation

$$d_{l_1} \times d_{s_1} \times d_{l_2} \times d_{s_2} \times \ldots\ldots \tag{84}$$

2. The Crystalline Field Perturbation. The crystalline field perturbation is represented by

$$H_{cryst} = \sum_{i=1}^{n} eV(\underline{r}_i). \tag{85}$$

The crystalline field does not act on the spin representations but splits the orbital representations of the individual electrons into irreducible representations of the symmetry group G_c of H_{cryst}

$$d_{l_i} = \sum_k \gamma_k. \tag{86}$$

An eigenstate of the system is defined by the eigenvalues of the operators

$$eV(\underline{r}_i),\ \gamma_1,\ \gamma_{1z},\ \underline{s_1}^2,\ \underline{s_{1z}};\ eV(\underline{r}_i),\ \gamma_2,\ \gamma_{2z},\ \underline{s_2}^2,\ \underline{s_{2z}};\ \ldots$$

$$[H_{cryst},\ \underline{S_z},\ (\underline{P})]. \tag{87}$$

The subscript z in γ_z indicates only that this operator plays a role similar to l_z. The operator P, in any case redundant, may be included in (87), if the crystalline symmetry has a center of inversion. It is

$$[H_{cryst},\ H_u] \neq 0; \tag{88}$$

therefore H_{cryst} may cause interaction between eigenstates of H_u with equal M_s, s_1 m_{s1}, s_2, m_{s2}, ...; parity has to be conserved if H_{cryst} is invariant under inversion. In effect the individual electron, because of the crystalline field, goes from a l, s, m_l, m_s to a γ, s, m_γ, m_s representation; the new elementary orbital eigenfunctions, which we call crystal orbitals, are linear combinations of the one electron orbital functions u_{nlm_l} and transform as the irreducible representations γ_i of G_c. Every possible distribution of electrons over the crystal orbitals may be called a crystal configuration. The Pauli principle has to be taken into account by not putting in an orbital γ_i of dimension r_i more than $2r_i$ electrons.

The relevant symmetry group at this stage is

$$G_{c_1} \times R_{sp_1} \times G_{c_2} \times R_{sp_2} \times \ldots\ldots \tag{89}$$

Degenerate eigenfunctions transform according to the representation

$$\gamma_1 \times d_{s_1} \times \gamma_2 \times d_{s_2} \times \ldots\ldots \tag{90}$$

3. The Electron-Electron Interaction. We examine now the effect of the electron-electron interaction by considering the Hamiltonian

$$H_o = \sum_{i=1}^{n} \left[\frac{p_i^2}{2m} - \frac{Ze^2}{r_i} + \sum_{j>i} \frac{e^2}{r_{ij}} \right]. \tag{91}$$

The electron-electron interaction "couples" the one-electron representations γ_i; the reduction follows in terms of irreducible representations of G_c:

$$\gamma_1 \times \gamma_2 \times \ldots\ldots \gamma_n = \sum_i \Gamma_i. \tag{92}$$

At this point the total spin S is introduced as good quantum number; this corresponds to the coupling of the spin representations d_{s_i}; however, since the spin-orbit interaction is not yet present, the relevant symmetry group for the spin part must be the full rotational group. This means that

$$d_{s_1} \times d_{s_2} \times \ldots\ldots = \sum_S D_S, \tag{93}$$

where D_S are irreducible representations of the full rotational group R_{sp}.

An eigenstate of the system is defined by the eigenvalues of the operators

$$H_o,\ \Gamma,\ \Gamma_z,\ S^2,\ \underline{S_z},\ (\underline{P}). \tag{94}$$

The subscript z in Γ_z indicates only that this operator plays a role similar to L_z, since it identifies an eigenfunction in the Γ manifold. It makes sense to include the operator P (which commutes with H_o) in (94) if it is also part of the set (87). It is

$$[H_o, H_{cryst}] \neq 0; \tag{95}$$

therefore H_o may cause interactions between eigenstates of H_{cryst} with the same M_S, i.e. crystal configuration interaction and electronic configuration interaction. Parity has to be conserved if H_{cryst} is invariant under inversion.

The relevant group at this stage is

$$R_{sp} \times G_c. \tag{96}$$

Degenerate eigenfunctions transform according to a representation

$$D_S \times \Gamma_i. \tag{97}$$

An energy level is represented by the symbol

$$^{2S+1}\Gamma_i. \tag{98}$$

4. The Spin-Orbit Interaction. The spin-orbit interaction is given by

$$H_{so} = \sum_{i=1}^{n} \xi(\underline{r}_i)\, \underline{l}_i \cdot \underline{s}_i. \tag{99}$$

The representation $D_S \times \Gamma_i$ of (97) reduces in terms of irreducible representations of G_c^S:

$$D_S \times \Gamma_i = (\sum_k \Gamma_k) \times \Gamma_i = \sum_j \Gamma_j. \tag{100}$$

An eigenstate of the system is defined by the eigenvalues of the operators

$$H_{so},\ \Gamma,\ \Gamma_z,\ (\underline{P}). \tag{101}$$

The subscript z in Γ_z indicates only that this operator plays a role similar to J_z, in that it identifies an eigenfunction in the Γ manifold. It makes sense to include the operator P (which commutes with H_{so}) in (101) if it is also part of (94). It is

$$[H_{so}, H_o] \neq 0;$$

therefore H_{so}, besides splitting the levels $^{2S+1}\Gamma_i$, may cause interaction between two different levels of this type, crystal configuration interaction and electronic configuration interaction.

The relevant group at this stage is G_c; degenerate eigenfunctions transform according to irreducible representations of G_c.

5. Example. Consider an ion with an electronic configuration (closed shell + d^3) in an octahedral crystalline field. This is the case of such ions as Cr^{3+} and V^{2+}. The d orbital splits into t_2 (γ_5) orbital and an e(γ_3) orbital. The configuration $t_2{}^3$ gives rise to the levels $^4A_2$, ^2E, $^2T_1$ and $^2T_2$, the configuration $t_2{}^2$ e to the levels $^4T_2$, $^4T_1$, etc. Finally, the spin-orbit interaction further splits the levels. This is illustrated in Figure 5.

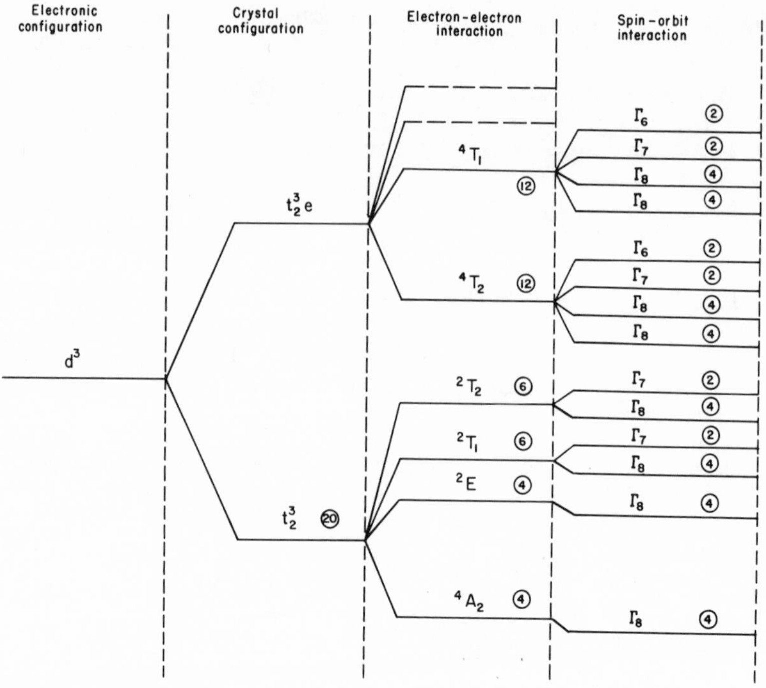

Figure 5. Splitting of the Energy Levels of a $3d^3$ Ion in an Octahedral Field.

IV.F. Kramers' Theorem

In evaluating the symmetry properties of the Hamiltonian we have taken into account only the coordinate transformations (proper and improper rotations) that leave the Hamiltonian invariant. We have mentioned the invariance of the Hamiltonian of an atom (or ion) with respect to the n! permutation operations that can be performed on the n electrons; because of the Pauli principle no complication is introduced into the problem by the permutation operations.

In considering the symmetry properties of the Hamiltonian all the operations that leave it invariant have to be considered; if the group one uses does not contain all these operations an "excess degeneracy" will result. An example of this type of degeneracy is the Kramers' degeneracy due to time reversal symmetry. Its effect can be expressed by the Kramers' theorem:

"If the number of electrons of an atomic system is odd, time invariant perturbations will not remove the degeneracy completely, as a twofold degeneracy of each level will remain."

We note here that, since the crystalline field interaction is time invariant, an ion with an odd number of electrons, when sitting at a lattice site, will present energy levels which have at least twofold degeneracies. We note also that the perturbation due to a homogeneous and constant magnetic H field is not time invariant; i.e. the H field removes even the Kramers' degeneracy.

V. EFFECT OF COVALENT BONDING

V.A. Why Covalent Bonding?

The crystalline field hypothesis is based on the assumption that the bond between the magnetic ion and the ligand ions is purely ionic. A need for relaxing this requirement and for allowing some mixing of the ion's and ligand's orbitals arises in several circumstances:

a) Such spectral parameters as Racah's B and C and spin-orbit interaction are found to be smaller in the solids than in the free ion; they are affected by the orbital mixing of ion and ligands.

b) First principle calculations of the crystalline field strength may require some covalence.

c) On the basis of the crystalline field hypothesis it is not possible to predict the occurrence of certain states generally high in energy which produce "charge transfer" absorption transitions of the order of hundreds or thousands times more probable than the crystalline field transitions.

IV.B. Molecular Orbitals

In a molecular complex <u>molecular orbitals</u> are the electronic wave functions that represent the one electron states. Consider the simple case of two atoms that come close to each other establishing a (covalent) bond; three types of molecular orbitals are relevant in establishing the bond: <u>σ orbitals</u>, which do not have any nodal plane containing the internuclear axis, <u>π orbitals</u>, which have one nodal plane containing the internuclear axis and δ orbitals which have two nodal planes containing the internuclear axis. The bonds that are established by these orbitals have different strengths: σ bond > π bond > δ bond.

The following atomic orbitals are in general available to produce a covalent bond:

Orbital	Normalized Angular Part
s	$\frac{1}{2\sqrt{\pi}}$
p_x	$\frac{\sqrt{3}}{2\sqrt{\pi}} \sin\theta \cos\phi$
p_y	$\frac{\sqrt{3}}{2\sqrt{\pi}} \sin\theta \sin\phi$
p_z	$\frac{\sqrt{3}}{2\sqrt{\pi}} \cos\theta$
d_{z^2}	$\frac{\sqrt{5}}{4\sqrt{\pi}} (3\cos^2\theta - 1)$
$d_{x^2-y^2}$	$\frac{\sqrt{15}}{4\sqrt{\pi}} (\sin^2\theta \cos 2\phi)$
d_{xy}	$\frac{\sqrt{15}}{4\sqrt{\pi}} (\sin^2\theta \sin 2\phi)$
d_{yz}	$\frac{\sqrt{15}}{2\sqrt{\pi}} (\sin\theta \cos\theta \sin\phi)$
d_{zx}	$\frac{\sqrt{15}}{2\sqrt{\pi}} (\sin\theta \cos\theta \cos\phi)$

The same orbitals are represented in Figure 6.

A bond is formed by the overlapping of two orbitals that have the same symmetry with respect to the internuclear axis. If we take the z direction as being along this axis, σ bonds can be produced by the overalpping of s-s, s-d_{z^2}, s-p_z, p_z-p_z, d_{z^2}-d_{z^2}, p_z-d_{z^2} orbitals, π bonds by the overlapping of p_x-p_x, p_y-p_y, d_{zx}-d_{zx}, d_{yz}-d_{yz} orbitals and δ bonds by the overlapping of d_{xy}-d_{xy}, $d_{x^2-y^2}$-$d_{x^2y^2}$ orbitals.

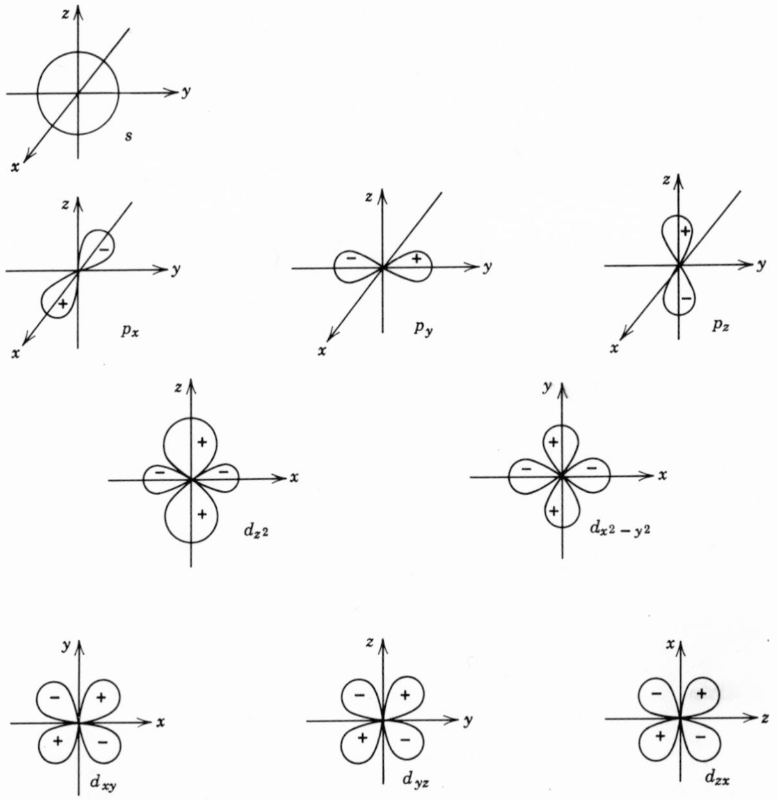

Figure 6. Relevant Orbitals for Chemical Bonding.

IV.C. Examples

Let us consider the formation of molecular orbitals in some simple molecules.

1. N_2 Molecule. The N atom has the following electronic configuration

$$N(Z=7): \underline{1s^2\ 2s^2}\ 2p_x^1\ 2p_y^1\ 2p_z^1.$$

The part underlined consists of the "core" electrons. A σ bond is established by the two atoms by using their $2p_z$ orbitals; two π bonds ($2p_x$-$2p_x$ and $2p_y$-$2p_y$) are also established. The N_2 molecule has the electronic configuration

$$N_2: \underline{1s_a^2\ 1s_b^2\ 2s_a^2\ 2s_b^2}\ \sigma_z^2\ \pi_x^2\ \pi_y^2.$$

2. HF Molecule. The F atom has the electronic configuration

$$F(Z=9):\ \underline{1s^2\ 2s^2}\ 2p_x^2\ 2p_y^2\ 2p_z.$$

The core part is considered. A σ bond is established by the F and H atoms by using a $2p_z$ and 1s orbitals, respectively. The remaining electrons $2p_x^2$ and $2p_y^2$ form two lone pairs. The electronic configuration of HF is

$$HF:\ \underline{1s^2\ 2s^2}\ \underline{\underline{2p_x^2\ \ 2p_y^2}}\ \sigma_z^2.$$

The lone pairs are underlined twice.

3. H_2O Molecule. The O atom has the electronic configuration

$$O(Z=8):\ \underline{1s^2\ 2s^2}\ 2p_x^1\ 2p_y^1\ 2p_z^2.$$

The core part is underlined. The following σ bonds are established by the O atom with two H atoms: $2p_x$-1s and $2p_y$-1s. The two electrons in $2p_z^2$ remain as a lone pair. The electronic configuration of the H_2O molecule is

$$H_2O:\ \underline{1s^2\ 2s^2}\ \underline{\underline{2p_z^2}}\ \sigma(2p_x\text{-}1s)^2\ \sigma(2p_y\text{-}1s)^2.$$

Experimentally it is found that the bond H-O-H angle is 105°, in good agreement with the value 90° predicted by this simple model; the difference of 15° may be accounted for by the repulsion of the two H atoms.

4. NH_3 Molecule. The N atom has the electronic configuration

$$N(Z=7):\ \underline{1s^2\ 2s^2}\ 2p_x\ 2p_y\ 2p_z.$$

The core part is underlined. The following σ bonds are established by the N atom with the three H atoms: $2p_x$-1s, $2p_y$-1s and $2p_z$-1s. No lone pair is present; the electronic configuration of NH_3 is

$$NH_3:\ \underline{1s^2\ 2s^2}\ \sigma(2p_x\text{-}1s)^2\ \sigma(2p_y\text{-}1s)^2\ \sigma(2p_z\text{-}1s)^2.$$

Experimentally it is found that the H-N-H bond angles are 106°, rather than the predicted 90°.

5. CH_4 Molecule; Hybridization. The C atom has the electronic

configuration

$$C(Z=6):\ 1s^2\ 2s^2\ 2p_x^1\ 2p_y^1.$$

The 2s and 2p orbitals are close in energy. In order to establish σ bonds with the four H atoms, the C atom "promotes" one electron from the 2s to the 2p shell to achieve the configuration

$$1s^2\ 2s^1\ 2p_x^1\ 2p_y^1\ 2p_z^1;$$

it then forms four new equivalent hybrid orbitals and directs them towards the corner of a tetrahadron. The hybrid orbitals are proper linear combinations of the one 2s and the three 2p orbitals. The four hybrid orbitals are used to establish σ bonds with the four H atoms.

In general the process of hybridization consists in the mixing of orbitals on a single atom; only orbitals of similar energy are mixed.

V.D. Octahedral Complexes

Let us consider the case of a (central) $3d^n$ ion in an octahedral coordination. In the strong field scheme the orbitals of the central ion change their properties due to the fact that the eigenfunctions representing the orbitals must form basis for some irreducible representation of the octahedral group O_h. Schematically this change can be indicated as follows

$$4p \rightarrow t_{1u}$$

$$4s \rightarrow a_{1g}$$

$$3d \rightarrow e_g + t_{2g}.$$

In effect the 4p and 4s levels will not split, the 3d level will split into a (lower) t_{2g} level and a (higher) e_g level. In physical terms one can indeed expect that the p orbitals are not split in an octahedral field for the electron charge density lobes of each p orbital are directed at two ligand ions. The inspection of the d orbitals in Figure 6 reveals that the d_{xy}, d_{yz} and d_{zx} orbitals have charge density lobes with maxima in a direction which is 45° away from the direction of the ligands. The lobes of the two d_{z^2} and $d_{x^2-y^2}$ orbitals are directed at the ligands; it is not obvious from this simple argument that they should have equal energy, but they do have equal energy, which is (and this is obvious) higher than the energy of the d_{xy}, d_{yz} and

d_{zx} orbitals.

Let us consider now the six ligands and let us assume that each of them has three orbitals available for bond formation as indicated in Figure 7. As the central ion, the ligands too will,

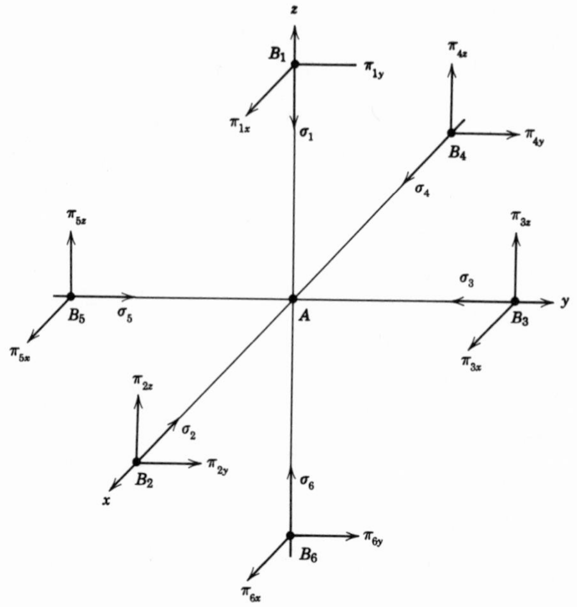

Figure 7. Ligand Orbitals in an Octahedral Complex.

in a manner of speaking, prepare themselves for the bonding. The σ bonds, being strong, will take precedence and the ligands will combine their σ_1 to σ_6 orbitals to create new ligand molecular orbitals compatible with the symmetry of the complex. These six new orbitals are reported in Table 3. In a similar manner the ligands will linearly combine the twelve π orbitals to create new ligand molecular orbitals compatible with the symmetry of the complex; these twelve new orbitals are also reported in Table 3.

In order to consider the molecular bonding of an AB_n complex, the following steps have to be taken:

1) Formation of the central ion orbitals.

2) Formation of proper combinations of ligand orbitals.

TABLE 3. ORBITALS FOR OCTAHEDRAL COMPLEXES

Representation	Ion Orbital	Ligand σ	Ligand π
A_{1g}	4s	$\frac{1}{\sqrt{6}}(\sigma_1+\sigma_2+\sigma_3+\sigma_4+\sigma_5+\sigma_6)$	
E_g	$3d_{z^2}$	$\frac{1}{\sqrt{3}}(2\sigma_1+2\sigma_6-\sigma_2-\sigma_3-\sigma_4-\sigma_5)$	
	$3d_{x^2-y^2}$	$\frac{1}{2}(\sigma_2-\sigma_3+\sigma_4-\sigma_5)$	
T_{1u}	$4p_x$	$\frac{1}{\sqrt{2}}(\sigma_2-\sigma_4)$	$\frac{1}{2}(\pi_{3x}+\pi_{1x}+\pi_{5x}+\pi_{6x})$
	$4p_y$	$\frac{1}{\sqrt{2}}(\sigma_3-\sigma_5)$	$\frac{1}{2}(\pi_{2y}+\pi_{1y}+\pi_{4y}+\pi_{6y})$
	$4p_z$	$\frac{1}{\sqrt{2}}(\sigma_1-\sigma_6)$	$\frac{1}{2}(\pi_{2z}+\pi_{3z}+\pi_{4z}+\pi_{5z})$
T_{2g}	$3d_{xz}$		$\frac{1}{2}(\pi_{2z}+\pi_{1x}-\pi_{4z}-\pi_{6x})$
	$3d_{yz}$		$\frac{1}{2}(\pi_{3z}+\pi_{1y}-\pi_{5z}-\pi_{6y})$
	$3d_{xy}$		$\frac{1}{2}(\pi_{2y}+\pi_{3x}-\pi_{4y}-\pi_{5x})$
T_{1g}			$\frac{1}{2}(\pi_{2z}-\pi_{1x}-\pi_{4z}+\pi_{6x})$
			$\frac{1}{2}(\pi_{3z}-\pi_{1y}-\pi_{5z}+\pi_{6y})$
			$\frac{1}{2}(\pi_{2y}-\pi_{3y}-\pi_{4y}+\pi_{5x})$
T_{2u}			$\frac{1}{2}(\pi_{3x}-\pi_{1x}+\pi_{5x}-\pi_{6x})$
			$\frac{1}{2}(\pi_{2y}-\pi_{1y}+\pi_{4y}-\pi_{6y})$
			$\frac{1}{2}(\pi_{2z}-\pi_{3z}+\pi_{4z}-\pi_{5z})$

3) Consideration of the interaction between the central ion and the ligands.

We have already elucidated the steps 1 and 2 above for an octahedral complex. It is now time to consider step 3.

In the crystalline field hypothesis the electrons of the central ion were considered to be entirely associated with this ion. The opposite extreme case is the one in which the electrons of the central ion and those of the ligands are so thoroughly mixed that there is practically equal probability of finding any electron on either the central ion or the ligands; this possibility would result in a completely covalent complex. The usefulness of the molecular orbital treatment is that, in effect, it can account not only for the extreme cases, but also for any intermediate case.

The interaction between the central ion and the ligands is clearly invariant under all the symmetry operations of the octahedral group, i.e. it transforms as the identity representation $\Gamma_{1g}(A_{1g})$ of the group O_h. The consequence of this is that a metal ion orbital and a ligand molecular orbital will interact if they belong to the same irreducible representation.

Taking now the central ion - ligand interaction into consideration, Table 3 suggests that the octahedral complex may have orbitals of three types:

1) Molecular orbitals with pure (non bonding) central metal ion character; for example t_{2g} in σ bonding.

2) Molecular orbitals with pure (non bonding) ligand ion character; for example t_{1g} and t_{2u}.

3) Molecular orbitals with mixed character arising from the interaction between central metal ion and ligand orbitals belonging to the same irreducible representation of O_h. Every time such interaction is possible a bonding and an antibonding levels will result.

We shall now examine first how σ bonds are established between the central ion and the ligands. The six ligand orbitals combinations transform according to A_{1g}, E_g and T_{1u}. The 3d ion orbitals transform according to T_{2g} and E_g (with T_{2g} lower in energy), the 4s orbital according to A_{1g} and the 4p orbitals according to T_{1u}. The A_{1g} ion orbital and the A_{1g} ligand orbital interact giving rise to an $a_{1g}(\sigma)$ bonding level and to an $a_{1g}(\sigma^*)$ antibonding level; in the same way the orbitals E_g give rise to the levels $e_g(\sigma)$ and $e_g(\sigma^*)$ and the orbitals T_{1u} give rise to the levels $t_{1u}(\sigma)$ and $t_{1u}(\sigma^*)$. On the other hand the three 3d ion orbitals $3d_{xy}$, $3d_{yz}$, and $3d_{zx}$ transforming according to T_{2g} are left unperturbed by the ligands and may be considered as pure central ion orbitals.

The resulting levels of the σ bond formation are shown in Figure 8, where the molecular orbital energy level diagram for the octahedral complex $[Fe(CN)_6]^{3-}$ is shown. In this complex 12 electrons from the six $(CN)^-$ ligands fill the three lower bonding levels and the five 3d electrons of Fe^{3+} occupy the non bonding

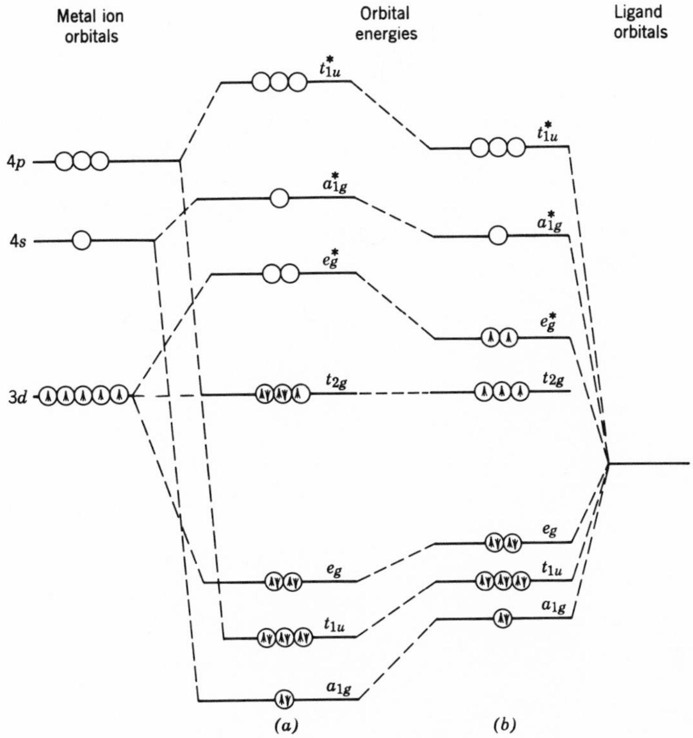

Figure 8. Molecular Orbital Energy Level Diagram of (a) $[Fe(CN)_6]^{3-}$, covalent, low spin; (b) $[FeF_6]^{3-}$, ionic, high spin.

t_{2g} level. In the same figure, on the right-hand side the molecular energy diagram of the $[FeF_6]^{3-}$ complex is shown. In this last case the smaller overlap between the F^- ligand σ orbitals and the Fe^{3+} orbitals produce a smaller covalent interaction; consequently, the $e_g(\sigma^*)$ level is closer to the non bonding t_{2g} level and the value of the crystalline field strength 10Dq(*) is smaller. The highly covalent case can be correlated to the low-spin strong field case of the crystalline field theory; the weakly covalent case can be correlated to the high-spin weak field case of the crystalline field theory. In the present σ-bonding scheme the crystalline field strength 10Dq is the difference in energy between the $e_g(\sigma^*)$ antibonding level and t_{2g} non bonding level.

Let us examine now the formation of π bonds in an AB_6 complex. The twelve ligand orbital combinations available for π bonding transform according to the representations T_{1u}, T_{2g}, T_{1g}, and T_{2u}. Now the T_{2g} ligand orbitals combine with the previously non bonding T_{2g} ion orbital and form the bonding $t_{2g}(\pi)$ and antibonding $t_{2g}(\pi^*)$ levels. The T_{1g} and T_{2u} ligand orbitals do not interact with any ion orbital and therefore preserve their ligand character. The ligand $T_{1u}(\pi)$ combinations interact with the $T_{1u}(4p)$ metal ion orbitals, which are already involved in the σ bonding scheme, and give rise to an additional level $t_{1u}(\pi)$. The t_{1u}^* antibonding level has, in this scheme, only a small ligand π character.

In the σ-π scheme, the crystal field strength 10Dq is the difference in energy between the $e_g(\sigma^*)$ level and the $t_{2g}(\pi^*)$ level. In Figure 9 we show the molecular orbital energy level scheme of an AB_6 σ-π bonded complex; we assume that each ligand participates in the bonding by using three orbitals and six electrons. The 36 electrons of the ligands are all located in the bonding and non bonding levels. The ion 3d electrons, not shown in the figure, are located in the $t_{2g}(\pi^*)$ and $e_g(\pi^*)$ levels.

The electronic configuration in the σ-π bonding scheme, when each ligand contributes six electrons and the metal ion has n electrons in the d shell, is given by

$$a_{1g}(\sigma)^2\, e_g(\sigma)^4\, t_{1u}(\sigma)^6\, t_{2g}(\pi)^6\, t_{1u}(\pi)^6\, t_{2u}(\pi)^6\, t_{1g}(\pi)^6$$

$$t_{2g}(\pi^*)^m\, e_g(\pi^*)^{n-m}$$

$$= {}^1A_{1g}\, t_{2g}(\pi^*)^m\, e_g(\pi^*)^{n-m}$$

where n electrons are distributed on the t_{2g} and e_g levels. An excited configuration is one in which an electron is brought from a lower to an upper level.

(*) 10Dq in crystalline field theory is the difference in energy between a t_{2g} orbital and an e_g orbital of a d^n ion in an octahedral symmetry site.(8)

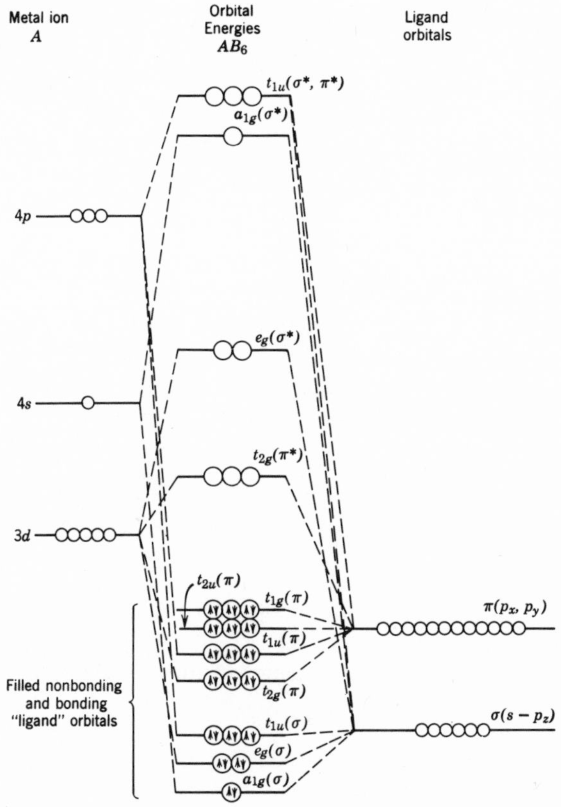

Figure 9. Molecular Orbital Energy Level Diagram of an AB_6 Octahedral Complex with σ and π Bonds (7) [reproduced from J. C. Ballhausen and Harry B. Gray, MOLECULAR ORBITAL THEORY: AN INTRODUCTORY LECTURE NOTE AND REPRINT VOLUME, copyright (c) 1965, W. A. Benjamin, Inc., New York and Amsterdam, by permission of the publisher].

V.E. Charge Transfer Spectra

Magnetic ions in crystals, in addition to the Laporte-forbidden transitions may present charge transfer spectra which can be classified as follows:

1. Ligand to metal transfer spectra, due to a transfer of an electron originally localized in a ligand orbital to the central ion.

2. Metal oxidation spectra, due to a transfer of an electron localized in the central ion to an excited ligand orbital that is not much mixed with orbitals of the central ion.

3. Rydberg spectra, due to a transfer of an electron localized in the central metal ion to an excited orbital that is not much mixed with the ligand orbitals.

4. Intraligand spectra, due to a transfer of an electron from a ligand orbital to another ligand orbital in a polyatomic ligand group. In this process only molecular orbitals of the ligands are involved.

The charge transfer spectra are, in general, Laporte-allowed, and therefore their intensity is much larger than the intensity of the intraconfiguration spectra. Most of the transitions have an f number of $\sim 10^{-1}$.

1. Ligand to Metal Transfer Spectra. In order to understand how these spectra come about, we refer to Figure 9 in which the molecular orbital energy level diagram of an octahedral AB_6 complex is represented.

A molecular orbital electronic configuration consists of a distribution of electrons on the various molecular orbital levels. Every electronic configuration produces a number of energy levels when the electrostatic interaction is taken into account. If all the bonding and non bonding levels are filled, the configuration $t_{2g}(\pi^*)^m\ e_g(\sigma^*)^{n-m}$ produces all the energy levels of crystalline field theory. Other electronic configurations not predicted by crystalline field theory with their accompanying energy levels are now possible. A charge transfer transition involves at least one level not predicted by the crystalline field theory. It is possible, for example, in principle to remove an electron from one of the bonding levels or from one of the pure ligand levels $t_{2u}(\pi)$ and $t_{1g}(\pi)$ and bring it to an antibonding level, such as $t_{2g}(\pi^*)$, if this is not completely filled, or to the level $e_g(\sigma^*)$. The ligand to metal transfer spectra are produced by transitions of this type.

We notice that selection rules are at work here, as for any other type of spectrum, and that the particular set of rules given by the specific symmetry of the complex is the valid one. Therefore $t_{2u}(\pi) \rightarrow t_{2g}(\pi^*)$, $e_g(\pi^*)$ transitions are allowed, for they connect states of different parities. If the metal ion contains six 3d electrons, the $t_{2g}(\pi^*)$ level is completely occupied, and the first relevant charge transfer transition is

$$\ldots t_{2u}(\pi)^6\, t_{1g}(\pi)^6\, t_{2g}(\pi^*)^6 \longrightarrow \ldots t_{2u}(\pi)^5\, t_{1g}(\pi)^6$$
$$t_{2g}(\pi^*)^6\, e_g(\sigma^*)^1.$$

The initial configuration gives a $^1A_{1g}$ level. In the final configuration we have to couple the two spins of the nonpaired electrons (S=0,1) and the two representations t_{2u} and e_g; this produces the energy levels $^1T_{1u}$, $^1T_{2u}$, $^3T_{1u}$, $^3T_{2u}$. Because the spin-selection rules are also at work and, for the O_h group the transitions $A_{1g} \rightarrow T_{1u}$, T_{2u} are both electric-dipole allowed, we expect two strong charge transfer bands due to these transitions. These bands appear in complexes like $[Co(NH_3)_5X]^{2+}$, where X = F^-, Cl^-, Br^-, I^-. We find that the energies of these bands decrease as we go from F^- to I^- because it is increasingly easier to remove an electron from the halogen ligand going from F^- to I^- (evidence of this is the decrease of electronegativity). When X = I^-, the charge transfer bands become so low that they overlap the crystalline field bands.

We notice that the transition $t_{1g}(\pi) \rightarrow e_g(\sigma^*)$ is Laporte-forbidden because it involves two states with the same parity. We expect a weaker band in correspondence to it.

If the metal ion has five electrons, the $t_{2g}(\pi^*)$ level is not completely filled, and Laporte-allowed transitions, such as $t_{2u}(\pi) \rightarrow t_{2g}(\pi^*)$ and $t_{1u}(\pi) \rightarrow t_{2g}(\pi^*)$, can take place; we expect these transitions to produce bands lower in energy than the $t_{2u}(\pi) \rightarrow e_g(\sigma^*)$ transition. It is found experimentally, for example, that the lowest allowed charge transfer band of $IrCl_6^{3-}$(Ir^{3+} is a $5d^6$ ion) is at about 45,000 cm^{-1} and is due to a $t_{1u}(\pi) \rightarrow e_g(\sigma^*)$ transition, whereas the lowest allowed charge transfer band, due to a $t_{1u}(\pi) \rightarrow t_{2g}(\pi^*)$ transition of $IrCl^{2-}$ (Ir^{2+} is a $5d^5$ ion), is at about 20,000 cm^{-1}. The difference in energy between the lowest allowed bands in the two cases should correspond to the crystalline field strength 10Dq.

For isoelectronic ions of the same series it is generally found that the higher the atomic number, the lower is the energy of the first charge transfer band. For isoelectric ions of different series the bands are found at higher energies going from the first to the second and third transition metal series; this is related to the fact that, as the principal quantum number of the d shell is increased, the metal ion is less stable towards oxidation; that is, more energy is required to produce a ligand-to-metal transition.

Tetrahedral complexes also present ligand-to-metal charge transfer spectra. The isoelectronic tetrahedral complexes MnO_4^-,

CrO_4^{2-}, VO_4^{4-} and TiO_4^{5-} have been examined experimentally by Teltow (9). The spectra of MnO_4^- and CrO_4^{2-} have also been studied by Wolfsberg and Helmholz (10), and two strong absorption bands in the visible and near ultraviolet have been attributed to ligand-to-metal ion electron transfer. As in the octahedral case, the energy of the first charge transfer band decreases with increasing atomic number of the metal ion and increases from the first to the second and third transition metal ion series.

2. Metal Oxidation Spectra. Spectra of this type may be observed when metal ions in low valence states are coupled with ligands with great electron affinity, as, for example, in the $Ru^{2+}(4d^6;\ t_{2g}^6)$ complex with pyridine as ligands; this complex presents an intense absorption band at $\sim$22,000 cm^{-1}, which is attributed to a transition $t_{2g}^6 \rightarrow t_{2g}^5\pi^*$, where π^* is an antibonding π orbital of the pyridine molecule. (Other lower and weaker bands are observed in this complex and attributed to $d \rightarrow d$ transitions; two very intense bands occuring at 35,000 to 45,000 cm^{-1} are attributed to ligand-to-ligand transitions.)

Similar spectra are observed in the Ru^{2+} complex with phenantroline as the ligand and also in $Ir^{3+}(5d^6;\ t_{2g}^6)$ complexes with pyridine and phenanthroline as ligands.

3. Rydberg Spectra. Rydberg spectra are found in lanthanides and actinides and are of the types $4f^n \rightarrow 4f^{n-1}5d$, $5f^{n-1}6d$, respectively. Ce^{3+} and Sm^{2+} present spectra of these types. Transition metal ions present $3d^n \rightarrow 3d^{n-1}\ 4p$ spectra in some square complexes such as $(PtCl_4)^{2-}$. The f numbers for these transitions are of the order 10^{-1}.

Divalent rare earth ions present strong absorption bands above groups of sharp lines. Such bands have been found by Butement (11) in divalent samarium, europium, and ytterbium and are explained by him as due to transitions from the $4f^n$ to the $4f^{n-1}5d$ configuration, whereas the sharp lines are assigned to transitions within the $4f^n$ configuration. This interpretation of the spectra is based on the fact that going from trivalent to divalent rare earths a lowering of the 5d orbital relative to the 4f orbital is expected. The oscillator strengths of the sharp lines are larger than in the trivalent ions, possibly because of a mixing of the $4f^n$ and $4f^{n-1}5d$ configurations due to asymmetric perturbations; for these transitions oscillator strengths with order of magnitude 10^{-4} have been found in $CaF_2:Sm^{2+}$.

4. Intraligand Spectra. If the ligands consist of polyatomic groups, strong absorption bands can be found in the spectra in correspondence to transitions involving only molecular orbitals with ligand character. These transitions are not, in general,

affected by the particular metal ion to which the ligands are bound. For this reason, in interpreting the spectra of metal ion complexes with polyatomic ligands, it is useful to compare them with the spectra of the ligands.

An example of a polyatomic ligand that produces this type of transition is the thiocyanate ion SCN^-.

References

1. E. P. Wigner, Group Theory and its Applications to Quantum Mechanics of Atomic Spectra, Academic Press, New York and London 1959.

2. V. Heine, Group Theory in Quantum Mechanics, Pergamon Press, New York, London, Oxford, Paris 1960.

3. M. Tinkham, Group Theory and Quantum Mechanics, McGraw Hill, New York - San Francisco - Toronto - London 1964.

4. C. J. Ballhausen, Introduction to Ligand Field Theory, McGraw Hill, London - New York 1962.

5. B. Di Bartolo, Optical Interactions in Solids, Wiley, New York, 1968.

6. H. L. Schlafer and G. Gliemann, Basic Principles of Ligand Field Theory, Wiley - Interscience, London, New York - Sydney - Toronto 1969.

7. C. J. Ballhausen and H. B. Gray, Molecular Orbital Theory, Benjamin, New York 1965.

8. D. S. McClure, "Electronic Spectra of Molecules and Ions in Crystals. Part II. Spectra of Ions in Crystals," in Solid State Physics, vol. 9, F. Seitz and D. Turnbull editors, Academic Press, New York 1959.

9. J. Teltow, "Das Liniehafte Absroptionsspektrum des Bichromations bei 20°K," Z. Physik Chem. B43, 375 (1939); also "Die Absorptionsspektrum des Permanganatchromat-, Vanadat- und Manganations in Kristallen," Z. Physik Chem. B43, 198 (1939).

10. M. Wolfsberg and L. Helmholz, "The Spectra and Electronic Structure of the Tetrahedral Ions MnO_4^-, CrO_4^{--} and ClO_4^-," J. Chem. Phys. 20, 837 (1952).

11. F. D. S. Butement, "Absorption and Fluorescence Spectra of Samarium, Europium and Ytterbium," Trans. Faraday Soc. 44, 617 (1948).

FUNDAMENTALS OF RADIATIVE TRANSITIONS OF IONS AND OF ION PAIRS IN SOLIDS

Ferd Williams

Department of Physics, University of Delaware

Newark, Delaware 19711

ABSTRACT

The interaction between radiation and ions in solids is analyzed semiclassically. The ions in the solids are treated quantum mechanically; the radiation, classically. Electronic transitions induced by time-dependent perturbations are considered generally and then the transition matrices for electromagnetic radiative transitions are determined. These transitions are related to the Einstein induced and spontaneous transition probabilities. Cooperative phenomena at high radiation densities are noted. Finally, the effects of irreversible phonon relaxation processes are considered.

I. INTRODUCTION

Our principal concern is with the interaction of atomic systems with electromagnetic radiation fields, specifically with the response of the electrons on ions in solids with such radiation fields. The analysis is semiclassical in the sense that the ions are treated quantum mechanically, whereas the electromagnetic radiation field is treated classically. After the more or less standard analysis for the interaction between radiation and matter and the calculation of transition matrices, we shall consider special effects which occur at high radiation densities and others due to the irreversible phonon relaxation processes which occur for ions in solids following radiative transitions.

II. TRANSITIONS INDUCED BY PERTURBATIONS PERIODIC IN TIME

Condensed matter consisting of ions in solids is herein approximated as a many-body system consisting of all the interacting electrons and nuclei. Before interaction with the radiation the system is assumed to be in the stationary state described by ψ_i, a solution to the time-independent Schrödinger equation. The wave function ψ_i is a function of the coordinates of all the electrons and all the nuclei, thus $\psi_i(\underline{r},\underline{R})$. The solution to the time-dependent Schrödinger equation for stationary states specifically the initial state is:

$$\Psi_i = \psi_i(\underline{r},\underline{R})e^{-(i/\hbar)E_i t} \tag{1}$$

where E_i is the eigenvalue of the Hamiltonian H of the many-body system.

After turning on the periodic perturbation H' the system is no longer in the stationary state described by eq. (1) but rather evolves in accordance with the time-dependent Schrödinger equation:

$$i\hbar \frac{\partial\Psi}{\partial t} = (H + H')\Psi \tag{2}$$

where Ψ is describable as a linear combination of the wave functions for the n stationary states of the system:

$$\Psi = \sum_n a_n(t)\Psi_n. \tag{3}$$

We substitute eq. (3) into eq. (2), use the orthonormality of the Ψ_n and find that the system transforms into the final stationary state Ψ_f as follows:

$$i\hbar \frac{da_f}{dt} = V_{fi}(t) \tag{4}$$

where $V_{fi}(t)$ is the matrix element connecting initial and final states:

$$V_{fi}(t) = \iint \Psi_f^* H' \Psi_i \underline{drdR} = V_{fi} e^{(i/\hbar)(E_f - E_i)t} \tag{5}$$

where $V_{fi} = \int \psi^* H' \psi \underline{drdR}$, containing only the explicit time-dependence of H'.

The radiation is assumed to be monochromatic and the periodic perturbation can therefore be written:

$$H' = Fe^{-i\nu t} + Ge^{i\nu t} . \tag{6}$$

We substitute eq. (6) into eq. (5), solve for $V_{fi}(t)$ which we then substitute into eq. (4) and obtain:

$$a_{fi} = \frac{F_{fi}\exp\{(i/\hbar)(E_f - E_i - h\nu)t\}}{(E_f - E_i - h\nu)} - \frac{F^*_{fi}\exp\{i/\hbar)(E_f - E_i + h\nu)t\}}{(E_f - E_i + h\nu)} \tag{7}$$

where the second subscript i indicates the state from which the system evolved. For the absorption of radiation the first term in eq. (7) dominates; for the emission, the second term. The squared modulus $|a_{fi}|^2$, which gives the probability of the transformation from ψ_i to ψ_f, can be shown to contain a factor which satisfies all the requirements of a Dirac delta function (1). For large t, that is $t >> 1/\nu$, the rate of the transition is:

$$\frac{d|a_{fi}|^2}{dt} = \frac{2\pi}{\hbar} |F_{fi}|^2 \delta(E_f - E_i \mp h\nu)\eta(\nu) \tag{8}$$

where $\mp$ refer to absorption and induced emission, respectively, and $\eta(\nu)$ is the density of final states. The transition obviously occurs only with energy conservation: $h\nu = h\nu_{fi}$, where ν_{fi} is of course the frequency of the radiation corresponding to the energy difference between the initial and final states.

III. TRANSITIONS INDUCED BY THE ELECTROMAGNETIC RADIATION FIELD

An electromagnetic field is described by the vector potential $\underline{A}$ and the scalar potential ϕ. For a pure radiation field with no local sources $\phi = 0$ and $\underline{\nabla}\cdot\underline{A} = 0$. From the equation of motion of the electrons of mass μ and charge e of our system the contribution to the Hamiltonian by the radiation field can be shown to be:

$$H' = -\frac{1}{2\mu}\left(\frac{e}{c}\underline{A}\cdot\underline{p} + \frac{e}{c}\underline{p}\cdot\underline{A} - \frac{e^2}{c^2}|\underline{A}|^2\right). \tag{9}$$

From the quantum-mechanical transformation of $\underline{p} \rightarrow -i\hbar\underline{\nabla}$ and $\underline{\nabla}\cdot\underline{A} = 0$, and neglecting the $|\underline{A}|^2$ term, we obtain the following approximate form for the matrix element F_{fi} of (8):

$$F_{fi} \approx \int \psi_f^* \left| -\frac{e}{\mu c} \underline{A}\cdot\underline{p} \right| \psi_i \, \underline{dr} \, \underline{dR} \,. \tag{10}$$

For radiation of wavelengths long compared to the dimensions of the absorbing or radiating system, the approximation of constant vector potential can be made, $\underline{A} = \underline{A}_o \exp\{ikr\} \approx \underline{A}_o$, and after transformation to spatial coordinates the well-known dipole approximation to the radiative transition rate is obtained,

$$\frac{d|a_{fi}|^2}{dt} = \frac{e^2\pi^2\nu_{fi}^2}{c^2\hbar^2} |\underline{A}_o|^2 |\underline{r}_{fi}|^2 \eta(\nu_{fi}) \,. \tag{11}$$

From electromagnetic theory it can be shown that the vector potential may be expressed in terms of the radiation density $\rho(\nu_{fi})$:

$$|\underline{A}_o(\nu_{fi})|^2 = \frac{2}{3} \frac{c^2}{\pi\nu_{fi}^2} \rho(\nu_{fi}) \tag{12}$$

and with this substitution in eq. (11) we recognize the Einstein induced transition B_{fi}, which is the radiative transition rate with unit radiation density:

$$B_{fi} = \frac{2}{3} \frac{e^2\pi}{\hbar^2} |\underline{r}_{fi}|^2 \eta(\nu_{fi}) \quad . \tag{13}$$

IV. SPONTANEOUS RADIATIVE TRANSITIONS

A system in an excited state can radiate in the absence of an electromagnetic field. Therefore, spontaneous emission must be included. A direct analysis of spontaneous emission requires quantization of the electromagnetic field. We shall apply the approach used by Einstein (2) in his classic work on induced and spontaneous radiative transitions to determine the spontaneous transition rate.

In the steady-state, the excitation rate must equal the de-excitation rate. For a two state system in a radiation field $\rho(\nu_{\ell m})$ with ℓ the higher and m the lower state, we have:

$$B_{m\ell}N_m\rho(\nu_{\ell m}) = B_{\ell m}N_\ell\rho(\nu_{\ell m}) + A_{\ell m}N_\ell \tag{14}$$

where $A_{\ell m}$ is the Einstein spontaneous transition coefficient and N_m and N_ℓ are the populations in the m and ℓ states respectively.

From the principle of detailed balance $B_{m\ell} = B_{\ell m}$, therefore,

$$\rho(\nu_{\ell m}) = \frac{A_{\ell m}/B_{\ell m}}{[(N_m/N_\ell) - 1]} \quad . \tag{15}$$

This formula applies generally to any steady-state (time-independent populations), therefore, it must apply to a body in equilibrium with its own thermal radiation. Thus:

$$N_\ell/N_m = e^{-h\nu_{\ell m}/kT} \quad . \tag{16}$$

Substitution of eq. (16) into eq. (15) yields:

$$\rho(\nu_{\ell m}) = \frac{A_{\ell m}/B_{\ell m}}{e^{h\nu_{\ell m}/kT} - 1} \tag{17}$$

and comparing eq. (17) with Planck's black body formula, we have:

$$A_{\ell m}/B_{\ell m} = 8\pi h\nu_{\ell m}^3/c^3 \quad . \tag{18}$$

Substitution of eq. (13) into eq. (18) gives:

$$A_{\ell m} = \frac{64\pi^4 e^2 \nu_{\ell m}^3}{3hc^3} |\underline{\mathbf{r}}_{fi}|^2 \eta(\nu_{\ell m}) \tag{19}$$

This result for spontaneous emission can be obtained using quantized field theory, as was originally done by Fermi (3).

From the preceding analysis the total radiative transition rate for emission is, therefore, the following:

$$1/\tau = B_{fi}\rho(\nu_{fi}) + A_{fi} \tag{20}$$

where τ is the radiative lifetime.

V. APPLICATION OF THE THEORY TO RADIATIVE TRANSITIONS OF IONS IN SOLIDS

The preceding analysis is strictly applicable to free ions and to zero-phonon transitions in solids, where the only interaction of the atomic system is with the radiation field. It is a good approximation for the very weak coupling case, such as trivalent rare earth ions, as discussed in a previous chapter.

For the weak field case, such as transition metal ions, and especially for the strong field case, there is coupling of the system with phonons, in addition to the coupling with the radiation field. This problem has been considered by Fowler and Dexter (4). From the adiabatic approximation, discussed in a previous chapter, there is an irreversible phonon relaxation process following optical absorption in polar materials, and therefore $\nu_{\ell m} \neq \nu_{m\ell}$. This is of course evident in the configuration coordinate model. Also, because of the adiabatic changes in the electronic wave function ψ^{el} accompanying this relaxation, $B_{\ell m} \neq B_{m\ell}$. In other words, the radiative de-excitation transition occurs for a different condition of lattice polarization, compared to the radiative excitation transition. Both the transition energy and the transition matrices are affected. This is, of course, responsible for one of the most striking characteristics of luminescence in solids. That is, the emission occurs in a spectral region different from the absorption, and the material is to a good approximation transparent to its own luminescent emission. This is, of course, the basis for Stokes' law and is a clear departure from Kirkoff's law relating emissivity to absorptivity. Luminescence of ions in solids can be interpreted by the preceeding analysis only approximately and with modifications that take account of coupling with phonons.

Under ordinary excitation intensities, the dominant term in the radiative transition rate for emission is according to eq. (20) the spontaneous transition. In this case, τ is the spontaneous radiative lifetime. With conditions of high radiation densities, such as occur with laser excitation, the induced transition rate for emission may dominate. Under these circumstances the term $(e^2/c^2)|\underline{A}|^2$ which was neglected in obtaining eq. (11) may be appreciable. Also, in the case of multiphoton processes, Lambropoulos (5) has noted that various channels via virtual and real intermediate states must be summed over to obtain the total transition rate between two stationary states. Finally, the approximation of a constant vector potential used in obtaining eq. (11) becomes less valid for shorter wavelength radiation interacting with impurities with large dimensions, such as donor-acceptor pairs with large interimpurity distances, that is, effective mass quasi-ion pairs.

REFERENCES

1. See L. D. Landau and E. M. Lifshitz, "Quantum Mechanics" (Pergamon Press, Oxford 1958),p. 146-147.

2. A. Einstein, Phys. Zeits. f. 18, 121 (1917).

3. E. Fermi, Rev. Mod. Phys. 4, 87 (1932).

4. W. B. Fowler and D. L. Dexter, Phys. Rev. 128, 2154 (1962); J. Chem. Phys. 43, 1768 (1965).

5. P. Lambropoulos, Bull. Am. Phys. Soc. 19II, 516 (1974).

ABSORPTION AND EMISSION SPECTRA

D. CURIE

Luminescence Laboratory

University of Paris, France

ABSTRACT

A general account of electron-phonon coupling effects on emission and absorption spectra (shape of the spectra and bandwidth) is presented. The purpose of this paper is mainly to show that the theoretical results which happen to be the most important ones for comparison with experiments (for instance the Gaussian spectra in the scope of the semi-classical approximation, the Huang and Rhys model and the detailed shape of the one-phonon replica band), which are generally obtained as a result of more or less highly sophisticated calculations, can indeed be accounted for in a very straightforward way if suitable hypotheses are accepted. There is also a discussion of some experimental results: large band spectra in II-VI compounds (the case of ZnS:Mn is specially examined), edge emission in CdS and R-line emission of ruby.

I. THE ONE-DIMENSIONAL CONFIGURATIONAL COORDINATE MODEL

I.A. Introduction

The use of one-dimensional configurational curves for describing absorption and emission phenomena in phosphors (Mott and Seitz) can merely lead to approximate results. However it is still useful in numerous cases.

Theoretically, this model is valid if and only if the interaction between electrons and phonons involves especially one vibrational mode. Typically, this situation can occur in two cases:

a) One local mode is highly predominant. For instance, let us consider the case of KCl:Tl which has been extensively studied by F. Williams and co-workers (1).

The electronic transitions occur between $^3P_1$ (or $^1P_1$) and $^1S_0$ states of the Tl^+ ion. This ion is placed in a substitutional K^+ site and, in the first approximation, only the "breathing mode" is taken into account; i.e., the six neighboring Cl^- ions are vibrating symmetrically around the Tl^+ center.

Let r be the value of the $Tl^+ - Cl^-$ distance, r_{eq} the equilibrium value of r in the fundamental state. The configurational coordinate is $Q = r - r_{eq}$.

b) No local mode occurs, but only the interaction with certain band mode phonons (i.e., in most cases, L.O. phonons) is taken into account. In such a case, the configurational coordinate Q is the normal coordinate associated with this vibrational mode, and it is not related to the position of one specific ion.

However, it is found that the model can still be used to provide satisfactory results even if the above hypotheses are not totally valid.

I.B. The Emission and Absorption Peaks (Fig. 1)

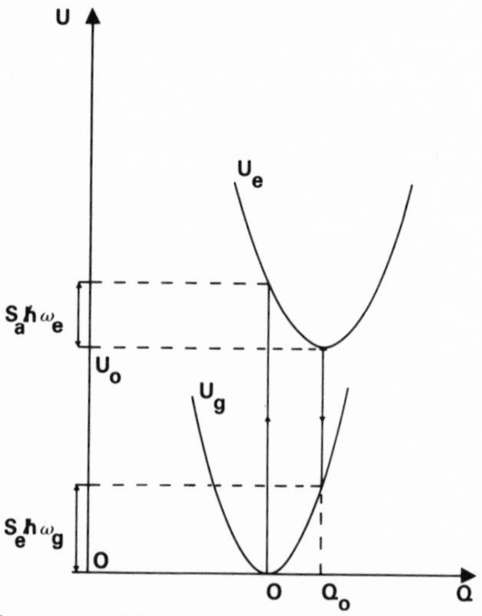

Figure 1 The one-dimensional configuration diagram.

Let us recall the usual notation. The potential energy curves are:

(a) for the ground state:

$$U_g(Q) = \tfrac{1}{2}k_g Q^2 \;;$$

(b) for the excited state:

$$U_e(Q) = \tfrac{1}{2}k_e\,(Q-Q_o)^2 + U_o.$$

U_o is the electronic energy difference between the excited state and the ground state, and Q_o is the change in equilibrium position of the ions between the two states.

We shall denote by $\hbar\omega_g$ and $\hbar\omega_e$ the quantum vibrational energies and by $h\nu$ the energy of emitted and absorbed photons.

The peak positions of emission and absorption bands are given by vertical transitions in the diagram (Franck-Condon principle):

peak position for emission : $h\nu_{oe} = U_o - \frac{1}{2} k_g Q_o^2$;

peak position for absorption : $h\nu_{oa} = U_o + \frac{1}{2} k_e Q_o^2$.

It will be useful to introduce the parameters S_e and S_a, defined by :

$$\frac{1}{2} k_g Q_o^2 = S_e \hbar\omega_g \qquad \frac{1}{2} k_e Q_o^2 = S_a \hbar\omega_e \tag{1}$$

Thus :

$$\begin{aligned} h\nu_{oe} &= U_o - S_e \hbar\omega_g \\ h\nu_{oa} &= U_o + S_a \hbar\omega_e \end{aligned} \tag{2}$$

whence the Stokes shift.

In this treatment, we have neglected zero-point energies in the initial and final states (See below paragraph I.D.2.). This assumption is valid for large values of the S's. S_a is the mean number of phonons emitted after the photon absorption, S_e the same after the photon emission.

I.C. The Shape of Emission and Absorption Spectra

This shape can be readily shown to be Gaussian if we accept both the Condon approximation and the semi-classical approximation. For the dipole case the emitted spectrum, for instance, is given by :

$$I(h\nu) = \frac{64\ \pi^4\ \nu^4}{3\ c^3} \sum \text{weighted}\ |M_{if}|^2 \ .$$

While performing this summation, it must be recalled that each possible initial state is weighted by an appropriate Boltzmann factor.

It will be convenient to introduce throughout the present paper the function $G(h\nu)$ defined by

$$G(h\nu) = \sum \text{ weighted } |M_{if}|^2$$

$$I(h\nu) = \frac{64\,\pi^4\,\nu^4}{3\,c^3}\,G(h\nu). \qquad (3)$$

For the sake of simplicity, $G(h\nu)$ is often denoted as the "shape of the spectrum". The ν^4 correction does not change the observed shape of the emission spectrum very much, however in order to make a convenient comparison between experimental and theoretical results, it is more correct to begin with the computation of $I(h\nu)/\ ^4$. In the semi-classical approximation, for instance, $G(h\nu)$ is Gaussian and $I(H\nu)$ is slightly skewed.

For the absorption cross-section $\sigma(h\nu)$ the correction consists only in dividing σ by the first power of ν ; the integrated cross-section is :

$$\int \sigma(h\nu)\,d(h\nu) = \frac{8\,\pi^3\,\nu}{3\,c}\,G(h\nu). \qquad (4)$$

If the vibrating frequencies in the ground state and the excited state are equal ($\hbar\omega_g = \hbar\omega_e$), then mirror symmetry occurs between the functions $G(h\nu)$ involved in absorption and emission processes.

<u>The Condon approximation</u> employs the assumption:

$$M_{if} = M_{eg} \int \psi^*_{en}\,(Q-Q_o)\,\psi_{gm}(Q)\,dQ\,.$$

M_{eg} is a matrix element depending only on the electronic states e and g, and the overlap integral is computed from the harmonic oscillator eigenfunctions in the respective m- and n-th vibrational states :

$$\psi_{gm}(Q) = N_m \exp\left[-\frac{1}{2}\left(\frac{Q}{a_g}\right)^2\right] H_m\left(\frac{Q}{a_g}\right)$$

$$\psi_{en}(Q-Q_o) = N_n \exp\left[-\frac{1}{2}\left(\frac{Q-Q_o}{a_e}\right)^2\right] H_n\left(\frac{Q-Q_o}{a_e}\right);$$

a_g and a_e are the classical amplitudes at zero-point energy level.

<u>The semi-classical approximation</u> consists in treating classically the final state, i.e. for the emission process the state ψ_{gm}. This approximation is valid for large values of S_e.

In this case $|\psi_{gm}(Q)|^2$ possesses a very sharp maximum around the classical amplitude Q_m given by

$$U_g(Q_m) = \frac{1}{2} k_g Q_m^2 = (m + \frac{1}{2}) \hbar \omega_g$$

and the square of the matrix element M_{if} is assumed to be proportional to $|\psi_{en}(Q_m-Q_o)|^2$.

At 0^oK temperature emission takes place from the state $n = 0$; ψ_{eo} is a Gaussian and so is $|M_{if}|^2$:

$$|M_{if}|^2 = \text{const}\cdot\exp\left[-\left(\frac{Q_m-Q_o}{a_e}\right)^2\right]. \qquad (5)$$

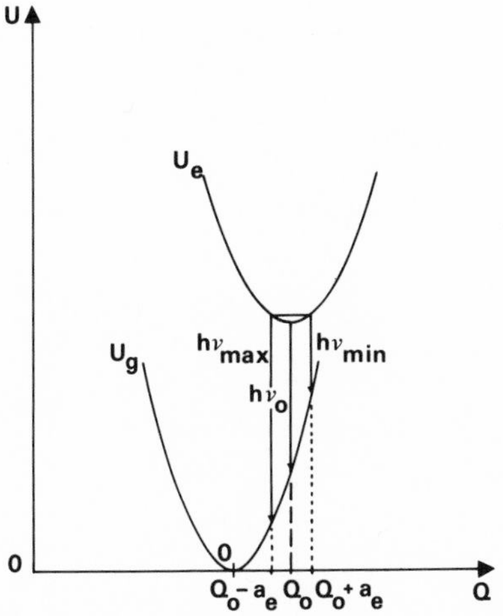

Figure 2 Transitions in the emission spectrum at 0^oK. a_e is the classical amplitude at zero-point energy, $h\nu_o$ the peak position of the emitted spectrum, and $h\nu_{min}$ and $h\nu_{max}$ the "tails" of this spectrum.

Fig.2 shows the transitions which occur respectively at $Q_m = Q_o + a_e$ and $Q_m = Q_o - a_e$. If moreover, $a_e << Q_o$, we may differentiate $U_g(Q)$ in the vicinity of $Q = Q_o$ (i.e. we replace the parabola by a straight line), and we have the energies of the emitted corresponding photons :

$$h\nu_{min} = h\nu_o - k_g Q_o a_e \qquad h\nu_{max} = h\nu_o + k_g Q_o a_e \ .$$

The emitted spectrum extends grossly from $h\nu_{min}$ to $h\nu_{max}$, with a peak for $h\nu = h\nu_o$. More accurately, the frequencies ν_{min} and ν_{max} are those for which the square of the matrix element M_{if} is reduced to 1/e of its maximum value (Fig.3).

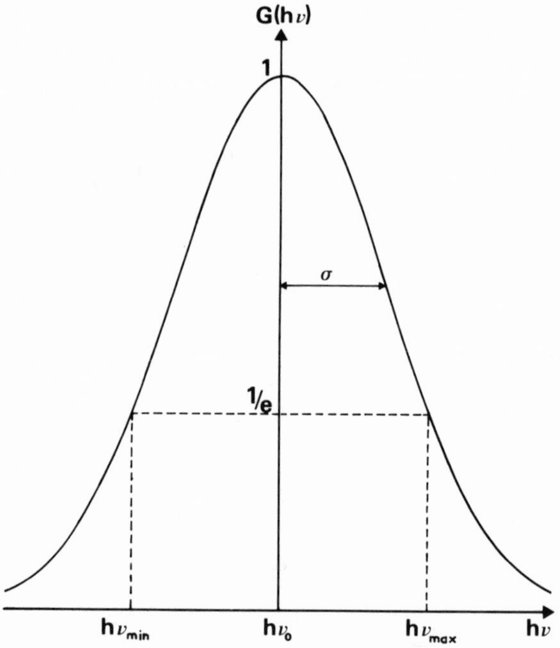

Figure 3 The shape of the spectrum.

The second moment σ_e is then shown to be given by :

$$2\,\sigma_e^2 = k_g^2\, Q_o^2\, a_e^2 \tag{6A}$$

or

$$\sigma_e^2 = S_e \frac{(\hbar\omega_g)^3}{\hbar\omega_e} \tag{6B}$$

The same calculation for absorption leads to :

$$2\,\sigma_a^2 = k_e^2\, Q_o^2\, a_g^2 \tag{7A}$$

$$\sigma_a^2 = S_a \frac{(\hbar\omega_e)^3}{\hbar\omega_g}\,. \tag{7B}$$

For higher temperatures emission can take place from any level $n \neq 0$. The calculation can be performed easily by means of Mehler's formula upon the Slater sums of the harmonic oscillator, and we shall not reproduce it (see for instance a fast derivation in (2)).

In the limits of the above approximations, emission and absorption spectra are still Gaussian, the peak positions remain unchanged and the bandwidth variation versus temperature are given by the well-known formulas

$$\sigma_e(T) = \sigma_e(0^oK) \left[\tanh \frac{\hbar\omega_e}{2\,kT}\right]^{-1/2} \tag{8}$$

$$\sigma_a(T) = \sigma_a(0^oK) \left[\tanh \frac{\hbar\omega_g}{2\,kT}\right]^{-1/2}\,. \tag{9}$$

These formulas are easily understood if we replace in (6A) and (7A) the amplitudes a_e and a_g relevant to zero-point energies by the mean-square roots $<a_e(T)>$ and $<a_g(T)>$:

$$< a_e(T) >^2 = \sum_{n=o}^{\infty} a_e^2(n) \frac{e^{-n\hbar\omega_e/kT}}{\sum_o^{\infty} e^{-n\hbar\omega_e/kT}} \;;$$

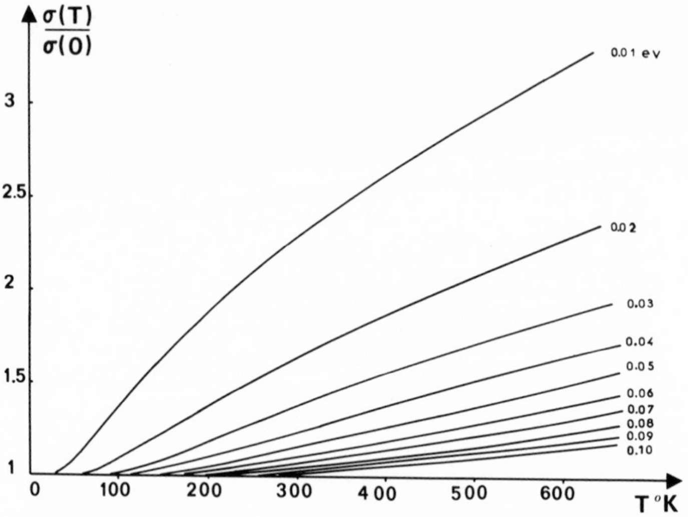

Figure 4 This net of curves gives by inspection the value of the vibrational quantum $\hbar\omega$ from experimental data of the variation of $\sigma(T)$ with temperature. The emission spectrum leads to ω_e, the absorption spectrum to ω_g.

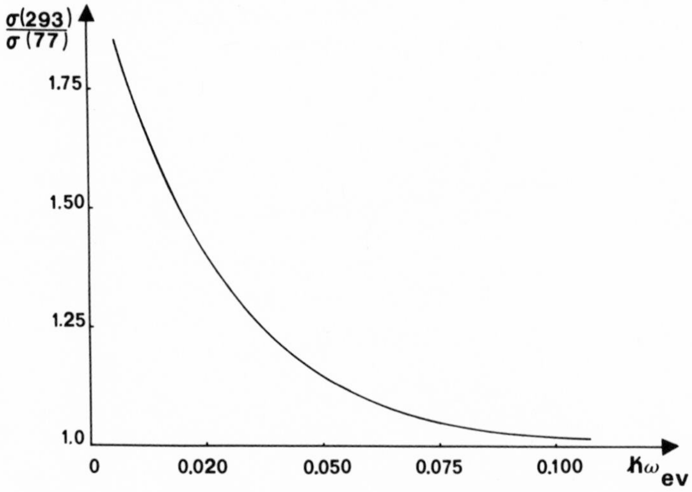

Figure 5 When experimental data of the bandwidth variation with T are not available at all temperature, $\hbar\omega$ can however be found from the ratio R of bandwidths measured respectively at room temperature and at liquid nitrogen temperature. But this procedure assumes formulas (8) or (9) to be valid; otherwise it leads to erroneous results.

$a_e(n)$ is the classical amplitude of the vibration in the n-th vibrational state :

$$\frac{1}{2} k_e a_e^2(n) = (n + \frac{1}{2}) \hbar\omega_e$$

The calculation is straightforward and gives :

$$< a_e(T) > = a_e \left[\tanh \frac{\hbar\omega_e}{2kT} \right]^{-1/2} \qquad a_e \equiv a_e(n = o) \quad ,$$

whence formula (8).

The formulas (2), (6) and (7), and (8), (9) allow us to compute the configurational diagram from the experimental emission and absorption spectra. Four parameters are needed in order to draw the configuration curves (U_o, Q_o, k_e and k_g). Therefore if emission and absorption (or excitation) spectra are known at several temperatures, the number of experimental data is redundant and we can not only obtain the diagram , but make sure it is valid (or, in other cases, find that it is not !). The value of the activation energy for thermal quenching can also be used for this purpose.

In some cases it happens that the number of accurately known parameters is too small to allow us to make such computations. In practice the formulas (8) and (9) are most useful, for they give immediately the values of the vibrational quanta $\hbar\omega_e$ and/or $\hbar\omega_g$ (Fig.4 and Fig.5).

I.D. A Study of Some Particular Cases

Two special cases allow the performance of calculations more exact than those based on the above approximations.

1. <u>The Case of Linear Coupling</u>. The vibrational frequencies in both excited and ground state are assumed to be equal:

$$\hbar\omega_e = \hbar\omega_g = \hbar\omega \ .$$

Such an approximation is perhaps particularly satisfactory for a shallow center interacting only with L.O. phonons ($\hbar\omega = \hbar\omega_{LO}$).

In this case the mean number of phonons emitted after optical emission and absorption are equal :

$$S_e = S_a = S$$

and formulas (2), (6B) and (7B) become, respectively :

$$h\nu_{oe} = U_o - S\,\hbar\omega \qquad\qquad h\nu_{oa} = U_o + S\,\hbar\omega$$

$$\boxed{\sigma_e^2 = \sigma_a^2 = S\,(\hbar\omega)^2 .} \tag{10}$$

The formulas valid in this special case have been well known for many years (Lax, Pekar, Kubo and many others).

Moreover, if we still accept the Condon approximation, the overlap integral

$$\int \psi^*_{en}\,(Q-Q_o)\,\psi_{gm}\,(Q)\,dQ = \Lambda^p_n\,(S)\;,$$

can be expressed in terms of the associated Laguerre functions

$$\Lambda^p_n\,(S) = \frac{n!\,e^{-S}S^p}{(n+p)!}\,L^p_n\,(S)\;,$$

without making use of the semi-classical approximation. Here $p = m-n$ and L^p_n is a Laguerre polynomial.

The summation over all emitting vibrational levels can be performed by means of the Myller-Lebedeff formula (2) and the result which is obtained in this way is identical to Huang's and Rhys' (see below, section II).

2. The Case of Quadratic Coupling. Pure quadratic coupling would occur if $Q_o = 0$ (whence $S = 0$) but $\hbar\omega_e \neq \hbar\omega_g$.

At low temperature, the emission spectrum would consist of a zero-phonon line ($m = n = 0$) whose energy $h\nu_o$ is given by

$$h\nu_o = U_o + \frac{1}{2}\,(\hbar\omega_e - \hbar\omega_g) \tag{11}$$

and a series of lines corresponding to the transitions from the level $n = 0$ to $m = 2, 4, \ldots, 2k, \ldots$ (for odd values of m the overlap integral vanishes for parity reasons). The energies of these lines are:

$$h\nu_{m=2k} = h\nu_o - 2k\,\hbar\omega_g \;.$$

In the same way, the absorption spectrum would consist of the above zero-phonon line (11) and the series of lines :

$$h\nu_{n=2k'} = h\nu_o + 2k'\,\hbar\omega_e \;.$$

If of course both conditions $Q_o = 0$ and $\hbar\omega_e = \hbar\omega_g$ would happen to be satisfied simultaneously, then the low temperature spectra would contain only the line (n = 0, m = 0). All other matrix elements would vanish as a result of the orthogonality of harmonic oscillator wave functions.

Experimentally, we are not aware of any case of purely quadratic coupling in luminescent phosphors. But the actual case ($Q_o \neq 0$, $\hbar\omega_e \neq \hbar\omega_g$) can be considered as a superposition of both linear and quadratic coupling.

Let us again consider the case T = 0°K. We turn back to formulas (1) and (2), which are apprxoimations valid for large values of S_e and S_a. More rigorously, if we still denote by S_e the mean number of phonons to be emitted after the peak optical transition $h\nu_{oe}$, we must write:

$$\frac{1}{2} k_g Q_o^2 = (S_e + \frac{1}{2}) \hbar\omega_g$$

and

$$h\nu_{oe} = U_o + \frac{1}{2} (\hbar\omega_e - \hbar\omega_g) - S_e \hbar\omega_g \cdot$$

We see, while applying (2), that we must add to U_o the same correction as in (11). The same applies for the absorption spectrum.

I.E. Simple Improvement of the Semi-Classical Approximation: H. Payen de la Garanderie's Law of the Rectilinear Diameter

We consider the general case ($Q_o \neq 0$, $\hbar\omega_e \neq \hbar\omega_g$) ; we still accept the Condon approximation and the semi-classical approximation as in I.C. Therefore we start from:

$$I(h\nu) = \frac{64 \pi^4 \nu^4}{3 c^3} G(h\nu) , \qquad (3)$$

while G(hν) is a Gaussian function of the configurational coordinate Q :

$$G(h\nu) = \text{const. } \exp \left[- \left(\frac{(Q-Q_o)}{<a_e(T)>} \right)^2 \right] , \qquad (12)$$

for emission. For T = 0°K, $<a_e(T)>$ equals the classical amplitude a_e at zero-point energy level.

In I.C., we accepted a further approximation, i.e., we differentiated $U_g(Q)$ at $Q = Q_o$: then the energy of emitted photons turns out to be a linear function of Q, and $G(h\nu)$ is therefore a Gaussian function of $h\nu$ itself.

H. Payen de la Garanderie (3) did not use this last approximation, but still used equation (12). Simple formulas are obtained in this way, which are valid for most non-Gaussian spectra.

$G(h\nu)$ has the same values for transitions occurring at

$$Q_1 = Q_o + \Delta Q \qquad \text{and} \qquad Q_2 = Q_o - \Delta Q \ ,$$

whatever ΔQ may be. Let $h\nu_1$ and $h\nu_2$ be the respective energies of these transitions:

$$h\nu_1 = U_o + \frac{1}{2} k_e \ (\Delta Q)^2 - \frac{1}{2} k_g \ (Q_o + \Delta Q)^2$$

$$h\nu_2 = U_o + \frac{1}{2} k_e \ (\Delta Q)^2 - \frac{1}{2} k_g \ (Q_o - \Delta Q)^2 \ .$$

We have from (12) :

$$G(h\nu_1) = G(h\nu_2) = G(h\nu_o) \exp \left[- \left(\frac{\Delta Q}{<a_e(T)>} \right)^2 \right]$$

From these premises, it is easy to show that the graph $\ell n\ G(h\nu)$ plotted versus $h\nu$ possesses a rectilinear diameter (Fig.6).

This means the middle M of points A_1 and A_2 has for abscissa :

$$\frac{h\nu_1 + h\nu_2}{2} = h\nu_o + \frac{1}{2} (k_e - k_g) \ (\Delta Q)^2$$

and for ordinate :

$$\ell n\ G(h\nu) = \ell n\ G(h\nu_o) - \left(\frac{(\Delta Q)}{<a_e(T)>} \right)^2 ,$$

and by eliminating $(\Delta Q)^2$ between these relations we see that the different positions of M which are obtained for all values of ΔQ lie on a striaght line whose slope is:

$$m = \tan\ \alpha = - 2 \ / \ (k_e - k_g) <a_e(T)>^2 \ .$$

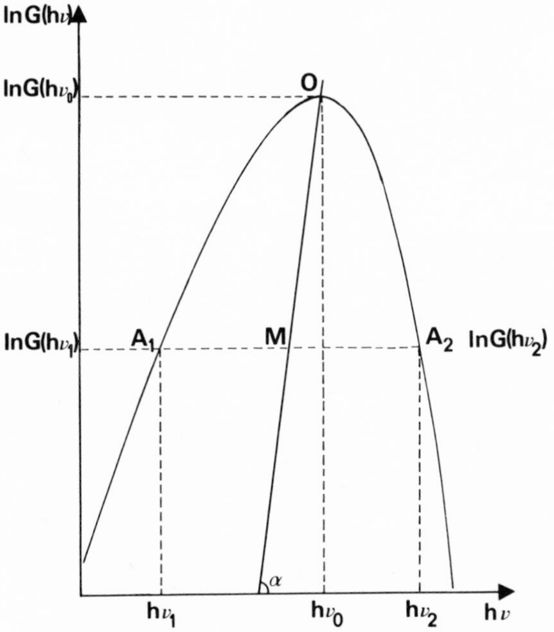

Figure 6 The rectilinear diameter (see text).

If $k_e < k_g$ this slope is positive, as is seen in Fig. 6. Both cases, m positive and m negative, are found experimentally.

When temperature changes, this slope would satisfy the following law :

$$m(T) = m(0^{\circ}K) \tanh \frac{\hbar\omega_e}{2kT} \quad . \qquad (13)$$

This formula and the value of m_0 can be of help in drawing the configurational diagram, especially in cases where data on the absorption or excitation spectra are missing.

However we must emphasize that many other assumptions can also lead to a rectilinear diameter : for instance, the Pekarian curve, i.e. Poisson's formula, may also result in a linear diameter (see II.D. below).

Experimentally, it is found that this "law of the rectilinear diameter" is satisfied in most cases on non-Gaussian spectra with a surprisingly good accuracy (Fig.7).

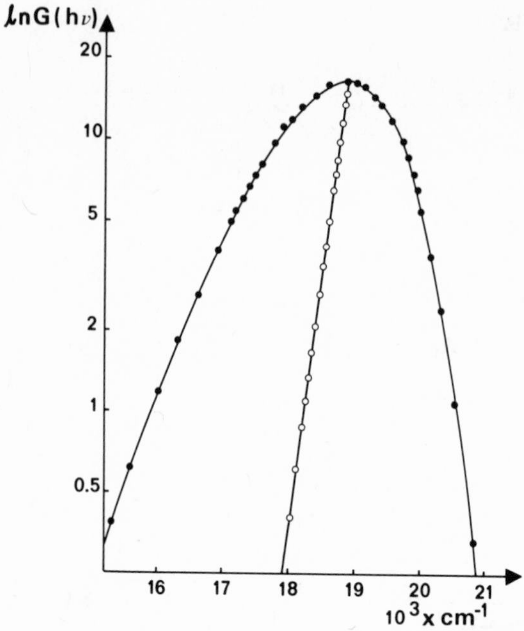

Figure 7 The rectilinear diameter : experimental data on the $MnCl_2$ + α - picoline complex.

On the contrary, formula (13) happens to be satisfied only in a few cases.

II. THE HUANG AND RHYS MODEL

Huang's and Rhys' calculations, originally performed for F-center absorption, give an accurate treatment of both absorption and emission phenomena in the case where it is assumed that all the lattice vibrational modes have the same frequency.

A straightforward derivation of the Huang and Rhys formula can be obtained on the basis of the following assumptions.

First, as mentioned above, we accept $\hbar\omega_e = \hbar\omega_g = \hbar\omega$. Then if U_o is the electronic energy difference between the excited state and the ground state (Fig.8), the emitted (or absorbed) photon energies are :

$$h\nu_{nm} = U_o + (n + \frac{1}{2})\,\hbar\omega - (m + \frac{1}{2})\,\hbar\omega \quad . \qquad (14)$$

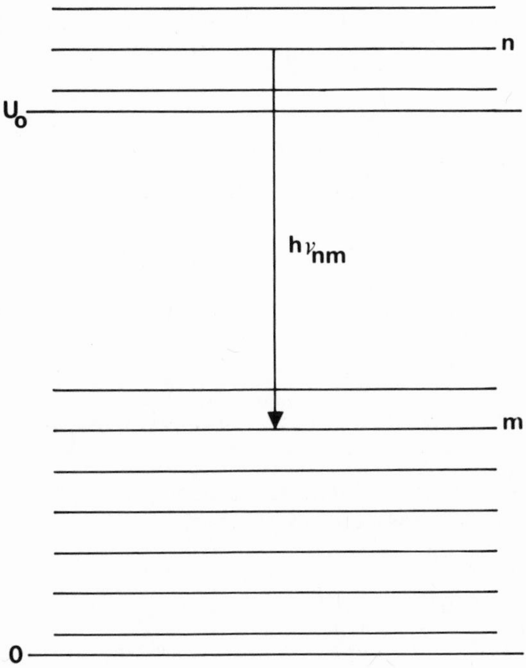

Figure 8 Transitions in the Huang and Rhys model. Note that in the n-th excited state level we have to deal with n vibrating oscillators, each of them corresponding to a phonon energy, while in the treatment of section I one harmonic oscillator is vibrating in the n-th quantum state.

II. A. The Emission Spectrum

In emission, m phonons are created, n phonons are annihilated; p = m - n is the gross number of created phonons and

$$h\nu_{nm} = U_o - p\,\hbar\omega \tag{15}$$

Let :

$$\pi_e = S\,(N + 1) \text{ and } \pi_a = S\,N \tag{16}$$

be the respective probabilities for emitting/absorbing one phonon at temperature T, and

$$N = \frac{1}{e^{\hbar\omega/kT} - 1} \cdot$$

S measures the strength of electron-phonon coupling and, as above, will be identified with the mean number of phonons emitted after or accompanying the optical transition.

We shall use as an hypothesis Poisson's distribution for the probability of emitting m phonons (proportional to $\pi_e^m/m!$) or absorbing n phonons (proportional to $\pi_a^n/n!$). Hence the probability of a process in which m phonons are created and n phonons are annihilated will be assumed to be :

$$e^{-(\pi_e + \pi_a)} \frac{\pi_e^m}{m!} \frac{\pi_a^n}{n!} .$$

The multiplicative factor $e^{-(\pi_e + \pi_a)} = e^{-S(2N + 1)}$ is introduced for normalization (the sum of the probabilities of all possible events must be equal to unity).

These assumptions mean, at low temperature, that the emission or absorption spectra are Poissonian (Pekarian) curves :

$$G(h\nu) = \text{const. } e^{-S} \frac{S^p}{p!} . \tag{17}$$

At 0°K indeed, emission takes place from the level n = 0. In addition, N = 0;therefore $\pi_e = S$ and $\pi_a = 0$. The only possible event is the emission of p = m phonons.

Now we can derive at once the emission spectrum at any temperature T . The square of the matrix element relevant to the n → m transition is :

$$|M_{if}|^2 = |M_{eg}|^2 \; e^{-(\pi_e + \pi_a)} \frac{\pi_a^n}{n!} \frac{\pi_e^m}{m!}$$

$$= |M_{eg}|^2 \; e^{-S(2N + 1)} \frac{(SN)^n}{n!} \frac{[S(N+1)]^m}{m!} .$$

The total emitted intensity at frequency ν is :

$$I(h\nu) = \sum_{n=0}^{\infty} I(h\nu_{nm}) \, \delta(U_o - p\,\hbar\omega - h\nu_{nm}) ,$$

i.e., apart from the multiplicative ν^4 factor which occurs in equation (3) :

$$G(h\nu) = |M_{eg}|^2 \, e^{-S(2N+1)} \sum_{n=o}^{\infty} \frac{S^{2n+p} \, N^n \, (N+1)^{n+p}}{n! \, (n+p)!} \, ,$$

(m = n+p). This expression may also be written:

$$G(h\nu) = |M_{eg}|^2 \, e^{-S(2N+1)} \left(\frac{N+1}{N}\right)^{p/2} I_p \left[2S\sqrt{N(N+1)}\right] \qquad (18)$$

in which

$$I_p\left[2S\sqrt{N(N+1)}\right] = \sum_{n=0}^{\infty} \frac{\left[S\sqrt{N(N+1)}\right]^{2n+p}}{n! \, (n+p)!} ,$$

is the modified Bessel function of complex argument $I_p(x) = i^{-p} J_p(ix)$. Formula (18) is just the well-known Huang and Rhys formula.

II.B. The Absorption Spectrum

Expression (14) still gives the positions of the absorption energy lines, but now m is the number of annihilated phonons and n is the number of created phonons. The gross number of created phonons in now p = n - m and instead of (15) we must write :

$$h\nu_{nm} = U_o + p\,\hbar\omega$$

When the definition of p is conveniently modified in this way, formula (18) still holds for the absorption spectrum. Mirror symmetry between emission and absorption spectrum is obvious.

II.C. A Remark

Pryce in his review paper (4) derives formulas for emission and absorption processes by first establishing the general result that the relative probability of a p phonon process is :

$$P(p) = e^{-S(T)} \frac{S(T)^p}{p!} , \qquad (19)$$

where:

$$S(T) = S(2N+1) = S \coth \frac{\hbar\omega}{2kT} . \qquad (20)$$

For T = 0 this result is the same as (17). But for $T \neq 0$ we must be careful, that, if one uses Pryce's vocabulary:

A one-phonon process is a process in which one phonon is either created (i.e. for photon emission the case m = 1, n = 0), or absorbed (i.e. the case m = 0, n = 1).

A two-phonon process is a process in which two phonons are involved, this means :

a) either two phonons are created (m = 2, n = 0)

b) or two phonons are annihilated (m = 0, n = 2)

c) or one phonon is created and one annihilated (m=1, n=1)

More generally, the p-phonon processes are (m = p, n = 0), (m = p-1, n = 1), (m = p-2, n = 2),..., (m = 0, n = p) whose total probability is according to our notation :

$$e^{-(\pi_e + \pi_a)} \sum_{\substack{n=0 \\ (m+n = p)}}^{p} \frac{\pi_e^{\,m}}{m!} \frac{\pi_a^{\,n}}{n!} .$$

The summation is straightforward and gives indeed :

$$e^{-S(2N+1)} \frac{S^p (2N+1)^p}{p!} .$$

We see a p-phonon process must not be confused with the p-term in the Huang and Rhys formula. This term indeed contains contributions of phonon processes of order p and of orders > p.

II.D. Shape and Asymmetry of Huang's and Rhys' and Pekarian Curves

1. The Moments. The expressions for the moments that are derived from (18) are found to be the same as those obtained by means of the semi-classical approximation in section I:

$$\langle p \rangle = S ,$$

$$\langle p^2 \rangle = S\,(2N+1) = S \coth \frac{\hbar\omega}{2kT} ,$$

$$\sigma^2 = \langle (h\nu - h\nu_o)^2 \rangle = S\,(\hbar\omega)^2 \coth \frac{\hbar\omega}{2kT} ,$$

and for large values of S the shape of the spectrum can be reasonably well approximated by a Gaussian for the values of p of physical interest (this means , around p = S). But the third moment is found to be different from zero :

in absorption $\langle (h\nu - h\nu_o)^3 \rangle = S\,(\hbar\omega)^3$;

in emission $\langle (h\nu - h\nu_o)^3 \rangle = -S\,(\hbar\omega)^3$,

thus proving that the shape of the spectrum is actually not symmetrical if we consider values of p situated at the tails of the emission or absorption curves.

2. A More Detailed Study of the Pekarian Curve. Markham and Konitzer (5) have shown from detailed experiments on F-center absorption that the shape of the spectrum does not change in a visible manner while temperature is increased from 4.5°K to 300°K. The width of the absorption band of course increases, but the asymmetry remains the same.

Therefore we shall examine only the Pekarian curves, theoretically valid at 0°K, not the most general Huang and Rhys distribution.

The Poisson distribution is non-symmetrical in the manner shown in Fig. 9.

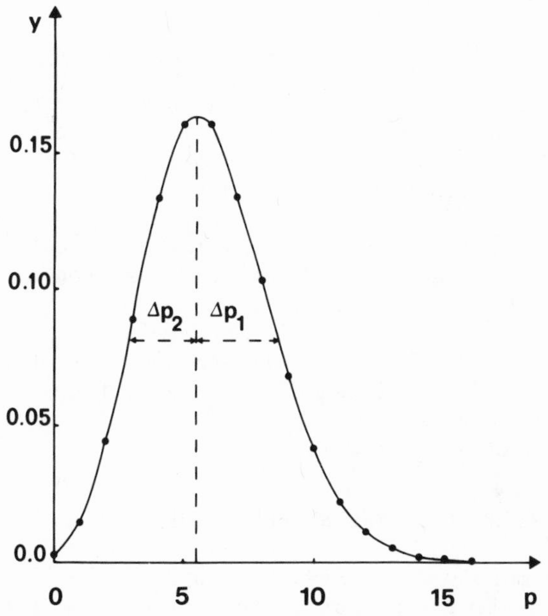

Figure 9 A Poissonian distribution (S = 6). $y = e^{-S}\,\dfrac{S^p}{p!}$.

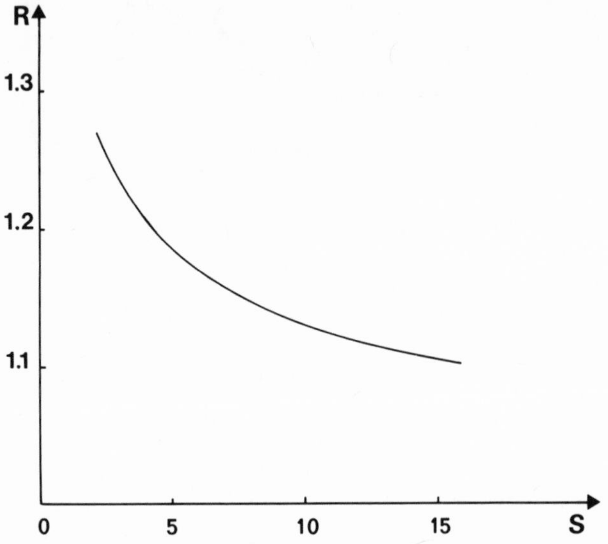

Figure 10 The asymmetry ratio R decreases and must reach unity for values of S $\to \infty$, but as a matter of fact it remains noticeably > 1 except for very large values of S.

Let us define the asymmetry by the ratio $R = \Delta p_1 / \Delta p_2$ of both sides of the band, taken at half-maximum. This ratio is typically of the order of 1.1 as observed by Markham ; it decreases while S increases (Fig. 10), although very slowly.

In addition, Fig. 11 shows that the "law of a rectilinear diameter" appears to be accurately satisfied as soon as $S \geqslant 5$.

However, we should remark that if Markham's result is correct, that the shape of the F-center absorption band does not change with temperature except for a modified energy scale, then the slope m of the rectilinear diameter is proportional to the reciprocal of the bandwidth instead of satisfying (13).

III. A DEEPER INSIGHT INTO THE SHAPE OF THE SPECTRA

Until now, we merely discussed the shape of the "envelope" curve, while in the limits of our approximations the spectra would consist of a series of narrow lines. In most cases, these lines join in a continuous spectrum, and only the zero-phonon line and a few replicas can be observed in some cases.

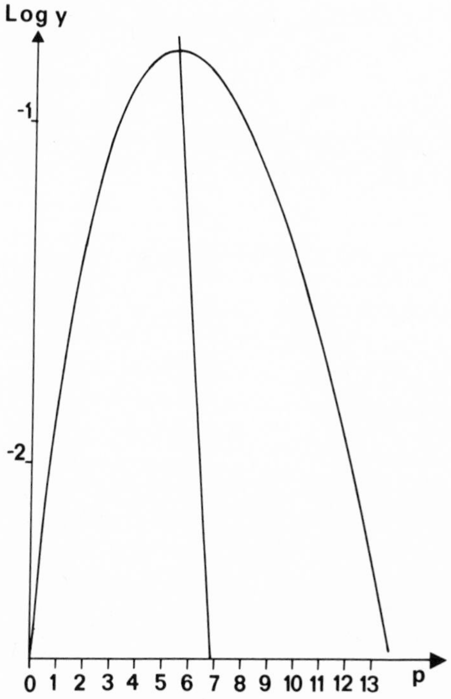

Figure 11 The Poisson distribution shown in Fig.9 (S = 6), redrawn in such a way as to show the occurrence of the linear diameter.

Other speakers, especially Rebane, will give a treatment of the electron-phonon interaction including many more details than I shall. I shall describe only some of the most important results.

III.A. Weak Coupling and Strong Coupling*

The parameter S (or S_e and S_a in section I) measures the intensity of electron-phonon coupling in the linear case.

Following Toyozawa, most people assume as a general rule:

(a) for $S \geqslant 5$ the zero-phonon line and its replicas are not observed and we are in the case of a continuous spectrum (strong coupling).

(b) for $S < 5$ at least the zero-phonon line may be observed (weak coupling) and for $S < 1$ this line dominates in the spectrum.

Some experimental examples will be described in the next section.

*For most "conventional" phosphors we are in the strong coupling case. But on the other hand weak coupling occurs in the vibronic spectra of most magnetic ions in crystals.

III.B. The Case of a Distribution of Phonons Interacting with the Emitting Center.

1. Selection Rules. The selection rules for determining the vibrational modes which interact strongly with the emitting centers are the same as those required for the possible occurrence of a Jahn-Teller effect, either in the ground state, or in the excited state.

This means for real representations:

Let us consider a vibrational mode belonging to the rep Γ ; the ground state belongs to Γ_g, the excited state to Γ_e. The direct sum :

$$\Gamma_g \otimes \Gamma_g + \Gamma_e \otimes \Gamma_e \quad *$$

must contain Γ in order to have a linear coupling.

For non-degenerate states, $\Gamma_g \otimes \Gamma_g$ and $\Gamma_e \otimes \Gamma_e \equiv \Gamma_1$, and among local vibrations around the center linear coupling occurs only with the breathing mode.

When the center site possesses an inversion symmetry (for instance this is the case if the site symmetry belongs to group O_h), only the even modes can interact linearly with the center, whatever the electronic states may be.

It is of interest to note that, on the contrary, the odd vibrational modes are in this case active only in infra-red transitions.

As an example, let us consider in KCl:Tl the local vibrations of the system made by the Tl^+ ion and its six neighbours Cl^-. 15 normal modes occur which can be classified as follows :

(a) the breathing mode A_{1g};
(b) two E_g vibrations occurring along the axes of the octahedron;
(c) three T_{2g} vibrations perpendicular to these axes;
(d) 9 odd modes ($2T_{1u}$ + T_{2u}).

The even modes A_{1g}, E_g, T_{2g} are all Raman active; the 2 T_{1u} modes are active in I.R., while the T_{2u} modes are inactive in both I.R. and Raman.

Kristoffel (6) has shown that the D_{4d} distortion produced by the E_g vibrations results in a Jahn-Teller effect. He also showed that the Condon approximation remains valid, the corrections introduced by taking into account the dependence of M_{eg} on the nuclear coordinates being insignificant. He ascribed this result to the fact that we have to deal with an allowed electric dipole transition ;

*The above selection rule (for a displacement of the equilibrium position of the ions) must not be confused with the selection rules for an allowed vibronic transition. See Wall's paper in this book.

on the contrary, for a forbidden transition which is forced by vibrations, the matrix element M_{eg} is expected to depend much stronger on the nuclear coordinates.

2. The Coupling Parameter S. We consider here the case of purely linear coupling, i.e., we neglect the dependence of the vibrational frequencies on the electronic state of the center.

Let us consider a vibrational mode α, whose angular frequency is ω_α, and this mode introduces a change $Q_{o\alpha}$ in the equilibrium value for the normal coordinate Q_α between the excited state and the ground state. We put in the same way as in section I:

$$S(\hbar\omega_\alpha)\,\hbar\omega_\alpha = \frac{1}{2}\,k_\alpha\,Q_{o\alpha}^2 .$$

Starting from the $S(\hbar\omega_\alpha)$ which are related to the coupling between one specific mode α and the center, we introduce the total coupling parameter which takes into account all interactions between all phonons and the center :

$$S(T) = \sum_\alpha (2N_\alpha + 1)\, S(\hbar\omega_\alpha) \;; \tag{21}$$

$$N_\alpha = \frac{1}{e^{\hbar\omega_\alpha/kT} - 1} .$$

At low temperatures, S(T) reduces to :

$$S = \sum_\alpha S(\hbar\omega_\alpha) . \tag{22}$$

If we consider a continuous distribution of interacting modes, of normalized density $\rho(\hbar\omega)$, then the summations in (21) and (22) are to be replaced by the integrals :

$$S(T) = \int S(\hbar\omega)\,\rho(\hbar\omega)\,\coth\frac{\hbar\omega}{2kT}\,d(\hbar\omega) \;; \tag{21a}$$

$$S = \int S(\hbar\omega)\,\rho(\hbar\omega)\,d(\hbar\omega) . \tag{22a}$$

It will be convenient to introduce the function $A(\hbar\omega)$:

$$A(\hbar\omega)\,d(\hbar\omega) = S(\hbar\omega)\,\rho(\hbar\omega)\,d(\hbar\omega)\,/\,S .$$

Then, after tedious calculations (4), it is possible to show that the normalized shape of the spectrum is given by :

$$G(h\nu) = e^{-S(T)} \sum_{p=0}^{\infty} \frac{S(T)^p}{p!} B_p(h\nu) \tag{23}$$

where the $B_p(h\nu)$ are appropriately normalized functions of $h\nu$. The p-term in this summation is the contribution of the p-phonon processes (see the remark made in II.C.).

We shall consider only the case of low temperatures. Then formula (23) reduces to :

$$G(h\nu) = e^{-S} \sum_{p=0}^{\infty} \frac{S^p}{p!} B_p(h\nu) \tag{24}$$

and a p-phonon process is merely a process in which p phonons are created (as phonon annihilation can now be neglected).

The term p = 0 is the zero-phonon line ; usually one puts

$$B_o(h\nu) = \delta(h\nu - U_o), \tag{25}$$

but more elaborate theories show that in perfect crystals B_o is indeed a narrow Lorentzian line.

$B_1(h\nu)$ is the normalized shape of the one-phonon replica. It reflects closely the shape of the density of modes $\rho(\hbar\omega)$. More accurately, if we assume the probability of creation of one phonon $\hbar\omega_\alpha$ to be $S(\hbar\omega_\alpha)$ per mode *, then:

$$B_1(h\nu) \text{ proportional to } S(\hbar\omega_\alpha)\ \rho(\hbar\omega) ,$$

and if normalizing :

$$B_1(h\nu) = A(\hbar\omega) ; \tag{26}$$

$$h\nu = U_o - \hbar\omega_\alpha , \quad \text{in emission ;}$$

$$h\nu = U_o + \hbar\omega_\alpha , \quad \text{in absorption.}$$

For higher order replicas the theory becomes more involved. A phonon process in which two phonons $\hbar\omega_\alpha$ and $\hbar\omega'_\alpha$ are created leads in emission to a photon energy :

---

* Remember eq.(16) . Here $N_\alpha = 0$.

$$h\nu = U_o - \hbar\omega_\alpha - \hbar\omega'_\alpha$$

and in absorption to :

$$h\nu = U_o + \hbar\omega_\alpha + \hbar\omega'_\alpha \cdot$$

The probability of this process is proportional to the product :

$$B_2(h\nu) = \int A(\hbar\omega)\ A(\hbar\omega')\ d(\hbar\omega') \cdot \qquad (27)$$

In performing the integral, one must write :

$$\hbar\omega = h\nu - U_o - \hbar\omega' \text{ , in absorption ;}$$

$$\hbar\omega = U_o - h\nu - \hbar\omega' \text{ , in emission .}$$

According to the discussion given in Pryce's paper (4) the $B_p(h\nu)$ of higher orders ($p \geqslant 3$) will look like Gaussian curves centered at :

$$h\nu_p = U_o \pm p\ \overline{\hbar\omega} \ ,$$

(+ in absorption, - in emission) ; $\overline{\hbar\omega}$ is given by :

$$\overline{\hbar\omega} = \int_0^\infty A(\hbar\omega)\ \hbar\omega\ d(\hbar\omega) \cdot \qquad (28)$$

The width of these Gaussians would be proportional to $\sqrt{p}$ and they must be normalized to unity in order to apply (24). We feel that this approximation may either be valid or not according to the shape of the phonon density $\rho(\hbar\omega)$.

3. The Moments of the Shape Function. No accurate general theoretical expression has been given until now for the shape of the spectrum at any temperature. The moments can be derived from (23).

The zeroth moment is of course unity :

$$\int_0^\infty G(h\nu)\ d(h\nu) = 1 \ .$$

This means, $G(h\nu)$ is normalized because $B_p(h\nu)$ is.

The first moment is found to be :

$$\int_0^\infty G(h\nu)\, h\nu \;\; d(h\nu) = h\nu_o = U_o \pm S\,\overline{\hbar\omega} \;, \qquad (29)$$

(+ in absorption, - in emission) where $\overline{\hbar\omega}$ is still defined by (28); i.e., S remains the mean value of the number of created phonons. This result does not depend on temperature.

The second moment which is grossly proportional to the band-width (but with a numerical coefficient which depends on the shape of the spectrum) is :

$$\sigma^2 = \langle (h\nu - h\nu_o)^2 \rangle = S \int_0^\infty A(\hbar\omega)(\hbar\omega)^2 \coth\frac{\hbar\omega}{2kT}\, d(\hbar\omega). \qquad (30)$$

The third moment which ought to be zero for a pure Gaussian is :

$$\langle (h\nu - h\nu_o)^3 \rangle = \pm S \int_0^\infty A(\hbar\omega) \;(\hbar\omega)^3 \; d(\hbar\omega). \qquad (31)$$

Corrections must be added to the above formulas for including quadratic coupling and also to go beyond the Condon approximation. These corrections predict a displacement of the $h\nu_o$ transition versus temperature. However, it happens that so many quite different effects can result in such a displacement that it is extremely difficult to ascribe this displacement, when experimentally observed, to these effects or to different phenomena (for instance, changes in the lattice constant when temperature increases).

IV. A STUDY OF SOME EXPERIMENTAL CASES

We shall study as an example a few cases, in which respectively S is rather large, S is about unity, S is much less than unity. We emphasize we do not plan to give a comprehensive review of all the experimental data which are available on the materials described below.

IV.A. Large Band Spectra of II-VI Compounds

1. ZnS:Cu. Cu^{++} incorporated in a substitutional Zn^{++} site in ZnS, in the absence of any other defect, produces an infra-red emission. This emission comes from transitions between the ^2E and $^2T_2$ states of the d^9 ion, which are split by the effect of spin-orbit coupling.

The usual green emission is ascribed to pairs of associated defects ; most probably in compensated ZnS:Cu,Cl we are dealing with a deep donor-acceptor pair (Cu^{++} donor, Cl^- acceptor). Other types of pairs may occur, in somes cases the Cl^- ion can be replaced by a sulphur vacancy, but in our opinion this second type of center occurs only if chlorine has been carefully eliminated or if the amount of oxygen in the material is most important.

Probable values for the peak positions of corresponding emission spectra are :

	Cubic samples	Hex. samples
ZnS:Cu,Cl	5350 Å (2.32 eV)	5150 Å (2.40 eV)
ZnS:Cu,|S|	about 0.05-0.07 eV larger	

As a result of the small differences between their masses, substitution of copper for zinc or of chlorine for sulphur results in a relatively small perturbation of the ZnS lattice. We expect the vibrational quantum $\hbar\omega_e = \hbar\omega_g$ to be equal or only slightly different from $\hbar\omega_{LO}$.

Experimentally, it is found for the emission bandwidths :

0.33 eV at room temperature (RT)
0.27 eV at liquid nitrogen temperature (LNT).

From Fig.5, i.e. by using the relation:

$$\sigma_e(T) = \sigma_e(0°K) \left[\tanh \frac{\hbar\omega_e}{2kT} \right]^{-1/2} \qquad (8)$$

one obtains (7), (8) :

$$\hbar\omega_e = 0.043 \text{ or } 0.044 \text{ eV}.$$

This last number is just the accepted value for zone center L.O. phonons. If, moreover, we accept $\hbar\omega_g$ to be also equal to $\hbar\omega_{LO}$, then it is possible to derive the configurational diagram from emission measurements only (9) ; and from this diagram one obtains $S_a = S_e = 7$, hence the computed peak of the absorption transition :

$$h\nu_{oa} = 3.00 \text{ eV } (4130 \text{ Å}).$$

This result is in good agreement with the occurrence of a blue band in the excitation spectrum of ZnS:Cu (10).

In the present state of things, we cannot say if $\hbar\omega_e = \hbar\omega_g$ is accurately equal to the energy of L.O. phonons in the perfect lattice, or if we are dealing with a local or pseudo-local mode which is vibrating with a neighbouring frequency. Much more detailed Raman studies than those presently available are necessary in order to solve this problem.

Other samples, possibly not well compensated, show different temperature variations. We did find in some cases $\hbar\omega_e < \hbar\omega_{LO}$, as well as $\hbar\omega_e > \hbar\omega_{LO}$. In addition, in more than a few cases, the temperature variation of the bandwidth is monotonic: $\sigma_e(4°K) > \sigma_e(77°K)$. These results support the assumption of different kinds of Cu centers in ZnS. However in all cases we found S = 6 or 7 and $h\nu_{oa}$ appeared in the blue region.

2. ZnSe:Cu. Typical values for the bandwidths relevant to the 6250 Å emission (which is homologous to the "green centers" in ZnS:Cu) are:

0.30 eV at RT
0.23 eV at LNT ,

whence $\hbar\omega_e$ = 0.033 eV while $\hbar\omega_{LO}$ (zone center) = 0.0314 eV. In this case, too, the perturbation introduced by copper is small.

3. CdS:Ag. In the case of cadmium sulphide, we expect that silver and not copper will produce a small perturbation in the lattice. We found for the 7200 Å emission the following data for the bandwidths (11):

0.34 eV at RT
0.27 eV at LNT

whence $\hbar\omega_e$ = 0.036 eV while $\hbar\omega_{LO}$ (z.c) = 0.038 eV. On the contrary, for the 6200 Å, $\hbar\omega_e$ seems to be much smaller , about 0.02 eV.

4. ZnS:Ag,Cl. The reported peak position of the main visible band of ZnS:Ag,Cl samples are about:

	Cubic samples	Hex. samples
ZnS:Ag,Cl	4500 Å	4370 Å

at RT and for usual excitation densities by means of an ordinary U.V. mercury lamp (Shionoya (12) has shown that the peak position of this band increases in energy for increasing excitation intensities).

The question arises: is this band analogous to the blue band of copper activated materials ? The answer is no; the blue band of ZnS:Ag,Cl is indeed the homologous emission of the green band of ZnS:Cu,Cl. The main evidence for this conclusion comes from polarization studies : in cubic crystal samples, both bands are polarized perpendicular to the [111] axis (13). But the temperature variation of bandwidths gives a supplementary proof : it is found that $\hbar\omega_e$ = 0.036 eV and, if in the absence of Raman studies upon silver-activated crystals we assume :

$$\hbar\omega\ (Ag) = \hbar\omega\ (Cu)\ \sqrt{M_{Cu}/M_{Ag}}$$

(M_{Cu} = 65, M_{Ag} = 108), we obtain $\hbar\omega$ (Ag) = 0.036 eV.

5. Self-Activated Materials. Contrary to all impurity band described above, S.A. emission shows systematically a much larger bandwidth (larger than 0.40 eV at RT), and in addition these bandwidths are more sensitive to temperature variations ($\hbar\omega_e$ is smaller). We have in some cases retained this argument for establishing the distinction between both types of emission, and this criterion proved to agree with the discrimination made from polarization studies, as established by Shionoya and co-workers (14).

IV.B. ZnS:Mn

Mn^{++} deserves a special study because the emission comes from an internal $^4T_1 \rightarrow {}^6A_1$ transition of the $3d^5$ ion, and not from a charge transfer between donor and acceptor levels, or between the conduction band and a localized level.

Langer and Ibuki (15) first observed the zero-phonon lines and the phonon structure in the absorption and emission spectra of ZnS:Mn. Previous work, more generally performed on powder samples, did not show this structure.

This emission raises many problems, which are far from being completely solved until now.

It is generally admitted that we have to deal with an electric dipole transition, "forced" by interaction with lattice vibrations. Consequently, the Condon approximation is not valid *.

In the author's opinion, a possible admixture of magnetic dipole transitions (especially in the zero-phonon line) can not be ruled out and this possibility has not been discussed enough presently.. Furthermore, in most commercial powder phosphors, most Mn^{++} ions are probably in perturbed sites which are situated in the neighbourhood of Cl^- ions or S vacancies.

In good cubic crystals with only a few stacking faults, on the other hand, most Mn^{++} ions seem to be actually in T_d sites (from e.p.r. studies). The position of the zero-phonon line is, in this case (cubic crystals), 17891 cm^{-1} for the $^4T_1 \rightarrow {}^6A_1$ transition.

---

* In a cubic (O_h) site, with an inversion center, "static" electric dipole transitions are forbidden for parity reasons, even if spin-orbit mixing of ^6S and quartet states is taken into account. Then the zero-phonon line, if observed, is necessarily magnetic dipole, while the broad absorption or emission bands arise from vibronic electric dipole transitions.

A detailed analysis of phonon replicas has been performed quite recently by Zigone, Beserman and Lambert (16). They rely on accurate Raman studies by Zigone, Beserman, Nusimovici and Balkanski (17), who did show that manganese in ZnS introduces :

(a) the "breathing mode" Γ_1 which vibrates at 298 cm^{-1} ;

(b) and resonant modes $\Gamma_1 + \Gamma_{15}$ at 311 and 335 cm^{-1}.

Well defined line peaks have been observed in emission by the above authors (16) at the following positions: (see Fig. 13):

17891 - 298, 311 and 335 cm^{-1} ;
- (298 + 298), (298 + 311), and (298 + 335) cm^{-1} ;
- (3 × 298) and - (2 × 298 + 333) cm^{-1}.

The breathing mode, 298 cm^{-1}, happens to be the only recurrent mode. This result has not yet received a detailed interpretation.

In addition, larger replicas are observed, which correspond possibly to interaction with non-localized modes of pure ZnS.

In cubic samples with numerous stacking faults, two other series of lines are found, which are ascribed to axial sites ; the corresponding zero-phonon lines are displaced by + 91 and + 118 cm^{-1} with respect to the cubic zero-phonon line. In pure wurtzite, the zero-phonon line is at 17891 + 208 cm^{-1}, but in most partially hexagonal materials the four lines are simultaneously observed.

The absorption transitions $^6A_1 \to {}^4T_2$ and $^6A_1 \to {}^4E$ show similar results.

If we turn now to the study of bandwidth variation versus temperature (17), formula (8) leads to $\hbar\omega_e$ = 0.037 eV, which is the energy value corresponding to 298 cm^{-1} phonons, but of course the accuracy of these determinations is not large enough for making the discrimination between these localized phonons and T.O. phonons at 0.0368 cm^{-1}. As a matter of fact, Langer and Ibuki were assuming that we have to deal with T.O. phonons.

IV.C. The R-Lines of Ruby

The R_1 and R_2 lines of Cr^{+++} in ruby must be considered as zero-phonon lines, because their position is the same in absorption and emission. The Stokes shift at 4°K, if it exists, is less than 0.002 cm^{-1} (18). These lines are electric dipole transitions, with possibly a small amount of magnetic dipole.*

---

* On the contrary, let us remember that in O_h sites of MgO:Cr^{+++}, Schawlow and co-workers (19) have observed only one R-line (because trigonal field is absent) and this line is of a magnetic dipole character.

The coupling of the excited state ^2E and the ground state $^4A_2$ with lattice vibrations is very weak ; however small intensity phonon replicas have been observed and discussed by Nelson and Sturge (18). Two one-phonon lines are observed (which reflect the $R_1 + R_2$ structure) at + 376.7 cm^{-1} in absorption and + 376.9 cm^{-1} in emission. These lines are vibronic electric dipole transitions. For polarization $\underline{E} \perp \underline{c}$, the integrated intensity of vibronic bands is about 17% the intensity of the R-lines. We may assume reasonably, that this value gives the order of magnitude of the coupling parameter S.

IV.D. Edge Emission in II-VI Compounds

Low temperature edge emission in CdS (Ewles-Kröger spectrum) was the first known example of a structure made of equally spaced emission narrow band:

$$h\nu_p = h\nu_o - p\,\hbar\omega$$

($\hbar\omega = \hbar\omega_{LO}$) whose relative intensities follow a Poisson's law (20) :

$$G(h\nu_p) = G(h\nu_o)\ \frac{S^p}{p!}\ .$$

Now two series of lines have been observed ; both are generally ascribed to donor-acceptor transitions. Both follow the above formulas.

Gutsche and Goede (21) have given an excellent review of this problem at Delaware International Conference on Luminescence. They obtained a very good agreement with the experimental shape of the spectrum by assuming for each component a Gaussian, i.e. :

$$G(h\nu) = e^{-S} \sum_{p=0}^{3} \frac{S^p}{p!} \exp\left[-\frac{(h\nu - h\nu_p)^2}{2\sigma^2}\right] \tag{32}$$

But instead of using the theoretical result σ proportional to $\sqrt{p}$, they put simply σ = constant; it appears that the fitting is quite good (Fig.12).

The ratio of intensities of the one-phonon line and the zero-phonon line is just equal to S. In CdS this parameter S is probably slightly < 1 : Hopfield gave S = 0.87, the above authors S = 0.97. In ZnS we have presumably S = 1 .

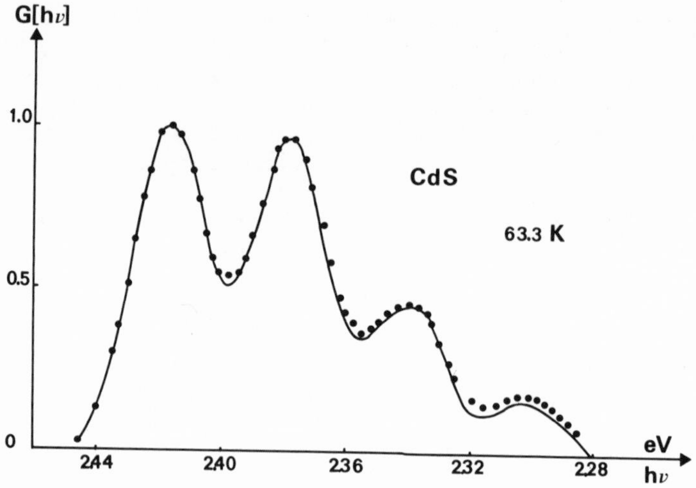

Figure 12 From Gutsche and Goede (21).
Spectral distribution of the intensity of the high energy Ewles-Kröger series. Points : experimental ; full line : theoretical curve from eq. (32).

IV.E. Stokes versus Anti-Stokes Emission

Near 0°K, the emission spectrum ceases abruptly at the zero-phonon line energy (Fig.13). This fact is observed experimentally in ruby and also in ZnS:Mn without stacking faults (the occurrence of some emission above 17891 cm^{-1} in a so-called "cubic" sample of ZnS:Mn, near 0°K, gives evidence that manganese sites do exist, in this sample, which are not actually cubic).

The part of the continuous emission spectrum above the zero-phonon line (sometimes called the anti-Stokes part) becomes detectable above 30°K in ZnS:Mn and at 77°K in ruby. Its intensity increases exponentially with temperature. For large values of S (semi-classical approximation), one obtains easily by integrating for $h\nu$ emitted > U_o :

$$\text{Fraction of the spectrum above } U_o \sim \exp\left[-\frac{S}{2}\tanh\frac{\hbar\omega}{2kT}\right].$$

More elaborate formulas have been derived for all values of S by Nelson and Sturge (18), but the experimental agreement is not very good.

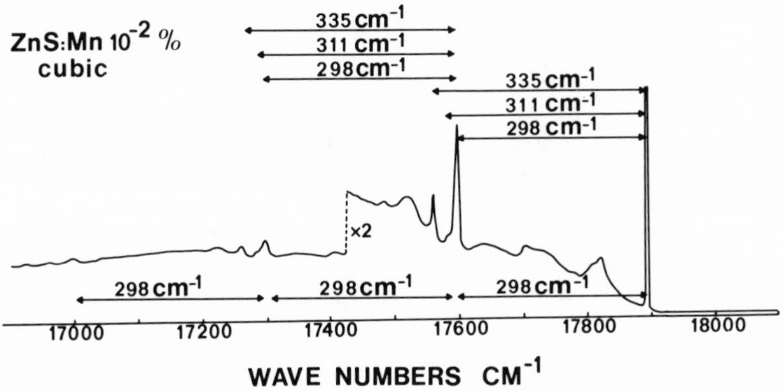

Figure 13 The high energy tail of the emission spectrum in cubic ZnS:Mn, at 4°K (16).

Now let us turn to the problem of excitation by using a radiation $h\nu_a < U_o$. We may call this situation anti-Stokes excitation.

Near 0°K, such an excitation is not possible, but it may be observed at "high" temperatures. A valuable criterion for "high" temperatures may be the occurrence of the anti-Stokes part of the emission spectrum described just above. For instance in ZnS:Mn, the 5682 Å line of a krypton laser (17599 cm^{-1}) is particularly suitable for this purpose (22) (let us remember that in cubic samples the zero-phonon line is situated at 17891 cm^{-1}).

Just a glance at the configuration diagram shows that excitation occurs, or not, according to the magnitude of the existing phonon density; if this phonon density is large enough for the excitation process to take place then the whole spectrum is emitted and the part $h\nu_e < h\nu_a$ is not emitted preferentially.

In the above case, it is found that the shape of the emitted spectrum is the same as by using ordinary Stokes excitation, and this proves that thermalization occurs in the excited state.

Erickson (23) has given a simple theory of this phenomenon. This theory may be summarized as follows : we start from the remark that the whole spectrum is emitted, or not, if the zero-phonon line itself is emitted or not. If excitation occurs at the energy $h\nu_a < U_o$, then we need the energy $W = U_o - h\nu_a$ to be taken out from absorbed phonons, and therefore the probability of the excitation process is proportional to exp(-W/kT).

Experimentally, it is found indeed that the emission intensity increases with temperature according to exp(-W/kT) (Fig.14). For the diamond H_3 line , a case where we have a line spectrum with well-defined phonon replicas , the expression $W = U_o - h\nu_a$ appears to be conveniently satisfied. On the other hand, in cases of large band spectra such as ZnS:Mn, the temperature variation is still exponential, but W is found to be smaller than $U_o - h\nu_a$.

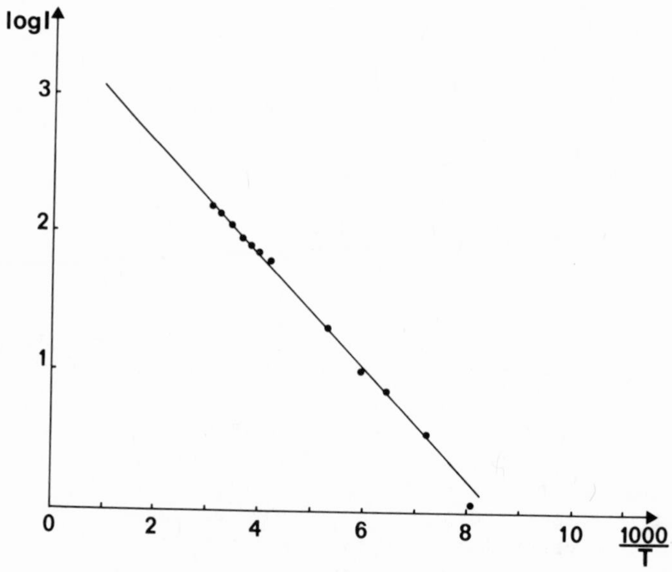

Figure 14 Anti-Stokes excitation of the diamond H_3 line (19868 cm^{-1}) excited by a Krypton laser 5208 Å line (19201 cm^{-1}) (22).

REFERENCES

1. F. Williams, J. Chem.Phys., 19, 457 (1951). P.D. Johnson and F. Williams, Phys. Rev., 117, 964 (1960).

2. D. Curie, "Luminescence in Crystals", Methuen (London) 1963, p.49, 54.

3. H. Payen de la Garanderie, Comptes-Rendus Académie des Sciences Paris, 260, 114 (1965).

4 . "Phonons in perfect lattices and in lattices with point imperfections", Scottish Universities Summer School 1965, Oliver and Boyd (Edinburgh), 1966, p.431.

5 . J.J. Markham, "F-centers in alkali-halides", Solid State Phys., Suppl.8, Academic Press 'New York) 1966, p.378.

6 . N.N. Kristofel, Optics and Spectr., 9, 324 (1960).

7 . H. Treptow, Phys.Stat.Sol. 6, 555 (1964).

8 . D. Curie, Acta Phys. Polonica, 26, 613 (1964).

9 . H. Payen de la Garanderie and D. Curie, J. Phys., 28, suppl. C 3, 124 (1967.

10. A. Halperin and H. Arbell, Phys.Rev., 113, 1216 (1959).

11. M.A. Dugué, Thesis, Paris 1972.

12. K. Era, S. Shionoya, Y. Washizawa, H. Ohmatsu. J. Phys.Chem. Sol., 29, 1827, 1843 (1968).

13. P. Jaszcyn-Kopec and S. Brandt, to be published in J. of Luminescence, 1974.

14. S. Shionoya, Y. Kobayashi and T. Koda, J. Phys.Soc.Japan, 20, 2046 (1965).

15. D. Langer and S. Ibuki, Phys.Rev., 138, A 809 (1965).

16. M. Zigone, R. Beserman and B. Lambert, to be published in J.of Luminescence, 1974.

17. R. Beserman, M.A. Nusimovici, and M. Balkanski, Phys.Stat.Sol., 34, 309 (1969).

18. D.F. Nelson and M.D. Sturge, Phys.Rev., 137, A 1117 (1965).

19. S. Sugano, A.L. Schawlow and F. Varsanyi, Phys.Rev., 120, 2045 (1960).

20. J.J. Hopfield, J.Phys.Chem.Solids, 10, 110 (1959).

21. E. Gutsche and O. Goede, J. of Luminescence, 1-2, 200 (1970).

22. R. Beserman, G. Curie and D. Curie, J. of Luminescence, 8, 326 (1974).

23. L.E. Erickson, J.of Luminescence, 5, 1 (1972).

QUANTUM THEORY OF LATTICE VIBRATIONS

R. Orbach*

Department of Physics, Tel Aviv University

Ramat Aviv, Tel Aviv, Israel

ABSTRACT

The quantum mechanical theory of lattice vibrations in solids is reviewed and summarized. The formalism is developed first in the classical manner, and the various symmetries of the normal mode eigenvalues are discussed. The transition to the quantum mechanical formalism is done by introducing operator forms for the appropriate physical (observable) quantities. As an extension of the quantum mechanical formalism, the thermodynamic properties of the lattice vibrations (entropy, free energy, etc.) are investigated. An extensive treatment of critical points in the phonon density of states is given; this includes a discussion of the so-called Van Hove singularities, and a cataloguing of the types of singularities involved. Finally, the lattice vibrational equation of state is formulated, and the effects of the boundary conditions on the frequency distribution are discussed.

I. LATTICE VIBRATIONS

I.A. Linear Diatomic Chain

1. Equations of Motion. This treatment of lattice vibrations deals with the theory of the departure of atoms from their equilibrium positions in solids. In the classical case, these departures are referred to as lattice vibrations and will be treated here as purely harmonic. Later in this section, the lattice vibrational modes will be quantized and dubbed "phonons". To

---

*Permanent address:
Department of Physics, University of California
Los Angeles, California 90024, U.S.A.

begin with, consider the simple case of a one-dimensional lattice composed of two kinds of atoms a distance a/2 apart, as in Figure 1. The bonds between atoms will be assumed to possess equal force constants μ. We shall assume the atoms of mass M are heavier than those of mass m, and the atomic positions will be labelled so that M,m lie on even, odd numbered sites respectively.

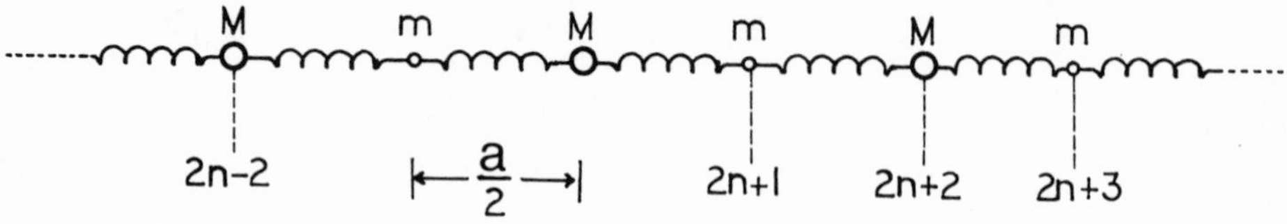

Figure 1 Linear Diatomic Chain

The chain is assumed to be composed of N masses of each type which are labelled by an index n = 0, 1, ... N-1. The linearized equations of motion for displacements along the x direction, assuming nearest neighbor restoring forces only, are then

$$m \frac{d^2u_{2n+1}}{dt^2} = -\mu[(u_{2n+1}-u_{2n}) - (u_{2n+2}-u_{2n+1})]$$

$$M \frac{d^2u_{2n}}{dt^2} = -\mu[(u_{2n}-u_{2n-1})-(u_{2n+1}-u_{2n})] \; . \tag{1}$$

Assuming periodic boundary conditions ($u_{2N+n} = u_n$), a solution of these equations may be written as

$$u_{2n+1} = A \exp \left\{ i[2\pi\ell(\frac{2n+1}{2N}) - \omega t] \right\}$$

$$u_{2n} = B \exp \left\{ i \; [2\pi\ell(\frac{2n}{2N}) - \omega t] \right\} \tag{2}$$

where ℓ is an integer. The length r along the line equals na and if we insist on a solution of the form $u \sim \exp(iK\cdot r)$, comparison with (2) shows

$$K = \frac{2\pi\ell}{Na} \quad . \tag{3}$$

We may substitute our solutions (2) into the original equations (1) to find

$$-m\omega^2 A \exp i[2\pi\ell \frac{2n+1}{2N} - \omega t]$$

$$= - \mu[(A \exp i[2\pi\ell \frac{2n+1}{2N} - \omega t]$$

$$- B \exp i[2\pi\ell \frac{2n}{2N} -\omega t]) - (B \exp i[2\pi\ell \frac{2n+2}{2N} -\omega t]$$

$$- A \exp i[2\pi\ell \frac{2n+1}{2N} - \omega t])]$$

or, simplifying,

$$(m\omega^2 - 2\mu)A + 2\mu \cos(2\pi\ell/2N)B = 0.$$

Similarly, from the other equation,

$$2\mu \cos(2\pi\ell/2N)A + (M\omega^2 - 2\mu)B = 0.$$

There exists a solution to these homogeneous equations only if the determinant of the coefficient of the amplitudes A and B vanishes

$$\begin{vmatrix} m\omega^2 - 2\mu & 2\mu \cos(2\pi\ell/2N) \\ 2\mu \cos(2\pi\ell/2N) & M\omega^2 - 2\mu \end{vmatrix} = 0 \, . \tag{4}$$

Solving for eigenvalue ω^2 yields

$$\omega^2 = \frac{\mu}{mM} [M+m \pm \sqrt{M^2+m^2+2Mm \cos(2\pi\ell/N)} \;] \, . \tag{5}$$

The eigenvectors yield amplitudes A and B in the ratio

$$\frac{A}{B} = \frac{2M\cos(2\pi\ell/2N)}{[M-m \mp \sqrt{M^2+m^2+2Mm\cos(2\pi\ell/N)}]} \quad . \tag{6}$$

2. One Dimensional Zone Scheme. The eigenvalue ω as a function of ℓ and the equivalent quantity $K = (2\pi\ell/Na)$ is is plotted in Figure 2.

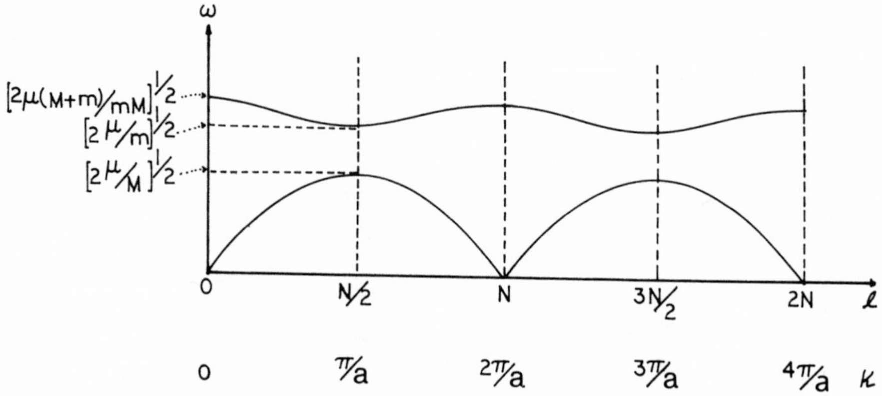

Figure 2 Values of the Eigenvalue ω.

It will be seen that for values of ℓ greater than N, the eigen-frequencies merely repeat. It turns out that the eigenvectors also repeat when ℓ exceeds N so that there exist distinct solutions for only N values of ℓ. Each solution is of course double valued so that a total of 2N distinct solutions have been found. Because we have allowed each of the 2N masses only one degree of freedom, this is of course the proper number.

The double valuedness of our solutions can be remedied by restricting our solutions as in Figure 3a. A more ususal way is to choose the domain for ℓ to lie between 0 and $\pm N$, rather than 0 and 2N. The range for K is then $\pm\ 2\pi/a$, and the allowed solutions look like those in Figure 3b. This is sometimes called the "extended zone" scheme.

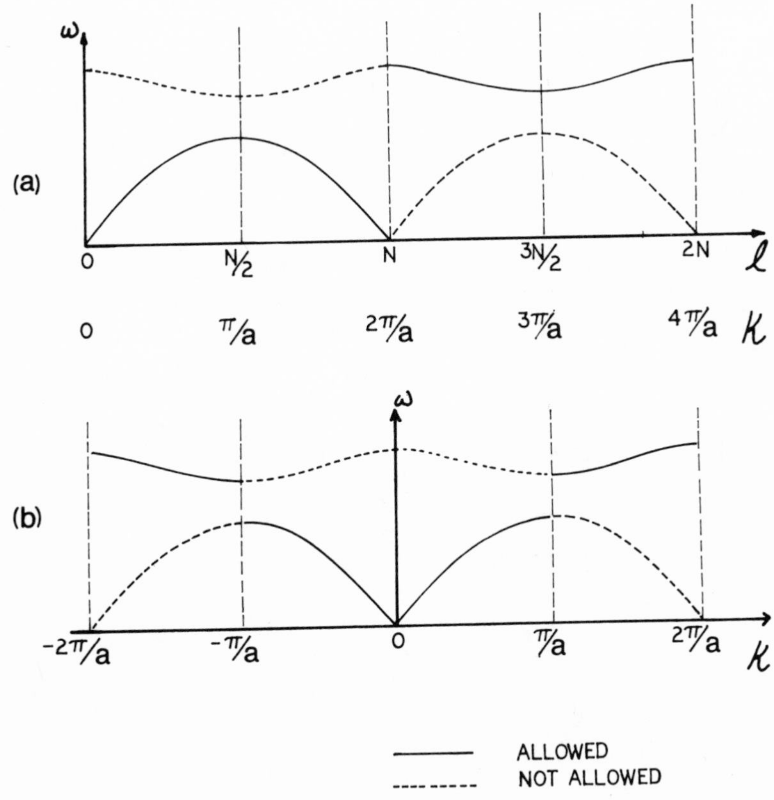

Figure 3 Restricted values for ω.

We may choose another scheme in which K goes from $0 \rightarrow 2\pi/a$ and accept the use of a multivalued function for $\omega(K)$ as in Figure 4a. Finally we may choose our multivalued function to lie in the limits $-\pi/a < K < \pi/a$ as in Figure 4b. This last choice is referred to as the "reduced zone" scheme.

(a)

(b)

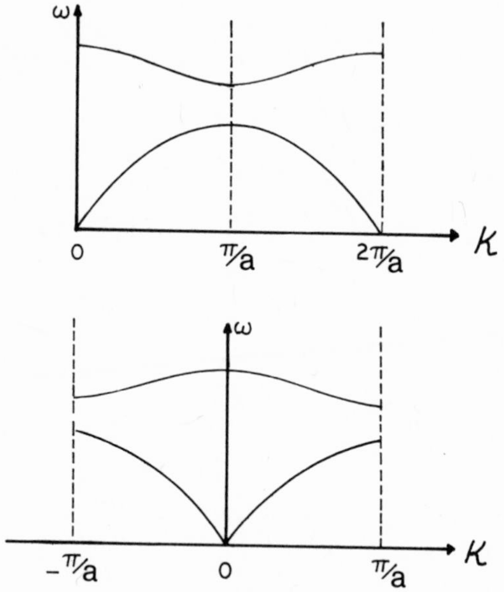

Figure 4 Zone schemes.

It should be clear that the choice of any particular scheme is quite arbitrary, and strictly a matter of convenience. For any K lying outside the chosen interval, the solutions merely repeat themselves and can be translated back by $4\pi/a$ in the single valued choices or by $2\pi/a$ in the multivalued choices. This should give some indication of the concept of a "zone" in which all allowed solutions are represented only once.

We may now examine the eigenvectors for various values of ℓ. The lower branch for $\ell < N/2$, or $K < \pi/a$, is specified by using the - sign in our expression (5) for ω, and the corresponding + sign in the expression (6) for A/B. This insures $A/B > 0$ and our wave is "acoustic" in character. The upper branch for $\ell < N/2$, $K < \pi/a$, may be selected by choosing the + sign in the expression (5) for ω, and the - sign in the expression (6) for A/B. Since the square root in the denominator $\geq$ M-m over this range, and the cosine in the numerator ≥ 0, we find that $A/B < 0$ for this branch. Our wave is then "optical" in character.

An interesting point is to set $\ell = N/2$, $K = \pi/a$, in the equation (6) for A/B. We find an indeterminancy and it is necessary to return to the original determinant. When $\ell = N/2$, one sees that

$$m\omega^2 = 2\mu$$

$$M\omega^2 = 2\mu$$

are allowed solutions with $A = 1$, $B = 0$ and $A = 0$, $B = 1$ respectively. Thus, one atom is stationary and the other vibrates at this peculiar point in K space.
The distinction between acoustic and optical fades as one moves away from $k = 0$ and disappears when $K = \pi/a$.

The concept of the zone should now be quite clear in the case of the diatomic chain. The extended zone scheme corresponds to a particular choice of sign for a given interval of K in our expression for ω; the reduced zone scheme chooses both signs and halves the interval in K space.

We note that the widths of the branches are given by

$$\begin{aligned} &(\text{optical}) && [2\mu(M+m)/mM]^{\frac{1}{2}} - (2\mu/m)^{\frac{1}{2}} \\ &(\text{acoustic}) && (2\mu/M)^{\frac{1}{2}} \end{aligned} \qquad (7)$$

If $m << M$, the top band width is

$$(2\mu/m)^{\frac{1}{2}}\,(1+\frac{m}{M})^{\frac{1}{2}} - (2\mu/m)^{\frac{1}{2}}$$

$$\cong (2\mu/m)^{\frac{1}{2}}\,[1+\frac{m}{2M} - 1] = (2\mu/M)^{\frac{1}{2}}\cdot\frac{1}{2}\,(\frac{m}{M})^{\frac{1}{2}} \quad . \qquad (8)$$

It is seen that the optical branch is narrower than the acoustic branch by $\frac{1}{2}\,(m/M)^{\frac{1}{2}}$. For salts with, say, one ion of mass m loosely coupled to a large number of other atoms which in turn are tightly coupled to one another and have a combined mass $M >> m$, one may regard the variation of ω with K for optical waves as slight in comparsion with the acoustic branch. We shall see the implications of this later when we discuss the Einstein and Debye approximations. Let it suffice to say that the optical mode group velocity, $\partial\omega/\partial K$, is small relative to acoustic velocities when $M >> m$.

3. Density of States. The number of modes of vibration per unit range of K is a constant for each branch because there are N vibrational modes distributed uniformly over ℓ. Noting that $K = 2\pi\ell/Na$, we see that there exists a separate solution for each branch every time K takes an increment of $2\pi/Na$. Then, the number of solutions for an interval between K and K+dK is given by $\frac{Na}{2\pi}\,dK$. Hence, if we define a distribution function $f(\omega)$ such that the number of modes in a particular branch between frequency ω and $\omega+d\omega$ is $f(\omega)d\omega$, we may make a one-to-one correspondence by setting

$$f(\omega)\,d\omega = \frac{Na}{2\pi}\;dK$$

We are thus led to the solution for $f(\omega)$

$$f(\omega) = (\frac{Na}{2\pi})\,(\frac{d\omega}{dK})^{-1} \qquad (9)$$

This function is plotted for the diatomic linear chain in Figure 5.

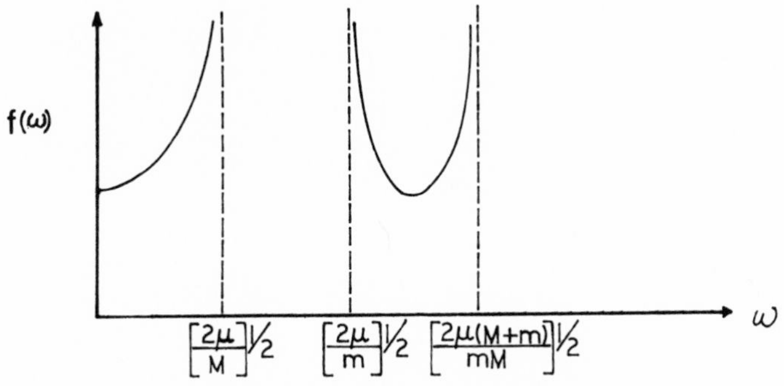

Figure 5 Density of States

Note that $f(\omega)$ diverges near the limits of the optical and acoustic branches. The behaviour of the distribution function is sometimes simulated by placing a delta function at the position of the optical branch equal in weight to N , the number of degrees of freedom. It is clear from the figure that this is a valid approximation when $M >> m$ and the optical band width shrinks in comparison to the acoustic width.

Clearly we get an additional solution to our secular equation each time we add a particle to the unit cell. Hence, if we want ω to be a single valued function of K, we must extend the domain of K by $2\pi/a$ for each particle added per unit cell in the extended zone scheme, or add another branch in the reduced zone scheme.

I.B. Three Dimensional Lattice. Classical Theory

1. Potential Energy. We now generalize our treatment to three dimensions, and arbitrary lattice structure. For simplicity we shall initially limit ourselves to one atom per unit cell, though this will be generalized at an appropriate later stage.

Let a static equilibrium lattice point be denoted by $\underline{X}^{\circ}(\underline{\ell})$. The displacement of this point can then be written as

$$\underline{X}(\underline{\ell}) = \underline{X}^{\circ}(\underline{\ell}) + \underline{u}_{\underline{\ell}} \; . \tag{10}$$

We may expand the potential energy in powers of the deviation from equilibrium

$$V - V_o = \sum_{\underline{\ell}\underline{\ell}'} \frac{1}{2} V_{\alpha\beta}(\underline{\ell},\underline{\ell}')\, u_{\underline{\ell}}^{\alpha}\, u_{\underline{\ell}'}^{\beta} +$$

$$+ \frac{1}{6} \sum_{\underline{\ell}\underline{\ell}'\underline{\ell}''} V_{\alpha\beta\gamma}(\underline{\ell},\underline{\ell}',\underline{\ell}'')\, u_{\underline{\ell}}^{\alpha}\, u_{\underline{\ell}'}^{\beta}\, u_{\underline{\ell}''}^{\gamma} + \ldots. \quad (11)$$

Here u_{ℓ}^{α} is the α component of the vector $\underline{u}_{\underline{\ell}}$ and one sums over repeated indices. There is no term linear in $u_{\underline{\ell}}$ because the set $\{\underline{X}^{o}(\underline{\ell})\}$ is supposed to represent a stable lattice.

2. Equations of Motion. The classical equation of motion is

$$M\, \ddot{u}_{\underline{\ell}}^{\alpha} = -\sum_{\underline{\ell}'} V_{\alpha\beta}(\underline{\ell},\underline{\ell}')\, u_{\underline{\ell}'}^{\beta} \quad (12)$$

where we have assumed the force is derivable from a potential. The necessary and sufficient condition for this to be true in a conservative field is

$$\frac{\partial F_{\underline{\ell}}^{\alpha}}{\partial u_{\underline{\ell}'}^{\beta}} = \frac{\partial F_{\underline{\ell}'}^{\beta}}{\partial u_{\underline{\ell}}^{\alpha}} \,. \quad (13)$$

From above, this implies

$$V_{\alpha\beta}(\underline{\ell},\underline{\ell}') = V_{\beta\alpha}(\underline{\ell}',\underline{\ell}) \,. \quad (13a)$$

If we assume that the V's depend only upon forces acting between atoms, but not necessarily central forces, we may write functionally

$$V_{\alpha\beta}(\underline{\ell},\underline{\ell}') = V_{\alpha\beta}(\underline{\ell}-\underline{\ell}') = V_{\beta\alpha}(\underline{\ell}'-\underline{\ell}) \,. \quad (14)$$

If a center of symmetry exists there is yet another relation upon the $V_{\alpha\beta}(\underline{\ell}-\underline{\ell}')$. Consider the set of points in Figure 6

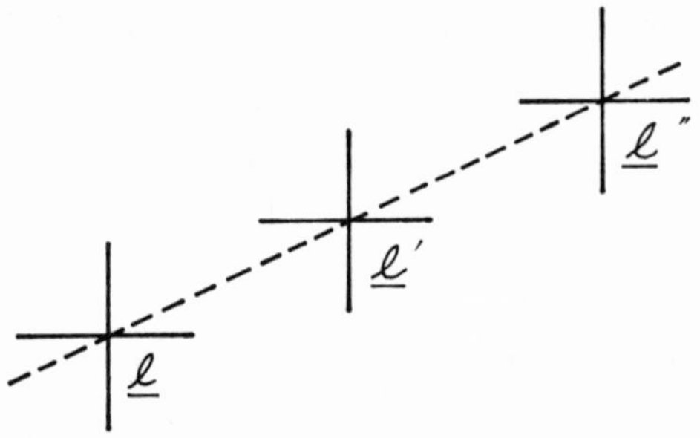

Figure 6 Set of Lattice Points

where $\underline{\ell}''$ can be derived from the inversion (or reflection) of $\underline{\ell}$ through $\underline{\ell}'$. Then by translational invariance,

$$V_{\alpha\beta}(\underline{\ell}-\underline{\ell}') = V_{\alpha\beta}(\underline{\ell}'-\underline{\ell}'')$$

and by inversion through $\underline{\ell}'$

$$= V_{\alpha\beta}(\underline{\ell}'-\underline{\ell}) \tag{15}$$

Thus, if we define $m = \underline{\ell}-\underline{\ell}'$

$$V_{\alpha\beta}(\underline{m}) = V_{\beta\alpha}(-\underline{m}) \tag{16}$$

in general and

$$V_{\alpha\beta}(\underline{m}) = V_{\alpha\beta}(-\underline{m}), \tag{17}$$

if a center of symmetry exists. We now return to the equation of motion (12):

$$M\,\ddot{u}_{\underline{\ell}}^{\,\alpha} = -\sum_{\underline{\ell}'} V_{\alpha\beta}(\underline{\ell},\underline{\ell}')u_{\underline{\ell}'}^{\,\beta} .$$

If all the $\underline{u}_{\underline{\ell}}$ were equal, we would merely be displacing the lattice as a whole and no acceleration would occur. Hence we have the further condition upon the $V_{\alpha\beta}(\underline{\ell},\underline{\ell}')$

$$\sum_{\underline{\ell}'} V_{\alpha\beta}(\underline{\ell},\underline{\ell}') = 0 \; . \tag{18}$$

3. Eigenvalue Spectrum and Eigenvectors. We look for a solution of eq. (12) of the form

$$\underline{u}_{\underline{\ell}} = \underline{e} \exp i\,[\underline{K}\cdot\underline{X}^{o}(\underline{\ell}) - \omega t]. \tag{19}$$

Then, inserting (19) in the equation of motion (12)

$$M\omega^2 e^{\alpha} = \sum_{\underline{\ell}'} V_{\alpha\beta}(\underline{\ell},\underline{\ell}')\, e^{i[\underline{K}\cdot(X^{o}(\underline{\ell}') - X^{o}(\underline{\ell})]}\, e^{\beta} =$$

$$= \sum_{\underline{m}} V_{\alpha\beta}(\underline{m})\, e^{-i\,\underline{K}\cdot\underline{X}^{o}(\underline{m})}\, e^{\beta} \tag{20}$$

where $\underline{X}^{o}(\underline{m}) = \underline{X}^{o}(\underline{\ell}) - X^{o}(\underline{\ell}')$. We set the right-hand side equal to

$$D_{\alpha\beta}(\underline{K})\, e^{\beta}$$

so that

$$D_{\alpha\beta}(\underline{K}) = \sum_{\underline{m}} V_{\alpha\beta}(\underline{m})\, e^{-i\,\underline{K}\cdot\underline{X}^{o}(\underline{m})} \; . \tag{21}$$

We call the matrix D the "dynamical matrix". Equation (20) can now be written

$$M\omega^2 e^{\alpha} = D_{\alpha\beta}(\underline{K})\, e^{\beta} \; . \tag{22}$$

This eigen-equation (where the $\underline{e}$ are the eigenvectors) has a solution only if

$$0 = \begin{vmatrix} [-M\omega^2+D_{XX}(\underline{K})] & D_{XY}(\underline{K}) & D_{XZ}(\underline{K}) \\ D_{YX}(\underline{K}) & [-M\omega^2+D_{YY}(\underline{K})] & D_{YZ}(\underline{K}) \\ D_{ZY}(\underline{K}) & D_{ZY}(\underline{K}) & [-M\omega^2+D_{ZZ}(\underline{K})] \end{vmatrix} . \qquad (23$$

The solution of the determinant will be a function of $\underline{K}$ and will be different in different parts of the zone. This has already been seen in the one dimensional case where the relative amplitudes of atomic motion (within a branch) changed as K went from 0 to π/a.

If there is more than one atom/unit cell, the eigen-equation (20) is modified to

$$M\omega^2\, e^{\alpha}(\underline{k}|\underline{K}) = \sum_{\underline{\ell}'\underline{k}'} V_{\alpha\beta}(\underline{\ell},\underline{k};\underline{\ell}',\underline{k}') \times$$
$$\times \exp\left\{i\, \underline{K}\cdot[\underline{X}^{\circ}(\underline{\ell}') - \underline{X}^{\circ}(\ell)]\right\} e^{\beta}(\underline{k}'|\underline{K}) \,. \qquad (24)$$

The indices $\underline{k}$ and $\underline{k}'$ label the atoms within the cell. The size of the determinant is thus the number of atoms/unit cell times the dimensionality of the crystal. For two atoms per unit cell in one dimension we have a 2 x 2; for two atoms per unit cell in 3 dimensions, a 6 x 6. Note that we have not taken the intra-cell phase change (change in $\underline{\ell}-\underline{\ell}'$ for different atoms in the same cell) into account in (24). This is because we may absorb the phases into the $\underline{e}(\underline{K})$ and work in a reduced zone scheme.

We return to the case of one atom per unit cell. We have shown

$$M\omega^2\, e^{\alpha} = D_{\alpha\beta}(\underline{K})\, e^{\beta} \qquad (25)$$

where

$$D_{\alpha\beta}(\underline{K}) = \sum_{\underline{m}} V_{\alpha\beta}(\underline{m})\, e^{-i\,\underline{K}\cdot\underline{X}^{\circ}(\underline{m})} .$$

Using

$$V_{\alpha\beta}(\underline{m}) = V_{\beta\alpha}(-m)$$

we find

$$D_{\alpha\beta}(\underline{K}) = \sum_{\underline{m}} V_{\beta\alpha}(-\underline{m})\, e^{-i\,\underline{K}\cdot\underline{X}^{o}(\underline{m})} =$$

$$= \sum_{\underline{m}} V_{\beta\alpha}(\underline{m})\, e^{i\,\underline{K}\cdot\underline{X}^{o}(\underline{m})} \tag{26}$$

because $\underline{m}$ is a dummy index. The $V_{\beta\alpha}(\underline{m})$ are real. If we reverse α and β we obtain

$$D_{\beta\alpha}(\underline{K}) = \sum_{\underline{m}} V_{\alpha\beta}(\underline{m})\, e^{i\,\underline{K}\cdot\underline{X}^{o}(\underline{m})} = [D_{\alpha\beta}(\underline{K})]^{*} \tag{27}$$

The dynamical matrix D is then Hermitian independently of the existence of a center of symmetry. There will exist three real solutions for ω^2 which we shall label

$$\omega(\underline{K},1)\ ;\ \omega(\underline{K},2)\ ;\ \omega(\underline{K},3) \tag{28}$$

corresponding to the eigenvectors

$$\underline{e}(\underline{K},1)\ ;\ \underline{e}(\underline{K},2)\ ;\ \underline{e}(\underline{K},3)\ . \tag{29}$$

We label the branches by an index j = 1,2,3 and pick the highest frequency branch (longitudinal) to be associated with j = 1, and the lower (transverse) with j = 2,3. In the strict sense, one has

$$\begin{aligned} &\underline{e}(\underline{K},1) \times \underline{K} = 0 \\ &\underline{e}(\underline{K},2) \cdot \underline{K} = \underline{e}(\underline{K},3) \cdot \underline{K} = 0\ . \end{aligned} \tag{30}$$

In a real crystal it so happens that these relations are obeyed only for certain symmetry directions of $\underline{K}$ and not for an arbitrary direction. For small $\underline{K}$, our solutions for the acoustic branch will reduce to

$$\omega_s = c_s\, K \tag{31}$$

where c_s is the usual low frequency velocity of sound. By construction $c_1 > c_2, c_3$, and in certain symmetry directions, $c_2 = c_3$. It is possible for a "transverse" velocity to exceed a "longitudinal" velocity, but this will be a rare occurrence.

The symmetry properties of the polarization vectors $\underline{e}(\underline{K},j)$ may be determined from (25):

$$M\omega^2 e^{\alpha} = D_{\alpha\beta}(\underline{K})\, e^{\beta}.$$

Taking the complex conjugate

$$M\omega^2(e^{\alpha})^* = [D_{\alpha\beta}(\underline{K})]^* (e^{\beta})^* =$$

$$= \sum_{\underline{m}} V_{\alpha\beta}(\underline{m})\, e^{i\,\underline{K}\cdot\underline{X}^{o}(\underline{m})} (e^{\beta})^* =$$

$$= D_{\alpha\beta}(-\underline{K})\, (e^{\beta})^*$$

But

$$M\omega^2\, e\,(-\underline{K}) = D_{\alpha\beta}(-\underline{K})\, e^{\beta}(-\underline{K})$$

so that

$$\underline{e}(-\underline{K},j) = \underline{e}^*(\underline{K},j). \qquad (32)$$

If a center of symmetry exists it is easy to show that

$$D_{\alpha\beta}(-\underline{K}) = D_{\alpha\beta}(\underline{K})$$

whence

$$\underline{e}(\underline{K},j) = \underline{e}^*(\underline{K},j) \qquad (33)$$

and the polarization vectors $\underline{e}$ are real. This, coupled with the condition (32), implies that if a spatial center of symmetry exists, there exists a $\underline{K}$ space inversion symmetry for the eigenvectors $\underline{e}(\underline{K},j)$.

We can show, however, that $\omega(\underline{K},j) = \omega(-\underline{K},j)$ even in the absence of a center of crystal symmetry. That is, there exists a $\underline{K}$ center of symmetry for ω even though a center of lattice symmetry does not exist. The proof is straightforward:

Consider the eigenvalue determinant (23). We know that ω^2 is real because $D_{\alpha\beta}(\underline{K})$ is Hermitean. We also know that $D_{\alpha\beta}(\underline{K}) = [D_{\alpha\beta}(-K)]^*$. Hence, taking the complex conjugate of the determinant (23):

$$0 = \begin{vmatrix} [-M\omega^2+D_{XX}{}^*(\underline{K})] & D_{XY}{}^*(\underline{K}) & D_{XZ}{}^*(\underline{K}) \\ D_{YX}{}^*(\underline{K}) & [-M\omega^2+D_{YY}(\underline{K})]^* & D_{YZ}{}^*(\underline{K}) \\ D^*_{ZX}(\underline{K}) & D^*_{ZY}(\underline{K}) & -M\omega^2+D^*_{ZZ}(\underline{K})] \end{vmatrix} =$$

$$= \begin{vmatrix} -M\omega^2+D_{XX}(-\underline{K}) & D_{XY}(-\underline{K}) & D_{XZ}(-\underline{K}) \\ D_{YX}(-\underline{K}) & [-M\omega^2+D_{YY}(-\underline{K}) & D_{YZ}(-\underline{K}) \\ D_{ZX}(-\underline{K}) & D_{ZY}(-\underline{K}) & [-M\omega^2+D_{ZZ}(-\underline{K}) \end{vmatrix} .$$

Comparing the above equation with (23) we see that $\omega^2(\underline{K},j)$ is a solution of the same equation as is $\omega^2(-\underline{K},j)$ and our statement is proven.

For complex $\underline{e}$, in general $D_{\alpha\beta}$ will be complex. The real part of $\underline{u}_{\ell}$ can be written

$$\begin{aligned} \text{Re } u_{\ell}^{\alpha} &= \text{Re } e^{\alpha} \exp i\,[\underline{K}\cdot\underline{X}^{o}(\underline{\ell})-\omega t] \\ &= a^{\alpha} \sin [\underline{K}\cdot\underline{X}^{o}(\underline{\ell})-\omega t] + \\ &+ b^{\alpha} \cos [\underline{K}\cdot\underline{X}^{o}(\underline{\ell})-\omega t] \end{aligned} \qquad (34)$$

where

$$\begin{aligned} a^{\alpha} &= -\text{ Im } e^{\alpha} \\ b^{\alpha} &= \text{Re } e^{\alpha} . \end{aligned} \qquad (35)$$

The ionic motion is then given by

$$\underline{u}_{\ell} = \underline{a} \sin [\underline{K}\cdot\underline{\ell}-\omega t] + \underline{b} \cos [\underline{K}\cdot\underline{\ell}-\omega t] \qquad (36)$$

where $\underline{\ell} \equiv \underline{X}^{o}(\underline{\ell})$. $\underline{a}$ and $\underline{b}$ may not be equal or orthogonal; the motion of the ion is then an elliptical path in the $\underline{a}\text{x}\underline{b}$ plane. As $\underline{K}$ has no special relationship to $\underline{e}$, this implies a sort of distorted "screw" polarization. We may achieve linear polarization when $\underline{a}$ is zero, or equivalently, when $\underline{e}$ is real, but these conditions require a center of symmetry or propagation along a symmetry direction.

4. Boundary Conditions and the Reciprocal Lattice. We now must determine those conditions on the wave vector which properly restrict the number of solutions. As we have done before for the linear chain we must first impose periodic boundary conditions. If, in a symmetry direction, there are N cells we set

$$\underline{u}_{\ell} = \underline{u}_{\ell+N} \; . \tag{37}$$

But $\underline{u}_{\ell} = \underline{e} \exp [i\,(\underline{K}\cdot\underline{\ell} - \omega t]$. The condition (37) can be satisfied if

$$e^{i\underline{K}\cdot\underline{N}} = 1 \; . \tag{38}$$

Let the crystal possess N_1, N_2, N_3, cells in the three translational symmetry directions. Then if τ_1, τ_2, τ_3 are the three primitive translation vectors of the unit cell (not necessarily orthogonal) we have from (38) the condition that

$$\begin{aligned} N_1\underline{K}\cdot\underline{\tau}_1 &= 2\pi p_1 \\ N_2\underline{K}\cdot\underline{\tau}_2 &= 2\pi p_2 \\ N_3\underline{K}\cdot\underline{\tau}_3 &= 2\pi p_3 \end{aligned} \tag{39}$$

where p_1, p_2, p_3 are integers. Consider the set of vectors

$$\underline{K} = 2\pi \left\{\frac{p_1}{N_1}\frac{\underline{\tau}_2 \times \underline{\tau}_3}{|\tau_1\tau_2\tau_3|} + \frac{p_2}{N_2}\frac{\underline{\tau}_3 \times \underline{\tau}_1}{|\tau_1\tau_2\tau_3|} + \frac{p_3}{N_3}\frac{\underline{\tau}_1 \times \underline{\tau}_2}{|\tau_1\tau_2\tau_3|}\right\} \tag{40}$$

where $|\tau_1\tau_2\tau_3| = \underline{\tau}_1\cdot(\underline{\tau}_2 \times \underline{\tau}_3) = \Omega_a$, the volume of the unit cell. It is clear that this choice of $\underline{K}$ will satisfy the periodicity requirements (39).

How many solutions are there? There are $N = N_1 \times N_2 \times N_3$ different ways to choose the p. For convenience we choose them to lie in the range $\pm \frac{N_1}{2}$ $\pm \frac{N_2}{2}$ $\pm \frac{N_3}{2}$ respectively so that $\underline{K}$ lies in the region

$$\underline{K} = \pi \left\{ \pm \frac{\underline{\tau}_2 \times \underline{\tau}_3}{|\tau_1\tau_2\tau_3|} \pm \frac{\underline{\tau}_3 \times \underline{\tau}_1}{|\tau_1\tau_2\tau_3|} \pm \frac{\underline{\tau}_1 \times \underline{\tau}_2}{|\tau_1\tau_2\tau_3|} \right\} . \tag{41}$$

This restriction upon $\underline{K}$ is equivalent to the symmetric reduced zone scheme discussed previously in one dimension. With this choice of limits upon the p, we have insured that the correct number N of solutions is obtained.

It is customary to define the primitive translations of the "reciprocal lattice" by

$$\underline{g} = \left(\frac{\underline{\tau}_2 \times \underline{\tau}_3}{|\tau_1\tau_2\tau_3|} , \frac{\underline{\tau}_3 \times \underline{\tau}_1}{|\tau_1\tau_2\tau_3|} , \frac{\underline{\tau}_1 \times \underline{\tau}_2}{|\tau_1\tau_2\tau_3|} \right) \tag{42}$$

where, again, the components of $\underline{g}$ are not necessarily orthogonal.

5. Brillouin Zones. We may now construct a solid figure in reciprocal lattice space which contains just the allowed number of degrees of freedom of the three dimensional real lattice. It is clear that the limits of $\frac{1}{2}\,\underline{g}$ define the bounding planes for this figure. The volume contained therein is called the "Brillouin zone," and the recipe for its construction is: Draw the reciprocal lattice. Bisect the line joining two points of the reciprocal lattice and construct a perpendicular plane. The volume enclosed by all these planes is the first Brillouin zone.

Examples: Consider the body-centered cubic lattice.

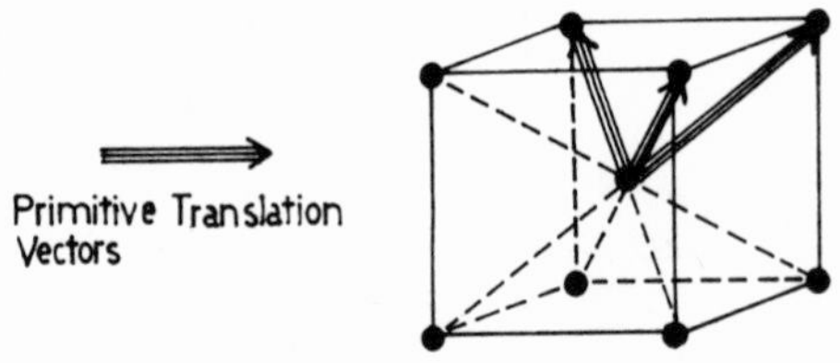

Figure 7 Body-Centered Cubic Lattice.

The primitive translation vectors are:

$$\frac{a}{2}\{(\underline{e}_X + \underline{e}_Y + \underline{e}_Z),(-\underline{e}_X + \underline{e}_Y + \underline{e}_Z),(-\underline{e}_X - \underline{e}_Y + \underline{e}_Z)\} \quad (43)$$

From (42), the reciprocal lattice is constructed from the three primitive translation vectors:

$$\underline{g} = \frac{1}{a}\{(\underline{e}_X + \underline{e}_Z),(-\underline{e}_X + \underline{e}_Y),(-\underline{e}_Y + \underline{e}_Z)\} \quad (44)$$

The resulting Brillouin Zone looks like a rhombic dodecahedron:

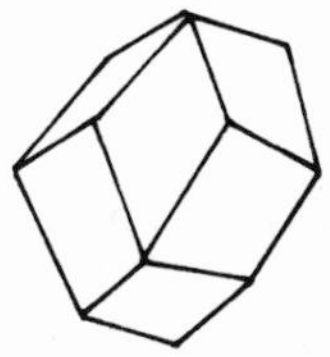

Figure 8 First Brillouin Zone for a Body-Centered Cubic Lattice.

The rhombic dodecahedron happens to be a primitive cell for the face-centered cubic lattice. This can be seen using the primitive translation vectors for the face-centered cubic lattice (note the similarity to (44))

$$\frac{a}{2}\{(\underline{e}_X + \underline{e}_Z),(-\underline{e}_X + \underline{e}_Y),(-\underline{e}_Y + \underline{e}_Z)\} \quad (45)$$

The reciprocal lattice vectors are

$$\underline{g} = \frac{1}{a}\{(\underline{e}_X + \underline{e}_Y + \underline{e}_Z),(-\underline{e}_X + \underline{e}_Y + \underline{e}_Z),(-\underline{e}_X - \underline{e}_Y + \underline{e}_Z)\} \quad (46)$$

which are, as stated above, the primitive translation vectors for the body-centered cubic lattice. The body-centered cubic and the face-centered cubic lattices are thus real-reciprocal lattice inverses of one another. The first face-centered cubic real lattice Brillouin zone looks like the shape in Figure 9.

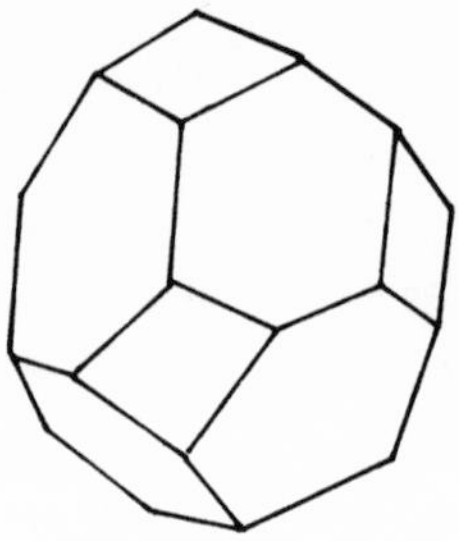

Figure 9 First Brillouin Zone for a Face-Centered Cubic Lattice.

The utility of the zones and/or lattices is clear. By construction, the termini of the wave-vectors $\underline{K}$ which give the correct number of solutions and which satisfy the periodic boundary conditions, make up the surface of the first Brillouin zone (times π). Rephrased, we have constructed a space using multiples of the wave vectors which were constructed under the condition of periodic boundary conditions. We have found what volume (and shape) of a solid figure formed from these wave vectors contains exactly the correct number of solutions ($N = N_1 \cdot N_2 \cdot N_3$). We have identified the figure with the first Brillouin zone and have given a recipe for its construction. From the point of view of the reciprocal lattice, it contains the volume associated with a single cell in reciprocal lattice space.

There are further repercussions. By construction

$$-\frac{N_1}{2} \lesssim p_1 < \frac{N_1}{2}$$

$$-\frac{N_2}{2} \lesssim p_2 < \frac{N_2}{2}$$

$$-\frac{N_3}{2} \lesssim p_3 < \frac{N_3}{2}$$

What would have happened if $\frac{N_1}{2} \lesssim p_1$? As we have seen before the solutions had been correctly counted when we set $p_1 < N_1/2$.

Thus, such a situation corresponds to a repetition of previous solutions, but with $\underline{K}$ in a zone other than the first. Because

$$\underline{u}_{\ell} = \underline{e} \exp [i(\underline{K}\cdot\underline{\ell} - \omega t)],$$

letting $p_1 \geqslant N_1/2$ (i.e. $\underline{K}_1 > \pi \dfrac{\underline{\tau}_2 \times \underline{\tau}_3}{|\tau_1\tau_2\tau_3|} \equiv \pi\underline{g}_1$) is entirely equivalent to $\underline{K}_1 > -\pi \dfrac{\underline{\tau}_2 \times \underline{\tau}_3}{|\tau_1\tau_2\tau_3|}$ by the same amount. That is, a shift by N_1 in p_1 changes K_1 by $2\pi \cdot \dfrac{\underline{\tau}_2 \times \underline{\tau}_3}{|\tau_1\tau_2\tau_3|}$ and thus changes $\underline{K}\cdot\underline{\ell}$ by $2\pi n_1$, (where n_1 = integer). The expression for $\underline{u}_{\ell}$ is then undisturbed.

Its solutions are periodic in p space with period ($p_1 = N_1$, $p_2 = N_2$, $p_3 = N_3$) or periodic in $\underline{K}$ space with period $2\pi\underline{g}$. Again, our result is that any solution for $\underline{K}$ outside the first zone may be re-expressed as a solution inside the first zone by a simple translation of 2π times a reciprocal lattice vector.

If there are η atoms per unit cell, then the "zone" is η times its one atom per unit cell volume. The difference between modes for which $\underline{K}$ is external rather than internal to the first zone is a matter of the relative phases of the different atoms within the unit cell. We may condense the larger volume into the first zone by absorbing these phases into $\underline{e}$ and introducing η sets of $\underline{e}$ instead of only one. These additional solutions are of course what has been referred to previously as "optical modes."

6. Energy Surface Theorem. It is possible to prove an interesting theorem using the properties of the Brillouin zone. Theorem: "If a center of symmetry exists in a crystal, then the directional derivative of $\omega(\underline{K},j)$ with respect to $\underline{K}$ vanishes at the center of a Brillouin zone face."

The proof is as follows. The secular equation determining $\omega(\underline{K},j)$ is

$$M\omega^2 e^{\alpha} = \sum_{\underline{m}} V_{\alpha\beta}(\underline{m})\, e^{-i\underline{K}\cdot\underline{m}}\, e^{\beta}.$$

Taking the directional derivative parallel to an arbitrary vector $\underline{K}'$ on both sides yields

$$2M\omega(\underline{K},j)e^{\alpha}\nabla_{\underline{K}'}\, \omega(\underline{K},j) = \sum_{\underline{m}} V_{\alpha\beta}(\underline{m}) \times \left(-\, i\, \frac{\underline{m}\cdot\underline{K}'}{|\underline{K}'|}\right) e^{-i\underline{K}\cdot\underline{m}}\, e^{\beta}. \qquad (47)$$

If a center of symmetry exists, then for every $\underline{m}$, there must also exist a $-\underline{m}$. Using $V_{\alpha\beta}(-\underline{m}) = V_{\alpha\beta}(\underline{m})$,

$$e^{\alpha}\, 2M\omega(\underline{K},j)\, \nabla_{\underline{K}'}\, \omega(\underline{K},j) =$$

$$= \sum_{|\underline{m}|} V_{\alpha\beta}(\underline{m}) \left(-i \frac{\underline{K}'}{|\underline{K}'|}\right)\cdot(\underline{m}\, e^{-i\underline{K}\cdot\underline{m}} - \underline{m}\, e^{i\underline{K}\cdot\underline{m}})\, e^{\beta} =$$

$$= - \sum_{|\underline{m}|} 2V_{\alpha\beta}(\underline{m}) \frac{\underline{K}'\cdot\underline{m}}{|\underline{K}'|} \sin(\underline{K}\cdot\underline{m})\, e^{\beta}. \tag{48}$$

At the center of the zone face it has already been shown that $\underline{K}$ is a multiple of

$$\pi\underline{g} = \pi \left\{ \pm \frac{\tau_2 \times \tau_3}{|\tau_1\tau_2\tau_3|} \pm \frac{\underline{\tau}_3 \times \underline{\tau}_1}{|\tau_1\tau_2\tau_3|} \pm \frac{\underline{\tau}_1 \times \underline{\tau}_2}{|\tau_1\tau_2\tau_3|} \right\} . \tag{49}$$

Because $\underline{m} = n_1 \underline{\tau}_1 + n_2\underline{\tau}_2 + n_3\underline{\tau}_3$
where n_1 , n_2 , n_3 are integers it is clear that the term

$$\sin(\underline{K}\cdot\underline{m}) = \sin(\pi\underline{g}\cdot\underline{m}) = 0$$

and either $\omega(\underline{K},j)$ or $\nabla_{\underline{K}'}\omega(\underline{K},j)$ must vanish. Thus, at the center of the zone, where $\underline{K}\cdot\underline{m} = 0$, we have shown that if $\omega(\underline{K},j)\neq 0$, $\nabla_{\underline{K}'}\omega(\underline{K},j) = 0$ for arbitrary $\underline{K}'$. The optical modes must thus come in flat at $\underline{K} = 0$ for arbitrary directions of $\underline{K}$. The same is true at the center of the zone faces for both acoustic and optical modes. It appears difficult to make any general statement concerning the value of $\nabla_{\underline{K}'}\omega(\underline{K},j)$ at the zone surface when a center of symmetry does not exist in the crystal. However, the following geometrical argument, in the absence of degeneracies, indicates those regions where the derivative of $\omega(\underline{K},j)$ normal to the surface of the Brillouin zone vanishes.

Consider a plane surface of an arbitrary Brillouin zone. Let $\omega(\underline{K}_s,j)$ be the eigenfrequency for the wave vector which terminates upon that face. By construction the wave vectors which terminate at the opposite face are $\underline{K}_s - 2\pi\underline{g}$ and $\omega(\underline{K}_s,j) = \omega(\underline{K}_s - 2\pi\underline{g},j)$. Construct the normal derivative to the zone face

$$\nabla_{\underline{K}'}(\underline{K},j)\Big|_{\substack{K=K_s \\ \underline{K}'\perp \text{ surface}}} .$$

By construction

$$\nabla_{\underline{K}'}\omega(\underline{K},j)\Big|_{\substack{\underline{K}=\underline{K}_s \\ \underline{K}'\perp \text{ surface}}} = \nabla_{\underline{K}'}\omega(\underline{K}-2\pi\underline{g},j)\Big|_{\substack{\underline{K}=\underline{K}_s \\ \underline{K}'\perp \text{ surface}}} \qquad (50)$$

If there exists a mirror plane of symmetry for the Brillouin zone which is at right angles to $\underline{K}'$, then

$$\nabla_{\underline{K}'}\omega(\underline{K},j)\Big|_{\substack{\underline{K}=\underline{K}_s \\ \underline{K}'\perp \text{ surface}}} = -\nabla_{\underline{K}'}\omega(-\underline{K},j)\Big|_{\substack{\underline{K}=\underline{K}_s \\ \underline{K}'\perp \text{ surface}}} \cdot \qquad (51)$$

These equations may be satisfied for nondegenerate energy surfaces only if the normal derivative vanishes. We have not had to resort to an explicit crystal structure, but we have had to assume a mirror plane of symmetry perpendicular to the direction of differentiation.

I.C.   Three Dimensional Lattice.   Quantum Theory

1.   <u>Introduction of Coordinates</u>.   We now quantize the theory of lattice vibrations.   Let us set

$$\underline{u}_{\underline{\ell}} = \sum_{\underline{K},j} q_{\underline{K},j}(t)\ \underline{\ell}(\underline{K},j)\ \exp(i\underline{k}\cdot\underline{\ell}) \cdot \qquad (52)$$

The $\underline{u}_{\underline{\ell}}$ are observable and hence are operators, as will be the $q_{\underline{K},j}(t)$.   We may write

$$q_{\underline{K},j}(t) = Q_{\underline{K},j}(0)\ \exp(-i\omega(\underline{K},j)t)$$

$$+ P_{\underline{K},j}(0)\ \exp(i\omega(\underline{K},j)t). \qquad (53)$$

This follows from the fact that, for any observable operator σ, the matrix element

$$(\dot{\sigma})_{nm} = i\omega_{nm}\ \sigma_{nm} = (i/\hbar)(E_n-E_m)\ \sigma_{nm} \qquad (54)$$

where $H\psi_n = E_n\psi_n$.   By introducing both $Q_{\underline{K},j}$ and $P_{\underline{K},j}$, we are allowing for $E_n-E_m < 0$ with a frequency greater than zero.   Thus, ω is to be regarded as a positive definite quantity.   Combining,

$$\underline{u}_\ell = \sum_{\underline{K},j} Q_{\underline{K},j}(0)\, \underline{e}\,(\underline{K},j)$$

$$\times \exp i\,(\underline{K}\cdot\underline{\ell} - \omega(\underline{K},j)t) + P_{\underline{K},j}\,(0)\, \underline{e}\,(\underline{K},j)$$

$$\times \exp i\,(\underline{K}\cdot\underline{\ell} + \omega(\underline{K},j)t). \tag{55}$$

We may now use the fact that $\underline{u}_\ell$ is an observable, and hence that $\underline{u}_\ell$ is Hermitian. Then the Hermitian adjoint of $\underline{u}_\ell$ is

$$\underline{u}_\ell^\dagger = \sum_{\underline{K},j} Q_{\underline{K},j}^\dagger\,(0)\, \underline{e}^*(\underline{K},j)$$

$$\times \exp i\,(-\,\underline{K}\cdot\underline{\ell} + \omega(\underline{K},j)t) + P_{\underline{K},j}^\dagger\,(0)\, \underline{e}^*(\underline{K},j)$$

$$\times \exp i\,(-\,\underline{K}\cdot\underline{\ell} - \omega(\underline{K},j)t). \tag{56}$$

But from (32) $\underline{e}^*(\underline{K},j) = \underline{e}\,(-\,\underline{K},j)$. Letting $\underline{K} \to -\,\underline{K}$ in the sum

$$\underline{u}_\ell^\dagger = \sum_{\underline{K},j} Q_{-\underline{K},j}^\dagger\,(0)\, \underline{e}\,(\underline{K},j)$$

$$\times \exp i\,(\underline{K}\cdot\underline{\ell} + \omega(\underline{K},j)t) + P_{-\underline{K},j}^\dagger\,(0)\, \underline{e}\,(\underline{K},j)$$

$$\times \exp i\,(\underline{K}\cdot\underline{\ell} - \omega(\underline{K},j)t) \tag{57}$$

where we have also used $\omega(\underline{K},j) = \omega(-\,\underline{K},j)$. We equate this result with $\underline{u}_\ell$ as given by (55) to see that

$$P_{-\underline{K},j}^\dagger\,(0) = Q_{\underline{K},j}\,(0)$$

$$P_{\underline{K},j}\,(0) = Q_{-\underline{K},j}^\dagger\,(0)\;. \tag{58}$$

Eliminating $P_{\underline{K},j}\,(0)$ we find

$$\underline{u}_{\ell} = \sum_{\underline{K},j} \{ Q_{\underline{K},j}(0)\, e(\underline{K},j) \exp i(\underline{K}\cdot\underline{\ell} - \omega(\underline{K},j)t)$$

$$+ Q^{\dagger}_{-\underline{K},j}(0)\, \underline{e}(\underline{K},j) \exp i(\underline{K}\cdot\underline{\ell} + \omega(\underline{K},j)t) \}, \quad (59)$$

or, letting $\underline{K} \to -\underline{K}$ in the second term,

$$\underline{u}_{\ell} = \sum_{\underline{K},j} \{ Q_{\underline{K},j}(0)\, \underline{e}(\underline{K},j) \exp i(\underline{K}\cdot\underline{\ell} - \omega(\underline{K},j)t)$$

$$+ Q^{\dagger}_{\underline{K},j}(0)\, \underline{e}^{*}(\underline{K},j) \exp i(-\underline{K}\cdot\underline{\ell} + \omega(\underline{K},j)t) \}. \quad (60)$$

We define

$$Q_{\underline{K},j}(t) = Q_{\underline{K},j}(0) \exp - i(\omega(\underline{K},j)t) \quad (61)$$

so that

$$\underline{u}_{\ell} = \sum_{\underline{K},j} \{ Q_{\underline{K},j}(t)\, \underline{e}(\underline{K},j) \exp i(\underline{K}\cdot\underline{\ell})$$

$$+ Q^{\dagger}_{\underline{K},j}(t)\, \underline{e}^{*}(\underline{K},j) \exp - i(\underline{K}\cdot\underline{\ell}) \}. \quad (62)$$

2. Commutation Relations. In order to study the Q, we invert by taking the dot product of $\underline{u}_{\ell}$ with

$$\underline{e}^{*}(\underline{K}',j') \exp - i(\underline{K}'\cdot\underline{\ell}).$$

We find,

$$\underline{u}_{\ell} \cdot \underline{e}^{*}(\underline{K}',j') \exp - i\,(\underline{K}'\cdot\underline{\ell}) = \sum_{\underline{K},j} \exp - i\,(\underline{K}'\cdot\underline{\ell})$$

$$\times\ \underline{e}^{*}(\underline{K}',j')\cdot[Q_{\underline{K},j}(t)\ \underline{e}\,(\underline{K},j) \exp i\,(\underline{K}\cdot\underline{\ell})$$

$$+\ Q_{\underline{K},j}^{\dagger}(t)\ \underline{e}^{*}(\underline{K},j) \exp - i\,(\underline{K}\cdot\underline{\ell})]. \qquad (63)$$

Summing over $\underline{\ell}$,

$$\sum_{\underline{\ell}} \underline{u}_{\ell} \cdot \underline{e}^{*}(\underline{K}',j') \exp[-i\,(\underline{K}'\cdot\underline{\ell})]$$

$$= \sum_{\underline{K},j} \sum_{\underline{\ell}} \exp i\,[(\underline{K}-\underline{K}')\cdot\underline{\ell}]\ Q_{\underline{K},j}(t)\ \underline{e}^{*}(\underline{K}',j')\cdot\underline{e}(\underline{K},j)$$

$$+ \sum_{\underline{K},j} \sum_{\underline{\ell}} \exp - i\,[(\underline{K}+\underline{K}')\cdot\underline{\ell}]Q_{\underline{K},j}^{\dagger}(t)\ \underline{e}^{*}(\underline{K}',j')\cdot\underline{e}^{*}(\underline{K},j). \qquad (64)$$

Using the definition of $\underline{\ell}$:

$$\underline{\ell} = n_1\underline{\tau}_1 + n_2\underline{\tau}_2 + n_3\underline{\tau}_3;\ n_1,n_2,n_3 \text{ integers},$$

we note

$$\sum_{\underline{\ell}} \exp[i\,(\underline{K}-\underline{K}')\cdot\underline{\ell}]$$

$$= \sum_{n_1,n_2,n_3} \exp[2\pi i\,(n_1/N_1)\,(p_1 - p_1')]$$

$$\times\ \exp[2\pi i\,(n_2/N_2)\,(p_2 - p_2')]$$

$$\times\ \exp[2\pi i\,(n_3/N_3)\,(p_3 - p_3')]. \qquad (65)$$

The sums are separable. Using the geometric progression formulae , we find:

$$\sum_{n_1=0}^{N-1} \exp\,[2\pi i\,(n_1/N_1)\,(p_1 - p_1')] = \frac{1 - \exp\,[2\pi i\,(p_1 - p_1')]}{1 - \exp\,[2\pi i\,(p_1 - p_1')/N_1]} \tag{66}$$

with similar expressions for n_2 and n_3. Because p and p' are integers, the numerator vanishes always. The sum differs from zero (perhaps) only when the denominator vanishes as well. This occurs for $p_1 = p_1'(p_1 < N_1/2)$, or for $p_1 = p_1' + N_1$. The latter equality implies a change in $\underline{K}$, of $2\pi\underline{g}$, and is referred to as Umklapp. In the absence of Umklapp, the sum equals N_1. Thus

$$\sum_{\underline{\ell}} \exp\,[i\,(\underline{K} - \underline{K}')\cdot\ell] = N_1 N_2 N_3\,\delta_{\underline{K},\underline{K}'}\;. \tag{67}$$

Setting the total number of atoms $N=N_1\times N_2\times N_3$, our equation becomes

$$\begin{aligned}\sum_{\underline{\ell}} \underline{u}_{\underline{\ell}}\cdot\,\underline{e}^*(\underline{K}',j')\,\exp - i\,(\underline{K}'\cdot\underline{\ell}) &= N\sum_j Q_{\underline{K}',j}\,(t)\;\underline{e}^*(\underline{K}',j')\cdot\,\underline{e}\,(\underline{K}',j)\\ &+ N\sum_j Q_{-\underline{K}',j}^{\dagger}\,(t)\;\underline{e}^*(\underline{K}',j')\cdot\,\underline{e}^*(-\underline{K}',j)\;.\end{aligned} \tag{68}$$

But $\underline{e}^*(\underline{K}',j')\cdot\,\underline{e}\,(\underline{K}',j) = \delta_{j,j'}$ because the eigenvectors $\underline{e}$ are orthonormal. Dividing through by N we find

$$\frac{1}{N}\sum_{\underline{\ell}} \exp\left\{- i\,(\underline{K}'\cdot\underline{\ell})\right\}\underline{u}_{\underline{\ell}}\;\cdot\;\underline{e}^*(\underline{K}',j') = Q_{\underline{K}',j'}\,(t) + Q_{-\underline{K}',j'}^{\dagger}\,(t). \tag{69}$$

Next, this result (69) is differentiated with respect to time:

$$\frac{1}{N}\sum_{\underline{\ell}} \exp\left\{-\, i\, (\underline{K}'\cdot\underline{\ell}\,\right\}\dot{\underline{u}}_{\underline{\ell}}\cdot \underline{e}^{*}(\underline{K}',j')$$

$$= +\, i\, \omega\, (Q^{\dagger}_{-\underline{K}',j'}\,(t) - Q_{\underline{K}',j'}\,(t))\,. \qquad (70)$$

Dropping the primes and adding the two equations:

$$\frac{1}{N}\sum_{\underline{\ell}} \exp\left\{-\, i\, (\underline{K}\cdot\underline{\ell})\right\}\underline{e}^{*}(\underline{K},j)\cdot(i\omega(\underline{K},j)\ \underline{u}_{\underline{\ell}} + \dot{\underline{u}}_{\underline{\ell}})$$

$$= 2\, i\, \omega\, (\underline{K},j)\ Q^{\dagger}_{-\underline{K},j}\,(t). \qquad (71)$$

Upon subtraction of the two equations:

$$-\, 2\, i\, \omega\, (\underline{K},j)\ Q_{\underline{K},j}\,(t) = \frac{1}{N}\sum_{\underline{\ell}} \exp\left\{-\, i\, (\underline{K}\cdot\underline{\ell})\right\}$$

$$\times\ \underline{e}^{*}(\underline{K},j)\cdot(-\, i\, \omega(\underline{K},j)\ \underline{u}_{\underline{\ell}} + \dot{\underline{u}}_{\underline{\ell}}). \qquad (72)$$

Thus, we arrive at the important relations for the Fourier amplitudes Q,

$$Q_{\underline{K},j}\,(t) = \frac{1}{2N}\sum_{\underline{\ell}} \exp\left\{-\, i\, (\underline{K}\cdot\underline{\ell})\right\}\underline{e}^{*}(\underline{K},j)\cdot$$

$$\cdot\ [\underline{u}_{\underline{\ell}} + i\, \dot{\underline{u}}_{\underline{\ell}}/\omega(\underline{K},j)], \qquad (73)$$

and

$$Q^{\dagger}_{\underline{K},j}\,(t) = \frac{1}{2N}\sum_{\underline{\ell}} \exp\, i\left\{(\underline{K}\cdot\underline{\ell})\right\}\underline{e}\ (\underline{K},j)\ \cdot$$

$$\cdot\ [\underline{u}_{\underline{\ell}} - i\, \dot{\underline{u}}_{\underline{\ell}}/\omega\, (\underline{K},j)]. \qquad (74)$$

We have succeeded in inverting our expression for $Q_{\underline{K},j}(t)$.

To find the commutation laws which they obey we note that $\underline{u}_{\underline{\ell}}$ are operators satisfying

$$[u_{\underline{\ell}'}^{\alpha}, u_{\underline{\ell}}^{\beta}] = 0, \quad [\dot{u}_{\underline{\ell}'}^{\alpha}, \dot{u}_{\underline{\ell}}^{\beta}] = 0$$

$$[\dot{u}_{\underline{\ell}'}^{\alpha}, u_{\underline{\ell}}^{\beta}] = (-i\hbar/M)\, \delta_{\underline{\ell},\underline{\ell}'}\delta_{\alpha,\beta} \,. \tag{75}$$

To make use of these expressions, we form the expression

$$[Q_{\underline{K},j}(t), Q^{\dagger}_{\underline{K}',j'}(t)] = (1/4\,N^2) \sum_{\underline{\ell}} \sum_{\underline{\ell}'}$$

$$\times \exp i\,(\underline{K}'\cdot\underline{\ell}') \exp\left\{- i\,(\underline{K}\cdot\underline{\ell})\right\} e^{\alpha\,*}(\underline{K},j)$$

$$\times e^{\beta}(\underline{K}',j')\,[(u_{\underline{\ell}}^{\alpha}(t) + i\,\dot{u}_{\underline{\ell}}^{\alpha}(t)/\omega(\underline{K},j))\,,$$

$$(u_{\underline{\ell}'}^{\beta}(t) - i\,\dot{u}_{\underline{\ell}'}^{\beta}(t)/\omega(\underline{K}',j'))] \,. \tag{76}$$

This expression is finite only when $\underline{\ell} = \underline{\ell}'$, $\alpha = \beta$ because of the commutation relations obeyed by the $\underline{u}_{\underline{\ell}}$. Thus

$$[Q_{\underline{K},j}(t), Q^{\dagger}_{\underline{K}',j'}(t)] = (1/4\,N^2) \sum_{\underline{\ell}}$$

$$\times \exp\,[i\,(\underline{K}'- \underline{K})\cdot\underline{\ell}]\; e^{\alpha\,*}(\underline{K},j)\; e^{\alpha}(\underline{K}',j')$$

$$\times \left\{ [u_{\underline{\ell}}^{\alpha}(t), - i\,\dot{u}_{\underline{\ell}}^{\alpha}(t)/\omega(\underline{K}',j')] + \right.$$

$$\left. + \quad [i\,\dot{u}_{\underline{\ell}}^{\alpha}(t)/\omega(\underline{K},j), u_{\underline{\ell}}^{\alpha}(t)] \right\} . \tag{77}$$

Using the commutation relations (75), we find

$$[Q_{\underline{K},j}(t), Q^{\dagger}_{\underline{K}',j'}(t)] = (1/4\,N^2)\sum_{\underline{\ell}}$$

$$\times \exp[i(\underline{K}' - \underline{K})\cdot\underline{\ell}]\, e^{\alpha\,*}(\underline{K},j) e^{\alpha}(\underline{K}',j') \times$$

$$\frac{\hbar}{M}\left(\frac{1}{\omega(\underline{K}',j')} + \frac{1}{\omega(\underline{K},j)}\right) . \tag{78}$$

The sum over $\underline{\ell}$ gives $N\delta_{\underline{K}',\underline{K}}$ and the sum over α then gives $\delta_{j',j}$. Finally

$$[Q_{\underline{K},j}(t), Q^{\dagger}_{\underline{K}',j'}(t)] = \frac{\hbar}{2NM\omega(\underline{K},j)}\,\delta_{\underline{K},\underline{K}'}\,\delta_{j,j'} . \tag{79}$$

3. Energy Matrix. We wish now to compute the total internal energy $T + (V - V_o)$ using the expressions for $\underline{u}_{\ell}$ we have derived above.

The kinetic energy equals

$$T = \frac{1}{2} M \sum_{\underline{\ell}} \dot{\underline{u}}_{\ell}^{\,2} . \tag{80}$$

Using (62), the time derivative of the displacement $\underline{u}_{\ell}$ is

$$\dot{\underline{u}}_{\ell} = \sum_{\underline{K},j} \{ - i\,\omega(\underline{K},j)\, Q_{\underline{K},j}(t)\, \underline{e}(\underline{K},j)$$

$$\times \exp i(\underline{K}\cdot\underline{\ell}) + i\,\omega(\underline{K},j)\, Q^{\dagger}_{\underline{K},j}(t)\, \underline{e}^{*}(\underline{K},j) \tag{81}$$

$$\times \exp - i(\underline{K}\cdot\underline{\ell}) \} .$$

This enables us to rewrite the kinetic energy (80) as

$$T = \sum_{\underline{\ell}} \sum_{\underline{K},j} \sum_{\underline{K}',j'} [\{ \tfrac{1}{2} M\omega(\underline{K},j)\omega(\underline{K}',j')\ \underline{e}\ (\underline{K},j)\cdot\ \underline{e}^*(\underline{K}',j')$$

$$\times \exp i\ (\underline{K} - \underline{K}')\cdot\underline{\ell}\ [Q_{\underline{K},j}\ (t)\ Q^{\dagger}_{\underline{K}',j'}\ (t) +$$

$$+ Q^{\dagger}_{\underline{K}',j'}\ (t)\ Q_{\underline{K},j}\ (t)]\} - \{ \tfrac{1}{2} M\omega(\underline{K},j)\omega(\underline{K}',j')$$

$$\times\ \underline{e}\ (\underline{K},j)\cdot\ \underline{e}\ (\underline{K}',j')\ \exp [i\ (\underline{K} + \underline{K}')\cdot\underline{\ell}]\ Q_{\underline{K},j}\ (t)$$

$$\times\ Q_{\underline{K}',j'}\ (t)\} - \{ \tfrac{1}{2} M\omega(\underline{K},j)\omega(\underline{K}',j')$$

$$\times\ \underline{e}^*(\underline{K},j)\cdot\ \underline{e}^*(\underline{K}'\cdot\ j')\ \exp [-\ i\ (\underline{K} + \underline{K}')\cdot\underline{\ell}]$$

$$\times\ Q^{\dagger}_{\underline{K},j}\ (t)\ Q^{\dagger}_{\underline{K}',j'}\ (t)\ \}]. \qquad (82)$$

We sum over $\underline{\ell}$ which results in a term $N\delta_{\underline{K}',\underline{K}}$ in the first curly bracket, $N\delta_{\underline{K}',-\underline{K}}$ in the second and third curly brackets, and $\delta_{j',j}$ in all three because of the orthonormality of the $\underline{e}$. (Note that $\underline{e}\ (\underline{K},j)\cdot\underline{e}^*(\underline{K}',j') \neq \delta_{j',j}$; this equality holds only when $\underline{K} = \underline{K}'$).

We find, upon simplifying,

$$T = \tfrac{1}{2} MN \sum_{\underline{K},j} \omega^2\ (\underline{K},j)\ \{ Q_{\underline{K},j}\ (t)\ Q^{\dagger}_{\underline{K},j}\ (t) +$$

$$+ Q^{\dagger}_{\underline{K},j}\ (t)\ Q_{\underline{K},j}\ (t)\} - \tfrac{1}{2} MN \sum_{\underline{K},j} \omega^2\ (\underline{K},j)$$

$$\times \{ Q_{\underline{K},j}\ (t)\ Q_{-\underline{K},j}\ (t) + Q^{\dagger}_{\underline{K},j}\ (t)\ Q^{\dagger}_{-\underline{K},j}\ (t)\ \}. \qquad (83)$$

The terms in the second curly bracket are troublesome. They are explicity time dependent, varying as $e^{\pm 2i\omega t}$. We shall show below that they cancel with similar terms in the potential energy. The total Hamiltonian will thus turn out to be stationary, though the separate kinetic and potential energy parts will not. The potential energy, $V-V_o$, is given by (11)

$$V-V_o = \sum_{\underline{\ell},\underline{\ell}'} \frac{1}{2} V_{\alpha\beta} (\underline{\ell},\underline{\ell}') \, u_{\underline{\ell}}^{\alpha} \, u_{\underline{\ell}'}^{\beta}$$

$$= \frac{1}{2} \sum_{\underline{\ell},\underline{\ell}'} V_{\alpha\beta} (\underline{\ell},\underline{\ell}') \sum_{\underline{K},j} \sum_{\underline{K}',j'} \{ Q_{\underline{K},j} (t) \, e^{\alpha} (\underline{K},j)$$

$$\times \exp\left(i \, \underline{K}\cdot\underline{\ell}\right) + Q_{\underline{K},j}^{\dagger} (t) \, e^{\alpha *}(\underline{K},j) \exp - i \, \underline{K}\cdot\underline{\ell} \}$$

$$\times \{ Q_{\underline{K}',j'} (t) \, e^{\beta} (\underline{K}',j') \exp i\left(\underline{K}'\cdot\underline{\ell}'\right) + Q_{\underline{K}',j'}^{\dagger}(t)$$

$$\times e^{\beta *}(\underline{K}',j') \exp - i \, \underline{K}'\cdot\underline{\ell}' \} . \tag{84}$$

We examine a typical term:

$$\frac{1}{2} \sum_{\underline{\ell},\underline{\ell}'} V_{\alpha\beta} (\underline{\ell},\underline{\ell}') \sum_{\underline{K},j} \sum_{\underline{K}',j'} \{ Q_{\underline{K},j} (t) \, e^{\alpha} (\underline{K},j)$$

$$\times \exp i \, (\underline{K}\cdot\underline{\ell}) \, Q_{\underline{K}',j'}^{\dagger} (t) \, e^{\beta *}(\underline{K}',j') \exp - i \, (\underline{K}'\cdot\underline{\ell}') \}$$

$$= \frac{1}{2} \sum_{\underline{\ell}'} \sum_{\underline{K},j} \sum_{K',j'} [\sum_{\underline{\ell}} V_{\beta\alpha} (\underline{\ell}',\underline{\ell}) \, e^{\alpha} (\underline{K},j)$$

$$\times \exp i \, \underline{K}\cdot(\underline{\ell} - \underline{\ell}')] \, Q_{\underline{K},j} (t) \, Q_{\underline{K}',j'}^{\dagger} (t) \, e^{\beta *}(\underline{K}',j')$$

$$\times \exp - i \, (\underline{K}' - \underline{K})\cdot\underline{\ell}' . \tag{85}$$

From (20), the term in square brackets is just $M\omega^2 (\underline{K},j)e^{\beta}(\underline{K},j)$. The right side of (85) can then be rewritten as

$$= \frac{1}{2} M \sum_{\underline{K},j} \sum_{\underline{K}',j'} \sum_{\underline{\ell}'} \omega^2 (\underline{K},j) \, e^{\beta} (\underline{K},j)$$

$$\times Q_{\underline{K},j} (t) \, Q_{\underline{K}',j'}^{\dagger} (t) \, e^{\beta *}(\underline{K}',j') \exp - i \, (\underline{K}' - \underline{K})\cdot\underline{\ell}' . \tag{86}$$

The sum over $\underline{\ell}'$ contributes a term $N\delta_{\underline{K}',\underline{K}}$, so that the expression can be simplified to

$$= \frac{1}{2} NM \sum_{\underline{K},j} \omega^2 (\underline{K},j) \, Q_{\underline{K},j} (t) \, Q^{\dagger}_{\underline{K},j} (t) \, . \tag{87}$$

The other terms develop similarly, so that we find finally,

$$V-V_o = \frac{1}{2} MN \sum_{\underline{K},j} \{ Q_{\underline{K},j} (t) \, Q^{\dagger}_{\underline{K},j} (t) +$$

$$+ Q^{\dagger}_{\underline{K},j} (t) \, Q_{\underline{K},j} (t) \} \, \omega^2 (\underline{K},j) +$$

$$+ \frac{1}{2} MN \sum_{\underline{K},j} \{ Q_{\underline{K},j} (t) \, Q_{-\underline{K},j} (t) +$$

$$+ Q^{\dagger}_{\underline{K},j} (t) \, Q^{\dagger}_{-\underline{K},j} (t) \} \, \omega^2 (\underline{K},j) \, . \tag{88}$$

The second curly brackets in $V-V_o$ just cancels the offending terms in T, as promised earlier. Adding, we find

$$H = T + \left(V-V_o\right) = NM \sum_{\underline{K},j} \omega^2 (\underline{K},j) \{ Q^{\dagger}_{\underline{K},j} (t) \, Q_{\underline{K},j} (t)$$

$$+ Q_{\underline{K},j} (t) \, Q^{\dagger}_{\underline{K},j} (t) \} \, , \tag{89}$$

a result which is of familiar form.

4. Eigenvalue Spectrum. We now wish to find the eigenvalue spectrum of H and the explicit matrix elements of Q. This information will enable us to compute most quantities of interest without resorting to an explicit expression for the energy eigenvectors. It is instructive to examine a particular term in H, that is, a term labelled by a specific $\underline{K}$,j. We will later sum the effects of all such terms independently. This is allowable because no cross terms in H appear between $\underline{K}$,j and $\underline{K}'$,j'. Our modes do not interact, and $\underline{K}$ and j separately are "good" quantum numbers. Consider the representative term (we drop the explicit $\underline{K}$,j labels temporarily)

$$H = NM \, \omega^2 \, [\, Q^{\dagger}Q + QQ^{\dagger}] \, . \tag{90}$$

We shall prove $\dot{Q} = - i \, \omega \, Q$, where ω is the frequency appearing in the Hamiltonian (90). In the Heisenberg representation

$$\dot{Q} = (i/\hbar)[H,Q].$$

Examining the commutator, we find

$$[H,Q] = NM\,\omega^2\,[Q^\dagger Q + QQ^\dagger,\,Q]$$

$$= NM\,\omega^2\,[Q^\dagger QQ - QQ^\dagger Q + QQ^\dagger Q - QQQ^\dagger]$$

$$= NM\,\omega^2\,\{\,[Q^\dagger,Q]\,Q + Q\,[Q^\dagger,Q]\,\} \tag{91}$$

$$= NM\,\omega^2\,\left(\frac{-2\hbar Q}{2NM\omega}\right) = -\,\hbar\,\omega\,Q$$

so that indeed,

$$\frac{i}{\hbar}\,[H,Q] = -\,i\,\omega\,Q = \dot{Q}\,. \tag{92}$$

Let us now assume that a set of states has been found which exactly diagonalizes H in (90) and are labelled by indices n,m:

$$H\,|n> = E_n\,|n> \tag{93}$$

where

$$<n|m> = \delta_{n,m} \tag{94}$$

Then from (92),

$$<m|Q|n> = -\,\frac{1}{\hbar\omega}\,<m|\,[H,Q]\,|n>\,. \tag{95}$$

However, using (93) and (94),

$$<m|\,[H,Q]\,|m> = <m|\,HQ - QH\,|n>$$

$$= (E_m - E_n)\,<m|Q|n>\,. \tag{95a}$$

Hence

$$<m|Q|n> = -\,\frac{E_m - E_n}{\hbar\omega}\,<m|Q|n>\,. \tag{96}$$

This can only be possible if either

$$E_n = \hbar\omega + E_m \tag{97}$$

or

$$< m|Q|n > = 0 . \tag{98}$$

Similarly, $< m|Q^\dagger|n > = 0$ unless $E_n = E_m - \hbar\omega$. We see that Q and $Q^\dagger$ can only connect states differing in energy by $\hbar\omega$, so that the accessible energy levels of our system must be placed at intervals of $\hbar\omega$. That is,

$$E_n = n\hbar\omega + E_0 \qquad n = 0, \pm 1, \pm 2, \ldots \tag{99}$$

The index n is an integer, and will now serve as a label for the eigenstate with energy $n\hbar\omega + E_0$. That is, the state $|n>$ now signifies a state with n units of excitation energy $\hbar\omega$. The operators Q and $Q^\dagger$ then lower and raise, respectively, the level of excitation by a single unit. For this reason the Q and $Q^\dagger$ are referred to as lowering and raising operators respectively. The Q operators will only have two finite matrix elements: $< n - 1|Q|n>$ and $<n + 1|Q|n >$. In order to evaluate these quantities, we return to

$$H = NM\omega^2 [Q^\dagger Q + QQ^\dagger] .$$

The commutation relations enable us to write

$$QQ^\dagger = [Q,Q^\dagger] + Q^\dagger Q$$

$$= (\hbar/2NM\omega) + Q^\dagger Q \tag{100}$$

so that

$$H = 2NM\omega^2 Q^\dagger Q + NM\omega^2 \hbar/2NM\omega$$

$$= 2NM\omega^2 Q^\dagger Q + \frac{1}{2}\hbar\omega . \tag{101}$$

The matrix element of H is

$$E_n = < n|H|n > = 2NM\omega^2 < n|Q^\dagger Q|n > + \frac{1}{2}\hbar\omega . \tag{102}$$

We insist that $< n|H|n >$ always be greater than 0, and, comparing with our previous result (99), we see it is possible to identify E_0 with $\frac{1}{2}\hbar\omega$, and that we must restrict $n \geq 0$. This <u>requires</u>

$$Q|o\rangle = 0, \tag{103}$$

because otherwise a negative energy state would be reached. Because we also have

$$H|o\rangle = \frac{1}{2}\hbar\omega|o\rangle = E_o|o\rangle , \tag{104}$$

we can finally write

$$E_n = (n + 1/2)\hbar\omega . \tag{105}$$

So far we have only discussed the eigenvalues of the Hamiltonian (89). It is also necessary to specify the value for the matrix element of the raising and lowering operators Q, as these will be of most use in later applications. We can find explicit values for terms like $\langle n|Q^\dagger|m\rangle$ by the following consideration. Using (102) the matrix element of the Hamiltonian H can be rewritten as

$$\langle n|H|n\rangle = 2NM\,\omega^2 \langle n|Q^\dagger Q|n\rangle + \frac{1}{2}\hbar\omega$$

$$= 2NM\,\omega^2 \sum_m \langle n|Q^\dagger|m\rangle \langle m|Q|n\rangle + \frac{1}{2}\hbar\omega \tag{106}$$

using the identity operator

$$\sum_m |m\rangle\langle m| = 1 . \tag{107}$$

From before, Q is a "lowering operator," that is

$$m = n - 1$$

and thus

$$E_n = \langle n|H|n\rangle = 2NM\,\omega^2 \langle n|Q^\dagger|n-1\rangle \langle n-1|Q|n\rangle + \frac{1}{2}\hbar\omega$$

$$= 2NM\,\omega^2 \langle n-1|Q|n\rangle^* \langle n-1|Q|n\rangle + \frac{1}{2}\hbar\omega$$

$$= 2NM\,\omega^2 |\langle n-1|Q|n\rangle|^2 + \frac{1}{2}\hbar\omega . \tag{108}$$

But $E_n = n\hbar\omega + \frac{1}{2}\hbar\omega$. Hence, we must have

$$|< n-1|Q|n >|^2 = \frac{n\hbar\omega}{2NM\omega^2} = \frac{n\hbar}{2NM\omega} \tag{109}$$

or, to within a phase factor,

$$< n-1|Q|n > = \sqrt{\frac{n\hbar}{2NM\omega}} = < n|Q^\dagger|n-1 > \tag{110}$$

and

$$< n+1|Q^\dagger|n > = \sqrt{\frac{(n+1)\hbar}{2NM\omega}} \tag{111}$$

which are the desired results. Re-introducing the subscripts $\underline{K}$,j, we have,

$$H = NM \sum_{\underline{K},j} \omega^2(\underline{K},j) \; [Q^\dagger_{\underline{K},j} Q_{\underline{K},j} + Q_{\underline{K},j} Q^\dagger_{\underline{K},j}]$$

$$= \sum_{\underline{K},j} \hbar\omega(\underline{K},j) \; (\hat{n}_{\underline{K},j} + \frac{1}{2}) \tag{112}$$

where $\hat{n}_{\underline{K},j}$ is the number operator

$$\hat{n}_{\underline{K},j} = [\frac{2NM\omega(\underline{K},j)}{\hbar}] \; Q^\dagger_{\underline{K},j} Q_{\underline{K},j} \; . \tag{113}$$

Using the lack of interference between the vibrational modes of different wave vector $\underline{K}$ and polarization index j, we may write the wave function for the entire system as

$$|\Psi_{vib}> = \prod_{\underline{K},j} |n_{\underline{K},j} > . \tag{114}$$

This wave function is an eigenvector of H in (112) with energy

$$E_{vib} = \sum_{\underline{K},j} \hbar\omega(\underline{K},j) \; (n_{\underline{K},j} + \frac{1}{2}) \tag{115}$$

where $n_{\underline{K},j}$ is the occupation number of each mode of vibration.

By construction

$$\hat{n}_{\underline{K},j}\,|\Psi_{vib}\rangle = n_{\underline{K},j}|\,\Psi_{vib}\rangle \tag{116}$$

and the number operator is diagonal in this representation. The matrix elements of the $Q_{\underline{K},j}$ are, from (110) and (111),

$$\langle n_{K,j}-1|Q_{\underline{K},j}|n_{\underline{K},j}\rangle = \left[\frac{\hbar n_{\underline{K},j}}{2NM\omega(\underline{K},j)}\right]^{\frac{1}{2}} ; \tag{117}$$

$$\langle n_{\underline{K},j}+1|Q^{\dagger}_{\underline{K},j}|n_{\underline{K},j}\rangle = \left[\frac{\hbar(n_{\underline{K},j}+1)}{2NM\omega(\underline{K},j)}\right]^{\frac{1}{2}} . \tag{118}$$

We label the quantized lattice vibrational modes, $|n_{\underline{K},j}\rangle$, "phonons", and use the index $n_{\underline{K},j}$ to label the degree of excitation in each phonon mode.

We can display an explicit matrix representation for Q and $Q^{\dagger}$ using a set of 3N state vectors, one for each $\underline{K}$,j phonon mode. Each state vector is an infinite column vector with zeroes everywhere, but in the $n_{\underline{K},j}$ + 1 place, where there is a 1. Thus

$$\begin{pmatrix}1\\0\\0\\ \cdot\\ \cdot\\ \cdot\end{pmatrix}_{\underline{K},j}$$ implies no phonons of type $\underline{K}$,j present;

$$\begin{pmatrix}0\\1\\0\\0\\ \cdot\\ \cdot\\ \cdot\end{pmatrix}_{\underline{K},j}$$ means one $\underline{K}$,j phonon present ;

$$\begin{pmatrix}0\\0\\0\\0\\1\\0\\ \cdot\\ \cdot\end{pmatrix}_{\underline{K},j}$$ means 4 $\underline{K}$,j phonons present.

In terms of these eigenvector matrices the Q which give the proper matrix elements and satisfy the commutation relations are

$$Q = \sqrt{\frac{\hbar}{2NM\omega}} \begin{pmatrix} 0 & \sqrt{1} & 0 & 0 & \cdots \\ 0 & 0 & \sqrt{2} & 0 & \cdots \\ 0 & 0 & 0 & \sqrt{3} & \cdots \\ 0 & 0 & 0 & 0 & \cdots \\ \vdots & \vdots & \vdots & \vdots & \ddots \end{pmatrix} ; \tag{119}$$

$$Q^{\dagger} = \sqrt{\frac{\hbar}{2NM\omega}} \begin{pmatrix} 0 & 0 & 0 & 0 & \cdots \\ \sqrt{1} & 0 & 0 & 0 & \cdots \\ 0 & \sqrt{2} & 0 & 0 & \cdots \\ 0 & 0 & \sqrt{3} & 0 & \cdots \\ 0 & 0 & 0 & \sqrt{4} & \cdots \\ \vdots & \vdots & \vdots & \vdots & \ddots \end{pmatrix} \tag{120}$$

The individual phonon state vectors can be put together to give us the overall state vector (compare (113) which is a simple product of all the $\underline{K},j$ column vectors. It is easy to show that this choice of matrices satisfies all the commutation laws.

II. THERMODYNAMIC PROPERTIES OF LATTICE VIBRATIONS

II.A. Construction and Use of the Partition Function

1. Distribution Function. We now investigate the thermodynamic behaviour of the lattice vibrations. In thermal equilibrium the average value of phonons in a state $\underline{K},j$ is symbolized by $\bar{n}_{\underline{K},j}$. The probability of finding $n_{\underline{K},j}$ is proportional to

$$\exp\left\{-[(n_{\underline{K},j}+1/2)\hbar\omega(\underline{K},j)/kT]\right\}. \qquad (121)$$

Then:

$$\bar{n}_{\underline{K},j} = \frac{\sum_{n_{\underline{K},j}} n_{\underline{K},j} \exp\left\{-[(n_{\underline{K},j}+\frac{1}{2})\hbar\omega(\underline{K},j)/kT]\right\}}{\sum_{n_{\underline{K},j}} \exp\left\{-[(n_{\underline{K},j}+\frac{1}{2})\hbar\omega(\underline{K},j)/kT]\right\}}$$

$$= \frac{1}{\exp(\hbar\omega(\underline{K},j)/kT)-1} \;. \qquad (122)$$

This is just the Einstein-Bose distribution. It has arisen solely from the eigenvalue spectrum of the quantized lattice vibrations which, in turn, is formed by the Hamiltonian and the commutation relations for the Q.

2. Summation of Terms. The partition function Z is

$$Z = \Sigma_{\text{all systems}} \exp(-E_{\text{system}}/kT)$$

$$= \sum_{\{n_{\underline{K},j}\}} \exp\left\{-[\sum_{\underline{K},j}(n_{\underline{K},j}+\frac{1}{2})\hbar\omega(\underline{K},j)/kT]\right\}$$

$$= \prod_{\underline{K},j}\sum_{n_{\underline{K},j}} \exp\left\{-[(n_{\underline{K},j}+\frac{1}{2})\hbar\omega(K,j)/kT]\right\}, \qquad (123)$$

where $\{n_{\underline{K},j}\}$ is a particular set of all the phonon occupation numbers $n_{\underline{K},j}$.

More algebra gives

$$Z = \prod_{\underline{K},j} \exp - [\hbar\omega(\underline{K},j)/2kT] \times$$

$$\times \sum_{n_{\underline{K},j}} \exp - [\hbar\omega(\underline{K},j) n_{\underline{K},j}/kT]$$

$$= \prod_{\underline{K},j} \frac{\exp - [\hbar\omega(\underline{K},j)/2kT]}{1 - \exp - [\hbar\omega(\underline{K},j)/kT]} \tag{124}$$

using the formula for the sum of a geometric progression. Defining

$$\Theta = \exp\,(-1/kT)$$

we arrive at

$$Z = \prod_{\underline{K},j} \frac{\Theta^{\hbar\omega(\underline{K},j)/2}}{1-\Theta^{\hbar\omega(\underline{K},j)}} . \tag{125}$$

3. <u>Internal Energy</u>. The average internal energy U is given by

$$U = kT^2(\partial/\partial T)\log Z. \tag{126}$$

Using $\partial/\partial T = (\partial\Theta/\partial T)\partial/\partial\Theta$ and $(\partial\Theta/\partial T) = \Theta/kT^2$ we find that (126) can be rewritten as

$$U = \Theta \frac{\partial}{\partial\Theta} \log Z. \tag{127}$$

The internal energy can then be evaluated using (125) and (127),

$$U = \Theta \frac{\partial}{\partial \Theta} \log \left[\prod_{\underline{K},j} \frac{\Theta^{\hbar\omega(\underline{K},j)/2}}{1-\Theta^{\hbar\omega(\underline{K},j)}} \right]$$

$$= \Theta \frac{\partial}{\partial \Theta} \sum_{\underline{K},j} \{\log [\Theta^{\hbar\omega(\underline{K},j)/2}] -$$

$$- \log [1 - \Theta^{\hbar\omega(\underline{K},j)}]\}$$

$$= \Theta \sum_{\underline{K},j} \left[\frac{(\hbar\omega(\underline{K},j)/2)\, \Theta^{[(\hbar\omega(\underline{K},j)/2)-1]}}{\Theta^{\hbar\omega(\underline{K},j)/2}} \right]$$

$$+ \Theta \sum_{\underline{K},j} \left[\frac{\hbar\omega(\underline{K},j)\ \Theta^{[\hbar\omega(\underline{K},j)-1]}}{1 - \Theta^{\hbar\omega(\underline{K},j)}} \right]$$

$$= \sum_{\underline{K},j} \left[\frac{\hbar\omega(\underline{K},j)}{2} + \frac{\hbar\omega(\underline{K},j)}{\Theta^{-[\hbar\omega(\underline{K},j)]}_{-1}} \right]$$

$$= \sum_{\underline{K},j} \left[\frac{\hbar\omega(\underline{K},j)}{2} + \frac{\hbar\omega(\underline{K},j)}{\exp\ [\hbar\omega(\underline{K},j)/kT]-1} \right] \tag{128}$$

This result could also have been arrived at using the eigenvalue (115) for the energy and the average value for the occupation numbers $n_{\underline{K},j}$ given by (122).

4. Free Energy. Next we compute the free energy, F.

$$F = - kT \ \ell n \ Z$$

$$= - kT \sum_{\underline{K},j} \ell n \left[\frac{\exp-(\hbar\omega(\underline{K},j)/2kT)}{1-\exp-(\hbar\omega(\underline{K},j)/kT)} \right]$$

$$= - kT \sum_{\underline{K},j} \left[- \frac{\hbar\omega(\underline{K},j)}{2kT} - \ell n \ (1-\exp-[\hbar\omega(\underline{K},j)/kT]) \right]$$

so that

$$F = \sum_{\underline{K},j} \left[\frac{\hbar\omega(\underline{K},j)}{2} + kT\, \ell n\, (1-\exp-[\hbar\omega(\underline{K},j)/kT]) \right] . \tag{129}$$

Note that the first term, the zero point energy, is an additive "constant."

5. Entropy. The entropy is derived from the usual relation

$$F = U - TS . \tag{130}$$

Thus

$$S = (U - F)/T =$$

$$\frac{1}{T} \sum_{\underline{K},j} \left\{ \frac{\hbar\omega(\underline{K},j)}{2} + kT\, \ell n\, (1 - \right.$$

$$- \exp - [\hbar\omega(\underline{K},j)/kT] - \frac{\hbar\omega(\underline{K},j)}{2} -$$

$$\left. - \frac{\hbar\omega(\underline{K},j)}{\exp\,[\hbar\omega(\underline{K},j)/kT]-1} \right\}$$

$$= - \frac{1}{T} \sum_{\underline{K},j} \left[kT\, \ell n\, (1-\exp-[\hbar\omega(\underline{K},j)/kT]\,) \right.$$

$$\left. - \frac{\hbar\omega(\underline{K},j)}{\exp\,[\hbar\omega(\underline{K},j)/kT]-1} \right] \tag{131}$$

It is a simple matter to transform this expression into the form

$$S = k \sum_{\underline{K},j} [(\bar{n}_{\underline{K},j} + 1)\, \ell n\, (\bar{n}_{\underline{K},j} + 1) -$$

$$- \bar{n}_{\underline{K},j}\, \ell n\, \bar{n}_{\underline{K},j}]. \tag{132}$$

This result will be quite useful when we compute transport properties. Note that the zero point energy explicitly cancelled when we computed the entropy. This is as it should be because the "occupation" of these states is constant regardless of temperature, order, etc.

6. Specific Heat. We may compute the specific heat using (128) for the internal energy U. We find

$$C_v = \frac{\partial U}{\partial T} =$$

$$= \frac{\partial}{\partial T} \sum_{\underline{K},j} \frac{\hbar\omega(\underline{K},j)}{2} + \frac{\hbar\omega(\underline{K},j)}{\exp\,[\hbar\omega(\underline{K},j)/kT]-1}$$

$$= \sum_{\underline{K},j} \frac{\hbar^2\omega^2(\underline{K},j)}{kT^2} \exp \hbar\omega(\underline{K},j)/kT$$

$$\times\,[\exp\,[\hbar\omega(\underline{K},j)/kT]-1\,]^{-2}$$

$$= k \sum_{\underline{K},j} \left[\frac{\hbar\omega(\underline{K},j)}{2kT}\right]^2 \Big/ \sinh^2\left[\frac{\hbar\omega(\underline{K},j)}{2kT}\right]. \qquad (133)$$

Let us first examine this expression when $T >> \hbar\omega(\underline{K},j)/k$ (the high temperature limit). Then,

$$C_v = k \sum_{\underline{K},j} \left\{ 1 - \frac{[\hbar\omega(\underline{K},j)]^2}{12k^2T^2} + \ldots \right\} . \qquad (134)$$

It is easily seen that only even powers of $\omega(\underline{K},j)$ will appear in the expansion (134) (1). In fact, it is not necessary to know the vibrational spectrum to find the specific heat in this limit. It can be shown that (134) can be rewritten as

$$C_v = 3Nk - \frac{N\hbar^2}{12MkT^2}\,[\,\sum_\alpha V_{\alpha\alpha}\,(0)]. \qquad (135)$$

At low temperatures however, there will always exist waves whose energy $< kT$, and any expansion scheme will break down. We must instead know the distribution of frequencies in order to determine C_v.

II.B. Density of States

1. Enumeration of States. We first examine the distribution of $\underline{K}$ values in 3 dimensions. We have previously defined $\underline{K}$ by the relation

$$\underline{K} = \frac{2\pi p_1}{N_1}\underline{g}_1 + \frac{2\pi p_2}{N_2}\underline{g}_2 + \frac{2\pi p_3}{N_3}\underline{g}_3 ,$$

where $\underline{g}_1$, $\underline{g}_2$, $\underline{g}_3$ are reciprocal lattice vectors. The faces of the Brillouin zone were constructed to lie normal to the tips of the vectors

$$\underline{K} = \pm\pi\underline{g}_1, \pm\pi\underline{g}_2, \pm\pi\underline{g}_3 .$$

The volume of the Brillouin zone is

$$\Omega_b = (2\pi)^3\underline{g}_1\cdot(\underline{g}_2\times\underline{g}_3) = (2\pi)^3\begin{vmatrix} g_{1x} & g_{1y} & g_{1z} \\ g_{2x} & g_{2y} & g_{2z} \\ g_{3x} & g_{3y} & g_{3z} \end{vmatrix} ,$$

whereas the volume of the primitive cell is,

$$\Omega_a = \underline{T}_1\cdot(\underline{T}_2\times\underline{T}_3) .$$

By construction, the product of the matrices is

$$(2\pi)^3\begin{pmatrix} g_{1x} & g_{1y} & g_{1z} \\ g_{2x} & g_{2y} & g_{2z} \\ g_{3x} & g_{3y} & g_{3z} \end{pmatrix}\begin{pmatrix} T_{1x} & T_{2x} & T_{3x} \\ T_{1y} & T_{2y} & T_{3y} \\ T_{1z} & T_{2z} & T_{3z} \end{pmatrix} = \begin{pmatrix} 2\pi & 0 & 0 \\ 0 & 2\pi & 0 \\ 0 & 0 & 2\pi \end{pmatrix} . \tag{136}$$

Taking the determinant of both sides of (136), we arrive at

$$\Omega_b\,\Omega_a = (2\pi)^3 . \tag{137}$$

However, by construction, the volume of the first Brillouin zone, Ω_b, contains N solutions. Hence, the number of solutions in the fractional volume element $\frac{d^3K}{\Omega_b}$ must be

$$\left(\frac{d^3K}{\Omega_b}\right) N = \frac{N\Omega_a}{(2\pi)^3} d^3K = \frac{V}{(2\pi)^3} d^3K , \tag{138}$$

where V is the volume of the entire crystal. Hence, the sum over wave vectors can be expressed as follows

$$\sum_{\underline{K}} = \sum_{p_1=-N_{1/2}}^{N_{1/2}} \sum_{p_2=-N_{2/2}}^{N_{2/2}} \sum_{p_3=-N_{3/2}}^{N_{3/2}} = \frac{V}{(2\pi)^3} \int_{BZ} d^3K \tag{139}$$

where the integral goes over only the volume of the first Brillouin zone.

The formula for the specific heat (133) becomes

$$C_v = \sum_j \frac{V}{(2\pi)^3} \int \frac{\hbar^2\omega^2(\underline{K},j)}{kT^2} \frac{\exp[\hbar\omega(\underline{K},j)/kT]}{[\exp[\hbar\omega(\underline{K},j)/kT]-1]^2} d^3K . \tag{140}$$

The integrand of (140) is given in terms of the frequency $\omega(\underline{K},j)$, and not of the wave vector $\underline{K}(\omega,j)$. Let $S(\omega)$ denote a surface of constant frequency ω within the Brillouin zone. The number of solutions contained within $S(\omega)$ is given by

$$\frac{V}{(2\pi)^3} \int_{S(\omega)} d^3K .$$

Let dS denote an incremental area of the surface $S(\omega)$. The element of volume with dS as base can be written as

$$\frac{dS\ d\omega}{|\mathrm{grad}_{\underline{K}}\omega(\underline{K},j)|} .$$

The number of solutions contained in this volume is from (138)

$$\frac{V}{(2\pi)^3} \frac{dS\ d\omega}{|\mathrm{grad}_{\underline{K}}\omega(\underline{K},j)|} \tag{141}$$

The total number of states in a frequency range $d\omega$ about the value ω is then given by

$$\sum_j \left[\frac{V}{(2\pi)^3} \int_{S(\omega)} \frac{dS}{|\mathrm{grad}_{\underline{K}}\,\omega(\underline{K},j)|} \right] d\omega \equiv \sum_j g_j\,(\omega)\,d\omega \;. \qquad (142)$$

The term in square brackets is known as the density of states $g_j(\omega)$ for the polarization mode j. The specific heat (140) can then be written as

$$C_v = \sum_j \int g_j(\omega)\, \frac{\hbar^2\omega^2}{kT^2}\, \frac{\exp(\hbar\omega/kT)}{[\exp(\hbar\omega/kT)-1]^2}\, d\omega \;. \qquad (143)$$

The density of states $g_j(\omega)$ is normalized to N by virtue of its definition:

$$\sum_{\underline{K}} = N = \int \frac{V}{(2\pi)^3}\, d\,K = \int g_j\,(\omega)\,d\omega \;. \qquad (144)$$

Because $dS \sim \omega^2$ at small $|\underline{K}|$, and $\partial\,\omega/\partial\,K \sim \partial\,/\partial K\,(cK) \sim c$, $g_j(\omega) \sim \omega^2$ for small ω. In general $g_j(\omega)$ may have many maxima or minima, and we may expand the solutions for $\omega(\underline{K},j)$ near one of these points in the following manner,

$$\omega^2 = \omega_o^{\,2} \pm c_j^{\,2}(\alpha K_1^{\,2} \pm \beta K_2^{\,2} \pm \gamma K_3^{\,2}), \qquad (145)$$

where ω_o is the extremum value of the $\omega(\underline{K},j)$ surface. The points ω_o are "special" because of the vanishing of the denominator in the expression (141) for the density of states $g_j(\omega)$ at that point. We shall consider four special cases: (1) the bottom of the acoustic branch, (2) other minima in the $\omega(\underline{K},j)$ surface, (3) maxima in the $\omega(\underline{K},j)$ surface and (4) saddle points.

2. <u>Critical Points, Small Wave Vector Limit. Debye Approximation</u>. At the bottom of the acoustic branch $\omega_o = 0$ and

$$\omega = c_j\,(\alpha K_1^{\,2} + \beta K_2^{\,2} + \gamma K_3^{\,2})^{\frac{1}{2}};\; \alpha,\beta,\gamma > 0. \qquad (146)$$

We assume spherical frequency surfaces so that

$$\omega = c_j K \, . \tag{147}$$

Then

$$|\mathrm{grad}_{\underline{K}}\omega| = c_j \ , \ dS = \frac{\omega^2}{c_j^{\,2}} \, d\,\Theta \, d\,\phi \, \sin\Theta \ ,$$

and

$$g_j(\omega) = \frac{4\pi V}{(2\pi)^3} \times \frac{\omega^2}{c_j^{\,3}} \, . \tag{148}$$

This is the Debye approximation. If we are so foolish as to demand that it should hold for all frequencies, we must set

$$\sum_j \int g_j(\omega) \, d\omega = 3N. \tag{149}$$

This approximation takes no note of the Brillouin zone boundaries, nor of anything else for large ω (i.e., large $\underline{K}$). Nevertheless we define a limiting frequency called the Debye frequency, $\omega_D^{(j)}$, by the condition, for <u>each</u> polarization j,

$$\int_0^{\omega_D^{(j)}} g_j(\omega) \, d\omega = N. \tag{150}$$

Integrating the expression for $g_j(\omega)$, one finds

$$\frac{4\pi V}{(2\pi)^3} \ \frac{1}{c_j^{\,3}} \ \frac{[\omega_D^{(j)}]^3}{3} = N,$$

or,

$$\omega_D^{(j)} = \left(\frac{6\pi^2 c_j^{\,3} N}{V} \right)^{1/3} \equiv \frac{k\Theta_D^{(j)}}{\hbar} \ , \tag{151}$$

where $\Theta_D^{(j)}$ is the Debye temperature. The specific heat becomes

$$C_V = \sum_j \int_0^{\omega_D^{(j)}} \frac{V}{(2\pi)^3} \frac{4\pi\hbar^2\omega^2\exp(\hbar\omega/kT)\omega^2}{kT^2[\exp(\hbar\omega/kT)-1]^2 c_j^3} d\omega$$

$$= \sum_j \frac{V}{2\pi^2} \frac{\hbar^2}{kT^2 c_j^3} \int_0^{\omega_D^{(j)}} \frac{\omega^4\exp(\hbar\omega/kT)}{[\exp(\hbar\omega/kT)-1]^2} d\omega$$

$$= \sum_j (V\hbar^2/2\pi^2 kT^2 c_j^3)(kT/\hbar)^5$$

$$\times \int_0^{x_{max}^{(j)}} \frac{x^4 \exp x \, dx}{(\exp x-1)^2} , \qquad (152)$$

where

$$x_{max}^{(j)} = \hbar\omega_D^{(j)}/kT .$$

Simplifying, using (151),

$$C_V = \sum_j 3Nk(T/\Theta_D^{(j)})^3 \int_0^{x_{max}^{(j)}} \frac{x^4\exp x \, dx}{(\exp x-1)^2} . \qquad (153)$$

We see that at low temperatures

$$x_{max}^{(j)} = \Theta_D^{(j)}/T >> 1$$

and the exponential behaviour of the integrand allows us to replace the upper limit by ∞ with little error. Hence,

$$\int_0^{x_{max}^{(j)}} \frac{x^4\exp x \, dx}{(\exp x-1)^2} \approx \int_0^{\infty} x^4 \exp(-x)dx = 4! ,$$

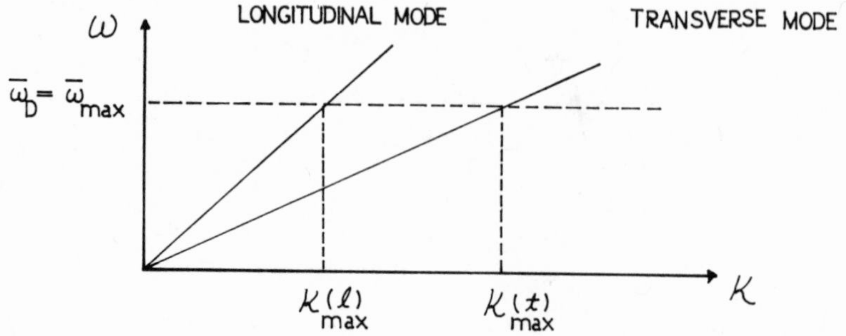

Figure 10 ω versus K Curves in the Debye Approximation.

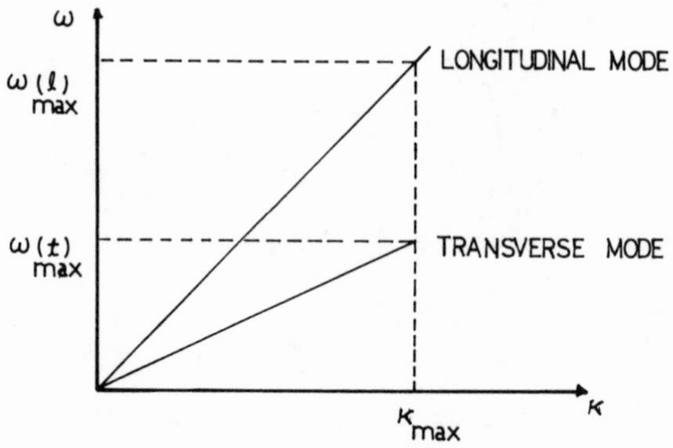

Figure 11 The Born Cut-off.

and

$$C_v = \sum_j 72\ Nk(T/\Theta_D^{(j)})^3\ , \tag{154}$$

at those temperatures such that $\Theta_D^{(j)}/T >> 1$. Debye defined an average velocity $\bar{c}$ by

$$\frac{3}{\bar{c}^3} = \frac{1}{c_\ell{}^3} + \frac{2}{c_t{}^3}\ , \tag{155}$$

where c_ℓ and c_t are the velocities for the longitudinal and transverse modes respectively in an isotropic solid. This leads to an average $\bar{\Theta}_D$ from the relation

$$\int_o^{\bar{\omega}_D} \bar{g}(\omega)\ d\omega = N, \tag{156}$$

where $\bar{\omega}_D$ and $\bar{g}(\omega)$ may be obtained from $\omega_D^{(j)}$ and $g_j(\omega)$ by replacing c_j everywhere by $\bar{c}$. Then

$$C_v = 9kN\left(\frac{T}{\bar{\Theta}_D}\right)^3 \int_o^{\bar{x}_{max}} \frac{x^4 \exp x\ dx}{(\exp x{-}1)^2}\ . \tag{157}$$

It is clear that the actual ω versus $\underline{K}$ curves have been approximated by the straight lines in Figure 10. Note that for $K_{max}^{(\ell)} < K < K_{max}^{(t)}$, the longitudinal modes have been excluded and only transverse modes have been counted. This is quite clearly in error as there must be N longitudinal modes and also N transverse modes of each type. Born attempted to remedy this deficiency and create a model which would more closely approximate the actual frequency distribution. A cut off on $\underline{K}$ rather than on ω is introduced.

This defines a "zone" sphere in $\underline{K}$ space by the condition that

$$\frac{4\pi V}{(2\pi)^3} \int_0^{K_{max}} K^2 \, dK = N,$$

or

$$|\underline{K}_{max}| = \left(\frac{6\pi^2 N}{V}\right)^{1/3} . \qquad (158)$$

One then finds the specific heat to be given by

$$C_v^{(Born)} = \sum_j 3Nk(T/\Theta_{Born}^{(j)})^2$$

$$\int_0^{x_{max}^{(j)}} \frac{x^4 \exp x \, dx}{(\exp x-1)^2} \qquad (159)$$

where $\Theta_{Born}^{(j)} = \hbar c_j K_{max}/k$ and $x_{max}^{(j)} = \Theta_{Born}^{(j)}/T$.

At high temperatures the Debye approximation gives the result

$$C_v = \sum_j 3Nk(T/\Theta_D^{(j)})^3 \int_0^{x_{max}^{(j)}} \frac{x^4 \exp x}{(\exp x-1)^2} dx$$

$$\cong 3Nk \sum_j (T/\Theta_D^{(j)})^3 \int_0^{x_{max}^{(j)}} (x^2 - \frac{x^4}{12}) \, dx,$$

when $x_{max}^{(j)} = \Theta_D^{(j)}/T \ll 1$. Integrating,

$$C_v = 3Nk \sum_j (T/\Theta_D^{(j)})^3 \;[\; \frac{1}{3}(\Theta_D^{(j)}/T)^3 - \frac{1}{60}(\Theta_D^{(j)}/T)^5 \;]$$

$$= 3Nk - \frac{Nk}{20T^2} \sum_j [\Theta_D^{(j)}]^2. \tag{160}$$

This has the correct high temperature form but the magnitude of the coefficient of the $1/T^2$ term does not in general compare well at all with the exact value $[-(N\hbar^2/12Mk) \sum_\alpha V_{\alpha\alpha}(o)]$ found previously in(135) for the coefficient of the $1/T^2$ term. It is clear that no special validity should be connected with the high temperature limit of the specific heat in the Debye approximation. It is at best a low temperature approximation and any extension to elevated temperatures is unwarranted.

We must treat in addition to the acoustic modes the intracell modes of vibration when there is more than a single atom per unit cell. One usually approximates these optical modes by Einstein terms, though some investigators insist upon tacking them onto the end of the Debye spectrum (i.e., defining

$$\int_o^{\omega_D} g(\omega)\, d\omega = nN$$

where there are n atoms/unit cell). This is misleading because ω is a relatively flat function of $\underline{K}$ for optical modes. Thus $|\text{grad}_{\underline{K}}\, \omega(\underline{K},j)|^{-1}$ is not unlike $\delta(\omega-\omega_{opt}^{(j)})$. Should such a form be assumed one finds,

$$C_v = \sum_j N \int \frac{\hbar^2\omega^2 \exp(\hbar\omega/kT)\,\delta(\omega-\omega_{opt}^{(j)})}{kT^2[\exp(\hbar\omega/kT)-1]^2}\, d\omega$$

$$= \sum_j N \frac{\hbar^2[\omega_{opt}^{(j)}]^2 \exp(\hbar\omega_{opt}^{(j)}/kT)}{kT^2[\exp(\hbar\omega_{opt}^{(j)}/kT)-1]^2} \;. \tag{161}$$

At low temperatures

$$C_v \sim (\hbar^2[\omega_{opt}^{(j)}]^2/kT^2)\exp-(\hbar\omega_{opt}^{(j)}/kT), \tag{162}$$

which is the familiar Einstein form for the specific heat. In practice one approximates the acoustic mode density of states by the Debye approximation and the optical mode density of states by the Einstein approximation. This leads to a total specific heat which is a sum of the two terms (157) and (161).

3. Critical Points. Minima. We shall consider now other minima in the $\omega(\underline{K},j)$ surface. The expansion of the frequency near the critical point ω_o is of the form

$$\omega^2 = \omega_o{}^2 + c_j{}^2\left(\alpha K_1{}^2 + \beta K_2{}^2 + \gamma K_3{}^2\right) . \tag{163}$$

Simplifying and expanding for small excursions of $\underline{K}$ near $\underline{K}_o$,

$$\omega = \omega_o + \frac{c_j{}^2}{2\omega_o}(\alpha K_1{}^2 + \beta K_2{}^2 + \gamma K_3{}^2). \tag{164}$$

For a spherical surface

$$K = \left[\frac{2\omega_o(\omega-\omega_o)}{c_j{}^2}\right]^{\frac{1}{2}} \tag{165}$$

and

$$\nabla_{\underline{K}}\,\omega = \frac{c_j{}^2 K}{\omega_o} = \left[\frac{2c_j{}^2\,\omega_o(\omega-\omega_o)}{\omega_o{}^2}\right]^{\frac{1}{2}} \tag{166}$$

$$= \sqrt{2}c_j\left(\frac{\omega-\omega_o}{\omega_o}\right)^{\frac{1}{2}}.$$

Using (142)

$$g_j(\omega) \propto \frac{K^2}{|\nabla_{\underline{K}}\omega|} = \frac{2\omega_o(\omega-\omega_o)}{c_j{}^2}\cdot\frac{(\omega_o)^{\frac{1}{2}}}{\sqrt{2}c_j(\omega-\omega_o)^{\frac{1}{2}}}$$

$$= \sqrt{2}\,\frac{\omega_o^{3/2}}{c_j^{3}}\,(\omega-\omega_o)^{\frac{1}{2}} \quad . \tag{167}$$

The density of states is thus zero for $\omega < \omega_o$, and rises vertically for $\omega \geq \omega_o$ as in Figure 12.

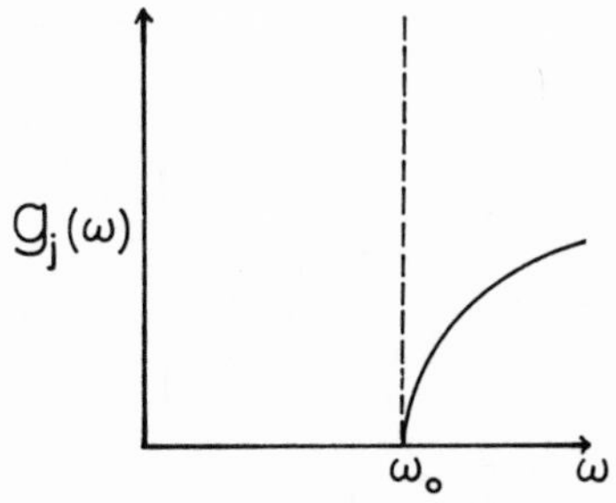

Figure 12 Density of States Due to a Minimum in the $\omega(\underline{K},j)$ Surface.

If we superimpose this behaviour upon a continuous Debye Background, we get a curve which looks like the one in Figure 13.

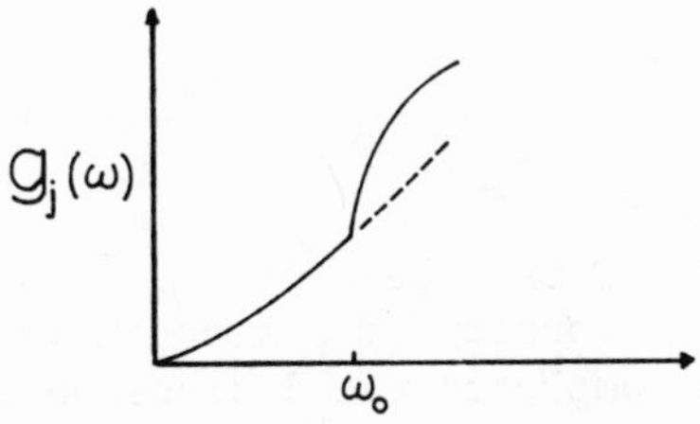

Figure 13 The Density of States of the Previous Figure Superimposed on a Debye Background.

4. Critical Points. Maxima. Let us consider now a maximum in the $\omega(\underline{K},j)$ surface. Here we choose the first minus sign in (145). The square root expanded near the critical point ω_o leads to

$$\omega = \omega_o - \frac{c_j^2}{2\omega_o}(\alpha K_1^2 + \beta K_2^2 + \gamma K_3^2). \tag{168}$$

Exactly as before, we find

$$g_j(\omega)\alpha(\omega_o-\omega)^{\frac{1}{2}} , \tag{169}$$

and behaves as in Figure 14.

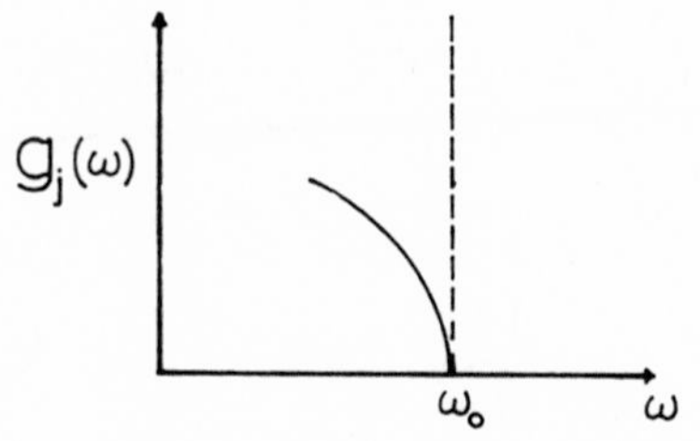

Figure 14. Density of States Due to a Maximum in the $\omega(\underline{K},j)$ Surface.

Again, the superposition of this behaviour upon a continuous background yields a total density of states curve like the one in Figure 15. This completes the discussion of the cases of absolute extrema. There do exist other types of surfaces which yield critical points in a real three dimensional crystal. The existence of these surfaces which are saddle points in the $\omega(\underline{K},j)$ surface was first pointed out by Van Hove (2). We now examine the density of states due to the two types of saddle points which are encountered in three dimensional crystals.

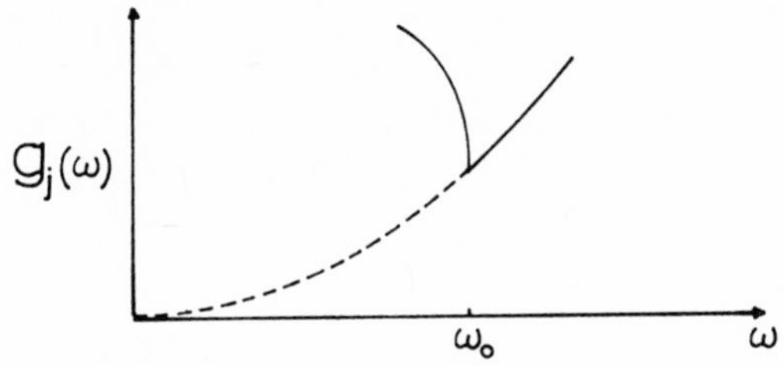

Figure 15 The Density of States of the Previous Figure Superimposed on a Continuous Background.

5. Critical Points. Saddle Points. The expansion of the frequency surface near the critical point is of the form

$$\omega = \omega_o \pm \frac{c_j^2}{2\omega_o}(K_z^2 - K_x^2 - K_y^2). \tag{170}$$

Consider first the positive sign in (170). Then, ω versus $\underline{K}$ is a minimum along the K_z direction but a maximum along the K_y and K_x directions.

It turns out that for a non-trivial spectrum, there always exists at least one saddle point of this type in a three dimensional lattice. This can be proven using topological arguments. We give a pictorial proof due to Wannier(3). Assume there are two maxima of $\omega(K,j)$ along a given direction in the reciprocal lattice as in Figure 16. Draw a series of equally spaced parallel planes through each maximum. On each plane there will also exist a maximum. Draw the line connecting all maxima on the intervening planes as in Figure 17. Somewhere along this path the maximum will be least; this is a saddle point.

It is obvious that there must exist at least one maximum of ω in the first Brillouin zone unless ω is trivially constant. Then, because $\omega(\underline{K},j) = \omega(-\underline{K},j)$, there will also exist another. Hence there will always be at least two maxima of $\omega(\underline{K},j)$ and concomitantly one saddle point.

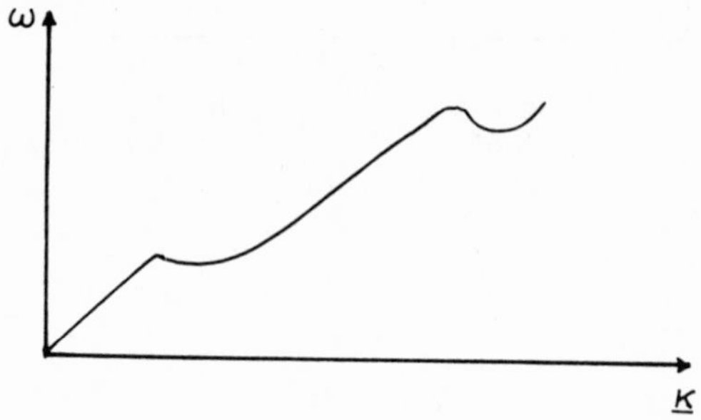

Figure 16 ω vs $\underline{K}$ Curve with Two Maxima.

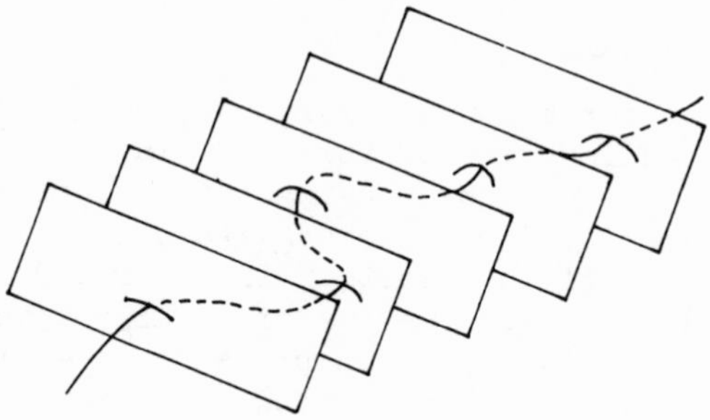

Figure 17 Line Connecting All Maxima.

Wannier shows that the form of the surface (170) with the positive sign leads to a density of states of the form

$$g_j(\omega) \sim [(c^2 + \omega-\omega_o)^{\frac{1}{2}}], \quad \omega < \omega_o$$

$$\sim [(c^2 + \omega-\omega_o)^{\frac{1}{2}} - 2(\omega-\omega_o)^{\frac{1}{2}}], \quad \omega > \omega_o \tag{171}$$

where c = const, or pictorially as in Figure 18. The surface (170) with the negative sign

$$\omega = \omega_o + \frac{c_j^2}{2\omega_o}(K_x^2 + K_y^2 - K_z^2), \tag{172}$$

gives rise to a density of states curve of the form in Figure 19.

6. Conditions on Number and Types of Critical Points. In summation we can describe the behaviour of the frequency surfaces near a critical point by letting

$$\omega = \omega_o + \frac{c_j^2}{2\omega_o}[\varepsilon_1 K_x^2 + \varepsilon_2 K_y^2 + \varepsilon_3 K_z^2]. \tag{173}$$

There then exist four distinct cases:

i = 0| $\varepsilon_1 = \varepsilon_2 = \varepsilon_3 = 1$, a minimum;

$$g_j(\omega) \sim (\omega-\omega_o)^{\frac{1}{2}} \quad \text{if} \quad \omega_o \neq 0:$$

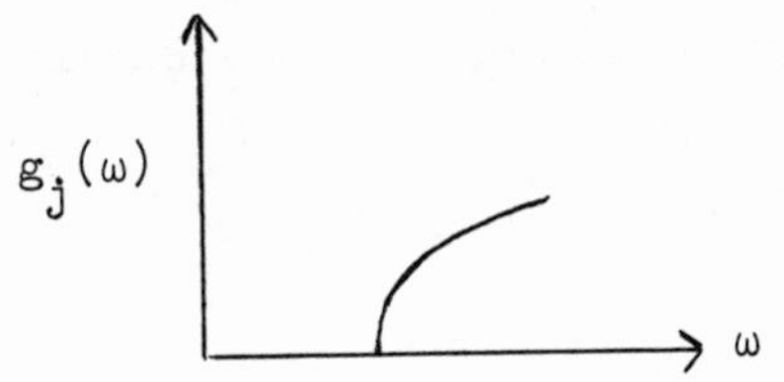

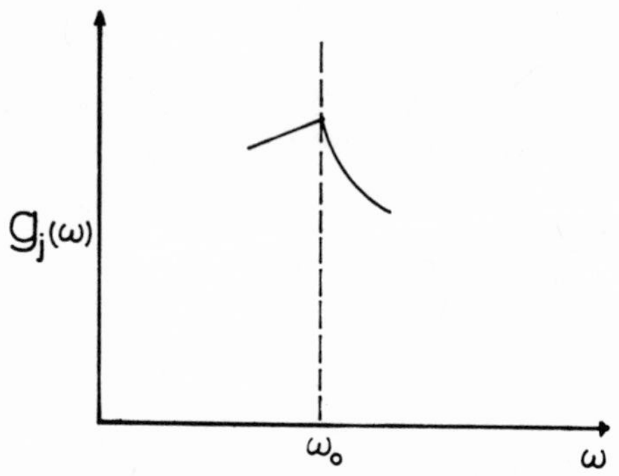

Figure 18 Density of States Due to a Saddle Point.

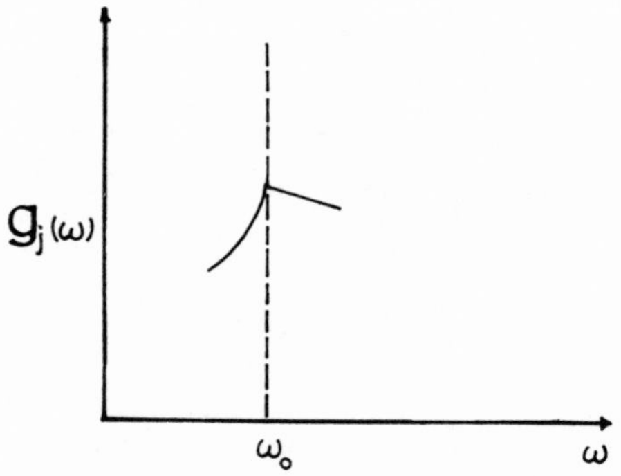

Figure 19 Density of States Due to a Saddle Point.

$g_j(\omega) \sim \omega^2$ if $\omega_o = 0$:

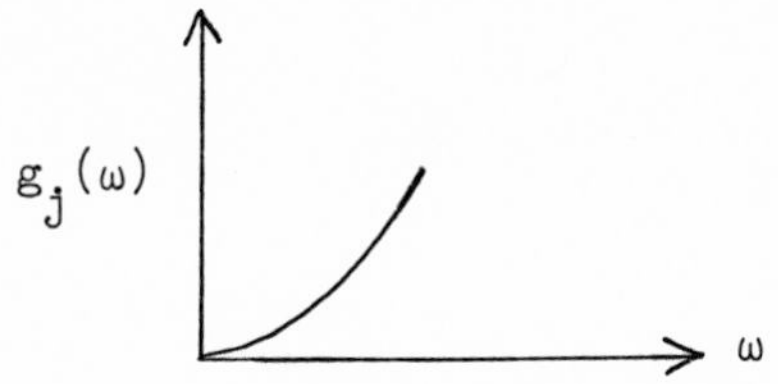

i = 3 | $\varepsilon_1 = \varepsilon_2 = \varepsilon_3 = -1$, a maximum;

$g_j(\omega) \sim (\omega_o - \omega)^{\frac{1}{2}}$:

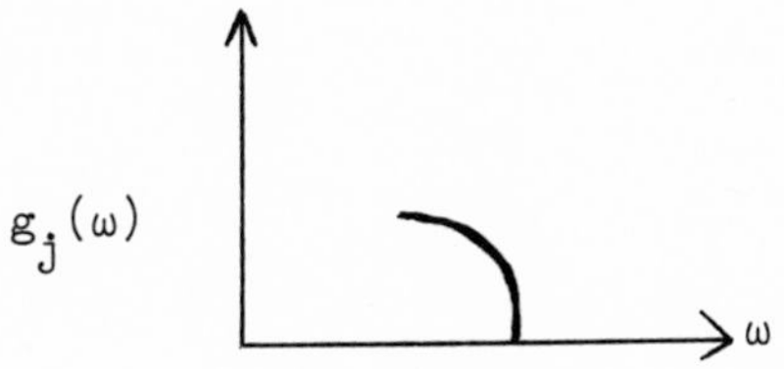

i = 1 | $\varepsilon_1 = \varepsilon_2 = -\varepsilon_3 = -1$, a saddle point

of the type $\omega = \omega_o + \frac{c_j^{\,2}}{2\omega_o}(K_z^{\,2} - K_x^{\,2} - K_y^{\,2})$

and $g_j(\omega) \sim [(c^2 + \omega - \omega_o)^{\frac{1}{2}}]$, $\omega < \omega_o$

$\sim [(c^2 + \omega - \omega_o)^{\frac{1}{2}} - 2(\omega - \omega_o)^{\frac{1}{2}}]$, $\omega > \omega_o$:

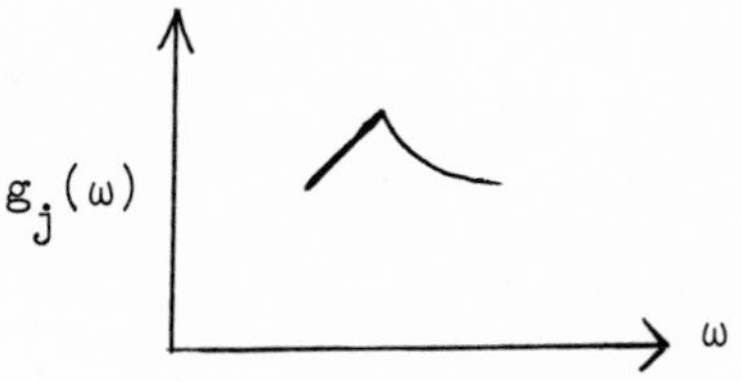

i = 2| $\varepsilon_1 = \varepsilon_2 = -\varepsilon_3 = 1$, a saddle point of the type

$$\omega = \omega_o + \frac{c_j^2}{2\omega_o}(K_x^2 + K_y^2 - K_z^2)$$ and

$$g_j(\omega) \sim [(c^2 + \omega_o - \omega)^{\frac{1}{2}} - 2(\omega_o - \omega)^{\frac{1}{2}}],\ \omega < \omega_o$$

$$\sim [(c^2 + \omega_o - \omega)^{\frac{1}{2}}],\ \omega > \omega_o:$$

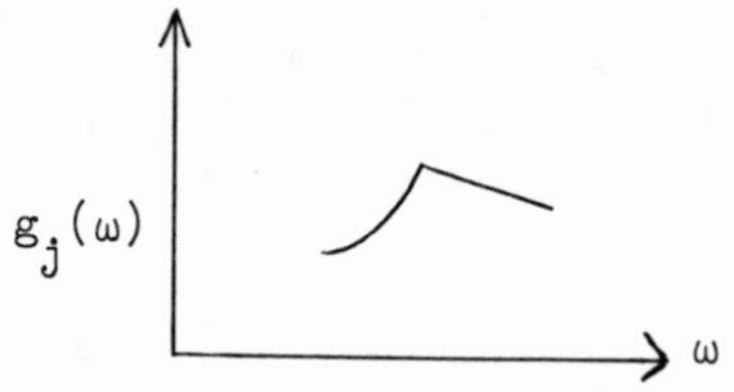

It can be shown using topological arguments due to Morse (see Van Hove's article(2)) that the number of times, N_i, a case i can occur in a three dimensional lattice is given by the set of equations

$$N_3 \geq 1;$$

$$N_1 - N_3 \geq 2;$$

$$N_2 - N_1 + N_3 \geq 1;$$

$$N_0 - N_2 + N_1 - N_3 = 0. \qquad (174)$$

The least number of "special" (or critical) points can be found by setting

$$N_3 = 1.$$

Then,

$$N_1 = 3,\ N_2 = 3,\ N_0 = 1.$$

We can have at <u>least</u> one minimum, one maximum, and three saddle points of each type which may, however, coalesce.

The "simplest" spectrum would thus look like the one in Figure 20 where a minimum exists at $\omega = 0$, a maximum exists at the zone boundary, and saddle points lie in between.

For a single polarization in a face-centered cubic lattice one can show the density of states will have the form in Figure 21.

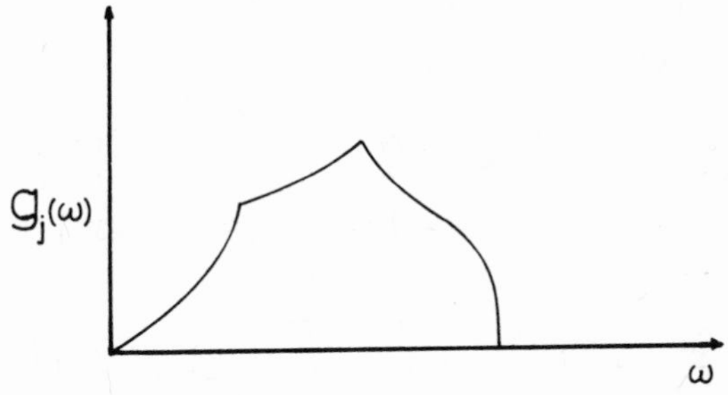

Figure 20 The Simplest Spectrum.

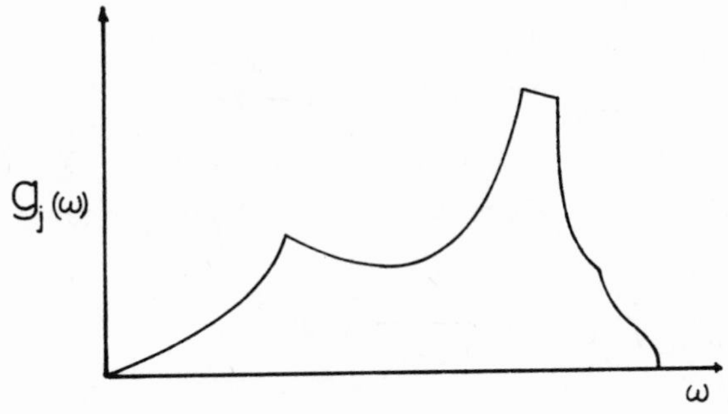

Figure 21 Density of States in a Face-Centered Cubic Lattice.

Using this form for the distribution function, the specific heat can be computed using the exact expression (143). In practice, one finds the number of critical points and their type from topological considerations, and then "moves" these points around until the best fit with experiment is obtained. This is considerably easier than computing $\omega(\underline{K},j)$ at selected points and differentiating to find $g_j(\omega)$. For an example of this technique, see J. C. Phillips[4] who applied Van Hove's method to find the lattice specific heat of aluminum. We now return to the thermodynamic treatment of lattice vibrations in order to consider in a simple way the effect of anharmonic terms which we have heretofore neglected.

II.C. Equation of State

1. Gruneisen Constant. We now consider the thermodynamics of lattice vibrations allowing for the effects of anharmonic forces. We have already found that the free energy F may be written, according to (129)

$$F = U_{eq} + kT \sum_{\underline{K},j} \log \{ 1 - \exp [-\hbar\omega(\underline{K},j)/kT] \}, \tag{175}$$

where U_{eq} is the static (equilibrium) lattice energy, and includes the zero-point energy for convenience. Using this expression, we may derive the equation of state connecting the parameters p, V, and T from the thermodynamic relation

$$p = -(\partial F/\partial V)_T . \tag{176}$$

Using (175) in (176), we find

$$p = -d U_{eq}/dV - \sum_{\underline{K},j} \{ [\exp (\hbar\omega(\underline{K},j)/kT)-1]^{-1} \}\hbar \frac{d\,\omega(\underline{K},j)}{dV} . \tag{177}$$

It is conventional to introduce the dimensionless quantity

$$\gamma_{\underline{K},j} \equiv - \frac{d\,\ell n\,\omega(\underline{K},j)}{d\ell n\,V} = - \frac{V}{\omega(\underline{K},j)} \frac{d\,\omega(\underline{K},j)}{dV} . \tag{178}$$

The $\gamma_{\underline{K},j}$ are referred to as the "Gruneisen" constants. The equation of state (177) becomes

$$p + \frac{d\,U_{eq}}{dV}$$

$$= \frac{1}{V} \sum_{\underline{K},j} \gamma_{\underline{K},j} \left\{ \frac{\hbar\omega(\underline{K},j)}{\exp(\hbar\omega(\underline{K},j)/kT)-1} \right\}$$

$$= \frac{1}{V} \sum_{\underline{K},j} \gamma_{\underline{K},j} \; U_{vib}(\underline{K},j), \tag{179}$$

where U_{vib} is the vibrational energy of the mode with wave vector $\underline{K}$ and polarization index j. We shall show that this microscopic definition of the Gruneisen constant will agree with the macroscopic definition (187) to be discussed next if all the $\gamma_{\underline{K},j}$ are set equal to a single constant γ.

To discuss the macroscopic equation of state, we note that c_v was a function only of (T/Θ_D). Hence $U_{vib} = TA(T/\Theta_D)$ where $A(T/\Theta_D)$ means a function depending only on (T/Θ_D). But, using the thermodynamic properties of the internal energy, we may also write $U_{vib} = B(V,T)$ where B is a function which depends only upon V, T. Evidently

$$TA(T/\Theta_D) = B(V,T). \tag{180}$$

To obtain a thermodynamic formula for γ, rather than (178) which proves to be difficult to evaluate, consider the derivative

$$\left(\frac{\partial A}{\partial \ln\Theta_D}\right)_T = \Theta_D \frac{\partial}{\partial \Theta_D} A(T/\Theta_D) = -\frac{T}{\Theta_D} A', \tag{181}$$

and the derivative

$$\left(\frac{\partial A}{\partial \ln T}\right)_V = \frac{T}{\Theta_D} A'. \tag{182}$$

Comparing (181) and (182) we see that

$$\left(\frac{\partial A}{\partial \ln\Theta_D}\right)_T = -\left(\frac{\partial A}{\partial \ln T}\right)_V. \tag{183}$$

The derivative

$$\left(\frac{\partial B(V,T)}{\partial V}\right)_T = \frac{1}{V}\left(\frac{\partial B(V,T)}{\partial \ell nV}\right)_T ,$$

may be rewritten

$$= \frac{1}{V}\left(\frac{\partial \ell n\Theta_D}{\partial \ell nV}\right)_T\left(\frac{\partial B(V,T)}{\partial \ell n\Theta_D}\right)_T \tag{184}$$

Taking the derivative with respect to $\ell n\ \Theta_D$ of the relation (180)

$$T\left(\frac{\partial A(T/\Theta_D)}{\partial \ell n\Theta_D}\right)_T = \left(\frac{\partial B(V,T)}{\partial \ell n\Theta_D}\right)_T$$

$$= -\ T\left(\frac{\partial A(T/\Theta_D)}{\partial \ell nT}\right)_V , \tag{185}$$

using (183). Comparing with (184), we finally arrive at the relation

$$\left(\frac{\partial B(V,T)}{\partial V}\right)_T = \frac{T}{V}\,\gamma\,\left(\frac{\partial A(T/\Theta_D)}{\partial \ell nT}\right)_V \tag{186}$$

where

$$\gamma = -\left(\frac{d\ \ell n\Theta_D}{d\ \ell nV}\right)_T . \tag{187}$$

Thus, this Gruneisen constant γ is related to the volume dependence of the Debye temperature. We shall relate this definition of γ to the $\gamma_{\underline{K},j}$ momentarily.

2. Lattice Vibrational Equation of State. Using the equality (180) we can rewrite the right side of equation (186) to find,

$$\left(\frac{\partial B(V,T)}{\partial V}\right)_T = \frac{\gamma}{V}\ [T\left(\frac{\partial B(V,T)}{\partial T}\right)_V - B(V,T)]. \tag{188}$$

Defining $F_{vib} = F - U_{eq}$, we see that, from our previous expression for F, F_{vib} has the same relative behaviour as a function of (V,T) and (T/Θ_D) that U_{vib} had. Thus we may write,

$$\left(\frac{\partial\{F-U_{eq}\}}{\partial V}\right)_T = \frac{\gamma}{V}\,[T(\frac{\partial F}{\partial T})_V - F + U_{eq}], \qquad (189)$$

because U_{eq} is only a function of V. From before,

$$F = U - TS$$

and we know $S = -(\frac{\partial F}{\partial T})_V$. Thus, transposing,

$$T\,(\frac{\partial F}{\partial T})_V - F = -U. \qquad (190)$$

Inserting (190) on the right side of (189),

$$-(\frac{\partial F}{\partial V})_T + (\frac{\partial U_{eq}}{\partial V})_T = \frac{\gamma}{V}(U-U_{eq}) = \frac{\gamma}{V}U_{vib}\,. \qquad (191)$$

Finally, using the relation (175), we find

$$p + \frac{dU_{eq}}{dV} = \frac{\gamma U_{vib}}{V}\,, \qquad (192)$$

which is the lattice vibrational equation of state. We see this equation is identical with our previous microscopic result (179) if and only if $\gamma_{\underline{K},j} = \gamma$ for all $\underline{K}$,j . Under those conditions, it follows that the definitions (178) and (187) are quite equivalent.

3. Relation Between Gruneisen Constant and Compressibility. In general, the effect of atmospheric pressure on the volume of our lattice is negligible, so that we may drop the term p in (192) to find

$$\frac{dU_{eq}}{dV} = \frac{\gamma U_{vib}}{V}\,. \qquad (193)$$

An additional relation is found by differentiating (192) with respect to volume V at constant T:

$$\left(\frac{\partial p}{\partial V}\right)_T + \frac{d^2U_{eq}}{dV^2} = \frac{\gamma}{V}\left(\frac{\partial U_{vib}}{\partial V}\right)_T - \frac{\gamma U_{vib}}{V^2} . \tag{194}$$

We may rewrite (188) as

$$\left(\frac{\partial U_{vib}}{\partial V}\right)_T = \frac{\gamma}{V}\left[T\left(\frac{\partial U_{vib}}{\partial T}\right)_V - U_{vib}\right]$$

$$= \frac{\gamma}{V}\left[Tc_v - U_{vib}\right]. \tag{195}$$

Using (193) and (195) in (194) we arrive at

$$\left(\frac{\partial p}{\partial V}\right)_T = -\frac{d^2U_{eq}}{dV^2} + \frac{\gamma^2}{V^2}\left(Tc_v - U_{vib}\right) - \frac{1}{V}\frac{dU_{eq}}{dV} . \tag{196}$$

To simplify this equation, we examine its equilibrium solution at T = 0. The quantity U_{eq} is expanded in a power series about its value at the equilibrium volume V_o:

$$U_{eq} = \left(U_{eq}\Big|_{V=V_o}\right) + (V-V_o)\frac{dU_{eq}}{dV}\Big|_{V=V_o}$$

$$+ \frac{1}{2}(V-V_o)^2 \frac{d^2U_{eq}}{dV^2}\Big|_{V=V_o} + \ldots . \tag{197}$$

In order that at T = 0, V_o represent the equilibrium volume, we must have

$$\frac{dU_{eq}}{dV}\Big|_{V=V_o} = 0,$$

and the last term on the right of (196) must therefore vanish under these conditions. At T = 0, the second term on the right also vanishes, so that we have left

$$\left(\frac{\partial p}{\partial V}\right)_{T=0} = -\frac{d^2U_{eq}}{dV^2} . \tag{198}$$

Multiplying by V_o and using $\frac{1}{\beta_o} = - V_o \left(\frac{\partial p}{\partial V}\right)_{T=0}$ we find

$$\beta_o = \left\{ V_o \frac{d^2U_{eq}}{dV^2} \right\}^{-1}. \tag{199}$$

We may solve for the term on the right by differentiating (197) with respect to the volume V. We find, using (193),

$$\frac{dU_{eq}}{dV} = \frac{\gamma U_{vib}}{V} = (V-V_o) \frac{d^2U_{eq}}{dV^2} . \tag{200}$$

Inserting into (199) we finally find the important result

$$\frac{V-V_o}{V_o} = \frac{\gamma U_{vib}}{V} \beta_o , \tag{201}$$

relating the volume expansivity to the Gruneisen constant, the vibrational energy, and the equilibrium compressibility at T=0. Note that γ must be a quantity caused by anharmonic forces, since lattice expansion, change of vibrational frequencies with volume etc., are all connected with the Gruneisen constant γ.

We may now directly relate the Gruneisen constant γ to known physical quantities. From (192),

$$p + \frac{dU_{eq}}{dV} = \gamma \frac{U_{vib}}{V} .$$

Differentiating with respect to T at constant V:

$$\left(\frac{\partial p}{\partial T}\right)_V = \frac{\gamma}{V} \left(\frac{\partial U_{vib}}{\partial T}\right)_V = \frac{\gamma}{V} c_V .$$

But,

$$\left(\frac{\partial p}{\partial T}\right)_V = - \frac{\frac{1}{V}\left(\frac{\partial V}{\partial T}\right)_p}{\frac{1}{V}\left(\frac{\partial V}{\partial p}\right)_T} ,$$

so that

$$\gamma = \left(\frac{V}{c_V}\right) \frac{\frac{1}{V}\left(\frac{\partial V}{\partial T}\right)_p}{\left(-\frac{1}{V}\right)\left(\frac{\partial V}{\partial p}\right)_T} = \frac{V}{c_V}\frac{\alpha}{\beta}$$

where $\alpha = (1/V)\,(\partial V/\partial T)_p$ is the thermal expansion coefficient and $\beta = -\,(1/V)\,(\partial V/\partial p)_T$ is the compressibility. We list below representative values of γ for the alkali halides.

Alkali Halide	γ	Alkali Halide	γ
LiF	1.99	KI	1.58
LiCl	1.54	RbF	1.28
NaF	1.57	RbCl	1.25
NaCl	1.43	RbBr	1.27
NaBr	1.55	RbI	1.50
KF	1.48	CsCl	1.97
KCl	1.34	CsBr	1.93
KBr	1.43	CsI	2.00

Over small temperature ranges, the agreement between the relation (201)

$$\frac{V-V_o}{V_o} = \frac{\gamma U_{vib}}{V}\,\beta_o$$

and the experimentally determined lattice expansion is reasonably good.

Alkali Halide	Extrapolated (theo) (0°K) β $10^{-12}cm^2dyne^{-1}$	Calculated $(V-V_o)/V$ at 293°K	Experimental $(V-V_o)/V$ at 293°K
NaCl	3.3	0.022	0.024
NaBr	3.9	0.025	0.027
NaI	(6.45)	0.031	0.030
KCl	4.8	0.021	0.021
KBr	5.5	0.025	0.027
KI	7.0	0.029	0.030

(An article "Equation of State of Certain Ideal Lattices," by Arenstein, Hatcher, and Neuberger(5), discusses in some detail the various definitions of γ and gives a numerical value for γ for a face-centered cubic lattice using a potential of the for $\lambda r^{-m} + Ae^{-\beta r}$).

II.D. Effect of the Boundary Conditions on the Frequency Distribution

We now complete the discussion of phonons by examining the effects of the assumption of periodic boundary conditions upon the distribution of frequencies $g_j(\omega)$. This treatment will follow that of R. E. Peierls whose theorem can be stated as follows (6): Consider a piece of the crystal similar in shape to the primitive cell and having a free surface; its frequency distribution is practically the same as that of a hypothetical lattice of the same shape with periodic (cyclic) boundary conditions imposed upon it.

The normal modes in both cases are of course not identical, and such is not claimed. What is claimed is that the distribution of the frequencies, the average number of solutions in an interval $\Delta\omega$ large enough to contain many frequencies, is the same in the two cases up to the leading order in N_o, where N_o is the number of cells along a crystal edge. We shall assume for simplicity that the crystal has cubic symmetry and that the shape of the primitive cell and of the entire crystal is also cubic; $N_o^{\,3} = N$ = total number of cells.

Consider a cube of N_o cells on a side so that

$$1 \leq n_1, n_2, n_3 \leq N_o .$$

As before, there exists a set of normal coordinates $Q_{\underline{K},j}(t)$ connected with the displacements $\underline{u}_{\underline{n}}$ by

$$\underline{u}_{\underline{n}} = \sum_{\underline{K},j} Q_{\underline{K},j}(t)\, \underline{e}\,(\underline{K},j) e^{i\underline{K}\cdot\underline{n}} + \text{c.c.} \tag{202}$$

In order to represent the frequency distribution we define the function,

$$D(\omega) = \begin{cases} 1 \text{ for } |\omega| < 1/2\, \Delta \\ 0 \text{ for } |\omega| > 1/2\, \Delta . \end{cases} \tag{203}$$

The number of frequencies in the interval $\left[\omega - \frac{1}{2}\Delta,\ \ \omega + \frac{1}{2}\Delta\right]$ is then given by

$$F(\omega) = \sum_{\underline{K},j} D(\omega(\underline{K},j) - \omega). \tag{204}$$

Because we are not interested in the exact values of the frequencies, we can replace this discontinuous function by a smooth function, also to be denoted by $D(\omega)$ and having the following properties:

$$\int_{-\infty}^{\infty} D(\omega)\, d\omega = \Delta,\ D(\omega) \ll 1, \text{ if } |\omega| \gg \Delta. \tag{205}$$

In addition we assume $D(\omega)$ sufficiently smooth so that its Fourier transform $g(\tau)$ rapidly converges. We write,

$$g(\tau) = \frac{1}{2\pi} \int_{-\infty}^{\infty} e^{-i\omega\tau}\, D(\omega)\, d\omega,$$

$$D(\omega) = \int_{-\infty}^{\infty} e^{i\omega\tau} g(\tau)\, d\tau \;. \tag{206}$$

The condition $g(\tau) << \Delta$ if $\tau >> 1/\Delta$ is consistent with $D(\omega) << 1$ if $\omega >> \Delta$. (One can see this directly from above, for

$\int D(\omega)\, d\omega\, e^{-i\omega\tau} \sim \Delta$ if $\tau \lesssim 1/\Delta$, but if $\tau >> 1/\Delta$, the exponential oscillates so rapidly that $\int D(\omega)\, d\omega\, e^{-i\omega\tau} << \Delta$).

A possible choice of functions which illustrate these properties is

$$D(\omega) = \frac{1}{\sqrt{2\pi}} e^{-\omega^2/2\Delta^2} \;;\; g(\tau) = \frac{\Delta}{2\pi} e^{-\Delta^2\tau^2/2} \;. \tag{207}$$

It will prove convenient to replace $F(\omega)$ by the symmetrical function

$$G(\omega) = F(\omega) + F(-\omega)$$

$$= \sum_{\underline{K},j} \{ D(\omega(\underline{K},j) - \omega) + D(\omega(\underline{K},j) + \omega) \}. \tag{208}$$

The error introduced by this replacement is small since the extra term $D(\omega(\underline{K},j) + \omega)$ is negligible for all but the lowest frequencies of the spectrum where $\omega \sim \Delta$.

We may write $G(\omega)$ in (208) in terms of $g(\tau)$ from (206):

$$G(\omega) = \sum_{\underline{K},j} \int_{-\infty}^{\infty} d\tau\, g(\tau)\, [e^{i\omega(\underline{K},j)\tau} e^{-i\omega\tau}$$

$$+ e^{i\omega(\underline{K},j)\tau} e^{i\omega\tau}]. \tag{209}$$

Combining the two terms,

$$G(\omega) = 2\int_{-\infty}^{\infty} d\tau\, g(\tau) \cos\omega\tau \sum_{\underline{K},j} e^{i\omega(\underline{K},j)\tau} \;. \tag{210}$$

However $D(\omega)$ is an even function of ω so that $g(\tau)$ is also an even function of τ. Thus (210) can be written as

$$G(\omega) = 4 \int_0^{\infty} d\tau\, g(\tau) \cos \omega\tau \sum_{\underline{K},j} \cos \omega(\underline{K},j)\tau. \tag{211}$$

Using the condition $g(\tau) \ll \Delta$ if $\tau \gg 1/\Delta$, it is only necessary to integrate (211) from 0 to $\sim \sigma/\Delta$ where σ is a numerical constant of order unity. Thus,

$$G(\omega) \cong 4 \int_0^{\sigma/\Delta} d\tau\, g(\tau) \cos \omega\tau \sum_{\underline{K},j} \cos \omega(\underline{K},j)\tau \quad . \tag{212}$$

We define a quantity $\Phi(\tau) = \sum_{\underline{K},j} \cos \omega(\underline{K},j)\tau, \tau < \sigma/\Delta$, so that

$$G(\omega) \cong 4 \int_0^{\sigma/\Delta} d\tau\, g(\tau) \cos(\omega\tau)\Phi(\tau) \, . \tag{213}$$

Let us examine the significance of $\Phi(\tau)$. Our fundamental solution for the vibrational amplitude $\underline{u}_{\underline{n}}$ is, because of (62)

$$\underline{u}_{\underline{n}} = \sum_{\underline{K},j} \{ Q_{\underline{K},j}(0)\, \underline{e}(\underline{K},j) \exp i\, (\underline{K}\cdot\underline{n}) \exp(-i\omega t)$$
$$+ Q^*_{\underline{K},j}(0)\, \underline{e}^*(\underline{K},j) \exp -i\, (\underline{K}\cdot\underline{n}) \exp(i\omega t)\} \, .$$

The quantities Q and Q^* are determined from the initial conditions [see (73)]

$$Q_{\underline{K},j}(0) = \frac{1}{2N} \sum_{\underline{n}} \exp -i\, (\underline{K}\cdot\underline{n})\, \underline{e}^*(\underline{K},j) \cdot$$

$$\cdot [\, \underline{u}_{\underline{n}}(0) + i\, \dot{\underline{u}}_{\underline{n}}(0)/\omega(\underline{K},j)],$$

with a similar expression for $Q^*_{\underline{K},j}(0)$. Consider the initial condition

$$u_{\underline{n}'}^{\alpha_0} = u, \; u_{\underline{n}}^{\beta} = 0, \; \underline{n} \neq \underline{n}', \; \beta \neq \alpha_0 .$$

That is, we displace the $\underline{n}'$ atom a distance u in the α_0 direction and release it without impulse at t = 0. We keep all the other atoms at their equilibrium positions until t = 0. Our solution for the displacement at the $\underline{n}$ site after a time t is then

$$u_{\underline{n}}^{\alpha}(t) = \sum_{\underline{K},j} \{[\frac{1}{2N} \exp -i (\underline{K}\cdot\underline{n}')$$

$$\times e^{\alpha_0 *}(\underline{K},j)\, u]\, e^{\alpha}(\underline{K},j) \exp i (\underline{K}\cdot\underline{n}) \exp [-i\omega(\underline{K},j)t]$$

$$+ [\frac{1}{2N} \exp i (\underline{K}\cdot\underline{n}')\, e^{\alpha_0}(\underline{K},j) u]$$

$$\times e^{\alpha *}(\underline{K},j) \exp -i (\underline{K}\cdot\underline{n}) \exp [i\omega(\underline{K},j)t]\} . \qquad (214)$$

Letting $\underline{K} \to -\underline{K}$ in the second expression, and using the symmetry relation (32)

$$u_{\underline{n}}^{\alpha}(t) = \sum_{\underline{K},j} \{ [\frac{1}{2N} \exp -i (\underline{K}\cdot\underline{n}')\, e^{\alpha_0 *}(\underline{K},j)\, u]$$

$$\times e^{\alpha}(\underline{K},j) \exp i (\underline{K}\cdot\underline{n}) \exp [-i\omega(\underline{K},j)t]$$

$$+ [\frac{1}{2N} \exp -i (\underline{K}\cdot\underline{n}')\, e^{\alpha_0 *}(\underline{K},j)\, u]\, e^{\alpha}(\underline{K},j)$$

$$\times \exp i (\underline{K}\cdot\underline{n}) \exp [i\omega(\underline{K},j)t] \} . \qquad (215)$$

Consider the above result for the displacement when $\alpha = \alpha_0$, $\underline{n} = \underline{n}'$ and $t = \tau$:

$$u_{\underline{n}'}^{\alpha_0} = \frac{u}{3N} \sum_{\underline{K},j} \cos \omega(\underline{K},j)\tau$$

$$\equiv \frac{u}{3N} \Phi(\tau) . \qquad (216)$$

We see that, apart from the constant factor u/N, $\Phi(\tau)$ represents the response after a time τ of the site at which our initial perturbation took place, in the direction of the initial displacement. We may use the identity

$$\sum_{\underline{n},\alpha} \left| e^{i\,\underline{K}\cdot\underline{n}}\, e^{\alpha}\,(\underline{K},j)\right|^2 = N \tag{217}$$

to rewrite (216) as

$$\frac{3N}{u}\, u^{\alpha_o}_{\underline{n}'} = \Phi(\tau) = \sum_{\underline{K},j} \left\{\frac{1}{N}\sum_{\underline{n},\alpha}\left| e^{i\underline{K}\cdot\underline{n}}\, e^{\alpha}\,(\underline{K},j)\right|^2\right\}$$

$$\times \cos\omega\,(\underline{K},j)\,\tau \tag{218}$$

$$= \frac{1}{N}\sum_{\underline{n},\alpha} \Phi^{\alpha}_{\underline{n}}\,(\tau),$$

where

$$\Phi^{\alpha}_{\underline{n}}\,(\tau) = \sum_{\underline{K},j}\left| e^{i\underline{K}\cdot\underline{n}}\, e^{\alpha}(\underline{K},j)\right|^2 \cos\omega(\underline{K},j)\,\tau. \tag{219}$$

The frequency distribution $F(\omega)$ in (208) is practically equivalent to $G(\omega)$ which is in turn determined by $\Phi(\tau)$ as in (213). We see from (213) that $g(\tau)\Phi(\tau)$ is the Fourier transform of $G(\omega)$ and from (218) that $\Phi(\tau)$ is composed of the additive contributions $\Phi^{\alpha}_{\underline{n}}(\tau)$ of the individual cells. The theorem which we are seeking to prove can now be stated in 2 parts:

I. Each function $\Phi^{\alpha}_{\underline{n}}(\tau)$ for a finite lattice is practically identical to that for an infinite lattice, $\Phi^{\alpha,\infty}_{\underline{n}}(\tau)$.

II. The sum $\Phi(\tau) = \frac{1}{N}\sum_{\underline{n},\alpha} \Phi^{\alpha}_{\underline{n}}(\tau)$ for the finite lattice is practically identical to the corresponding sum $\Phi^{\infty}(\tau) = \frac{1}{N}\sum_{\underline{n},\alpha} \Phi^{\alpha,\infty}_{\underline{n}}(\tau)$ for the infinite lattice. If we can prove these propositions, then it is clear that the effect of introducing periodic boundary conditions is negligible since their imposition upon an infinite lattice cannot reflect any change in the physical properties of the lattice.

Demonstration of I: Let $d(\underline{n})$ denote the distance of the cell $\underline{n}$ from the nearest boundary of the finite crystal. According to (216) and (218) $\Phi^{\alpha}_{\underline{n}}(\tau)$ represents the action of the cell upon itself. This action will be almost independent of the boundary for a time $t < 2d(\underline{n})/c$ because the effect of the boundary could not have been transmitted to the cell $\underline{n}$ in a smaller time. This is because $\Phi^{\alpha}_{\underline{n}}(\tau)$ represents the subsequent motion of a cell which at t=0 was perturbed while all the surrounding cells were initially not disturbed. The subsequent disturbance can propagate at best with a velocity c. It takes a time $d(\underline{n})/c$ for the disturbance to reach the surface, and $d(\underline{n})/c$ to get back to the original cell. Hence, the cell at $\underline{n}$ can only feel the effect of the surface after a time $2d(\underline{n})/c$. Our argument that $\Phi^{\alpha}_{\underline{n}}(\tau)$ is roughly independent of the surface for times $\tau \lesssim 2d(\underline{n})/c$ is therefore justified. However, τ is restricted by (212) for the given frequency interval Δ, to be less than σ/Δ, where σ is of order unity. Hence only waves from those cells for which

$$d(\underline{n}) < c/2\Delta \tag{220}$$

are strongly influenced by the wall. In other words, the functions $\Phi^{\alpha}_{\underline{n}}(\tau)$ for which $d(\underline{n}) > c/2\Delta$ are practically the same as the functions $\Phi^{\alpha,\infty}_{\underline{n}}(\tau)$ for an infinite crystal.

Demonstration of II: We divide the crystal into an interior and a surface part, by construction of a surface at a distance $d = c/2\Delta$ from the boundary, where Δ is the fixed frequency interval determining the accuracy of the frequency measurement. We take Δ to be a given fraction β of the maximum vibrational frequency of the crystal, ω_{max}. Let d be a certain fraction δ of the side $N_o a$ of the cube (a = lattice constant). By definition

$$\Delta = \beta\omega_{max}; \; d = \delta N_o a. \tag{221}$$

Then,

$$\delta\beta = \frac{\Delta}{\omega_{max}} \cdot \frac{d}{N_o a} = \frac{c}{2\omega_{max} N_o a} \equiv \frac{\lambda_{min}}{4\pi N_o a}$$

where $\lambda_{min} = 2\pi c/\omega_{max}$ is the corresponding minimum wavelength of the crystal. This wavelength is of the order the lattice constant a (i.e., $\lambda_{min} \sim 2\pi a$) so that roughly

$$\delta\beta = \frac{1}{2N_o} \quad . \tag{223}$$

The two fractions δ and β are thereby reciprocal: the smaller the frequency interval Δ, the thicker the boundary layer d. The ratio of the number of atoms in the boundary layer to those in the interior of the finite crystal is given by

$$\varepsilon = \frac{6dN_o^{\,2}a^2}{N_o^{\,3}a^3} = 6\,\delta \tag{224}$$

so that $\varepsilon\beta \sim 3/N_o$. For a cube one cm on a side one has $N_o \sim 10^8$, so that ε may be small even if β is small. Thus, for $\beta \sim 10^{-4}$, one finds $\varepsilon \sim 3 \times 10^{-4}$, indicating that the number of "surface" atoms is negligibly small in comparison with the number of atoms in the interior of the cube. This will cause the contribution of the surface layer to the sum $\Phi(\tau) = \sum_{\underline{n},\alpha} \Phi^{\alpha}_{\underline{n}}(\tau)$ to be negligible, proving statement II.

Statements I and II are however, limited. If a high accuracy of frequency measurement is desired (i.e., if β is very small) then ε cannot be too small, for the "surface" atoms will then contribute significantly to $\Phi(\tau)$, causing it to differ from $\Phi^{\infty}(\tau)$ considerably (7).

What is the limiting accuracy of the frequency measurement? It must occur for $\varepsilon \sim 1$ whence $\beta \sim 1/N_o$. This yields a frequency accuracy

$$\Delta \sim \omega_{max}/N_o \sim 10^{-8}\,\omega_{max} \sim 10^6 \text{ sec}^{-1} \tag{225}$$

for a crystal 1 cm on a side. Put another way, cyclic boundary conditions fail when the time to traverse the crystal, τ_t becomes comparable to the inverse of the desired accuracy of frequency measurement, $1/\Delta$. They are valid only when

$$\Delta\tau_t >> 1 \quad . \tag{226}$$

REFERENCES

1. E. W. Montroll, J. Chem. Phys. **11**, 481 (1943).

2. L. Van Hove, Phys. Rev. 84, 1189 (1953).

3. G. H. Wannier, Elements of Solid State Theory, Cambridge University Press, London, 1959, p. 73.

4. J. C. Phillips, Phys. Rev. 104, 1263 (1956); Phys. Rev. 105, 1933 (1957).

5. M. Arenstein, R.D. Hatcher and J. Neuberger, Phys. Rev. 131, 2087 (1963).

6. R. E. Peierls, Proc. National Institute of Science of India 20, 121 (1954).

7. For effects of finite size, see M. Hass, Phys. Rev. Lett. 13, 429 (1974).

THEORY OF VIBRONIC SPECTRA

William A. Wall

US Army LWL, Aberdeen Proving Gnd., Md. 21005

Present Address: US Naval EODF, Indian Hd., Md. 20640

ABSTRACT

Prominent sidebands often appear in the absorption and fluorescence spectra of systems consisting of a small quantity of magnetic ions in an ionic host lattice. The sidebands are called vibronic, and result from a dynamic coupling of the magnetic ion to the motion of other lattice constituents. These sidebands may reflect a frequency distribution for the perfect or perturbed lattice modes. An interpretation of the vibronic structure for the perfect lattice case, as first presented by R. A. Satten, is discussed.

A vibronic analysis requires a correlation of available information on lattice modes with optical data, and so the first portion of the discussion reviews infrared absorption, Raman scattering, and neutron scattering theory. A simple model for vibronic processes is then presented. Selection rules for vibronic transitions are formulated from group-theoretical considerations regarding the uncoupled electronic and vibrational systems. The role of critical points and their effect upon the frequency distribution of lattice modes is also taken into account. Finally, the system MgO:V^{2+} is treated in some detail.

I. INTRODUCTION

The fluorescent spectra of ionic crystals containing small quantities of magnetic ion impurities are strongly dependent upon temperature. Typically, the spectra present narrow lines accompanied by wide bands whose intensity increases with increasing temperature. These sidebands have been successfully analyzed in terms of a simultaneous interaction of the magnetic ion with a radiation field and with thermal vibrations of a localized or lattice type, via a dynamic crystal field. The second order processes resulting from such an interaction are called vibronic and involve the emission of a photon and the emission or absorption of one or more phonons. Vibrational frequencies appropriate to the magnetic ion and its surroundings modulate the frequency associated with an electronic transition and are reflected in the sideband structure.

In analyzing vibronic transitions, it is necessary to consider whether the modulating frequencies for a given system are those of lattice modes or of local modes. The presence of the impurity destroys the translational invariance of the crystal. The relative importance of perturbed and unperturbed modes is determined essentially by the nature and range of the interaction. In the treatment of vibronic processes given below, it is assumed that the relevant modes are those of the perfect lattice and that perturbation effects may be neglected. There is no a priori restriction on the $\underline{K}$-vectors of lattice modes which may be vibronically active. In a simple model it is found that the sidebands reflect a "frequency distribution" of those lattice modes allowed to participate by selection rules, with this distribution modified by a frequency-dependent weighting function.

Interpretation of a vibronic sideband requires at least a partial knowledge of the vibrational characteristics of the host crystal. In the first half of this article, three principal sources of such information are considered, namely, infrared absorption, Raman scattering, and thermal neutron scattering measurements. The basic theory relevant to each technique is sketched, and the obtainable information regarding lattice modes and the limitations of each are discussed. In the second half, a simple model for vibronic transitions is presented. The interaction of the ion with its surroundings is assumed to be electrostatic in origin; this model is not generally valid. The expression obtained for the vibronic matrix element allows a formulation of selection rules based on symmetry. Normal modes are characterized by irreducible representations of the host crystal's space group. A critical point analysis of the dispersion curves indicates possible sources of prominent structure in the frequency distribution and, therefore, in the vibronic sideband. Data from infrared absorption, Raman scattering, and neutron scattering measurements indicate at least some critical point frequencies and may specify both the nature of the critical points and the type of structure they can produce. The correlation of such information with vibronic selection rules may then allow the assignment of observed spectral features to particular regions of the Brillouin Zone or, conversely, the construction of approximate frequency distributions. The article concludes with an example, an interpretation of the vibronic sideband of the $^2E_g \rightarrow {}^4A_{2g}$ transition of $MgO{:}V^{2+}$.

II. INFRARED ABSORPTION

II.A. Characteristics of Infrared Absorption in Crystals

The interaction of electromagnetic radiation with a crystal can provide information about the crystal's vibrational and electronic spectrum, corresponding to a coupling of the radiation field with nuclear and electronic motion, respectively. Of interest here are the processes involved in the absorption of near infrared light by the crystal and the information that may be obtained about vibrational frequencies of the crystal.

The basic mechanism of infrared absorption is the creation of a net dipole moment in the unit cell which is driven by the radiation field. The simplest example of such a system is an ionic crystal with two ions per unit cell. At $\underline{K} = 0$ the optical modes describe a motion in which the center of mass of the cell is stationary, so that a fluctuating dipole moment of frequency $\omega_j(\underline{K} = 0)$ results. Other processes such as the deformation of electronic clouds and charge transfer can result in an "effective charge" on atoms and, thereby, in an oscillating dipole moment for the unit cell. If the radiation field has components with the frequency of the oscillating dipole, then coupling can take place and the radiation field may supply energy to the dipole. From a quantum mechanical viewpoint this energy exchange corresponds to the destruction of a photon and the creation of one or more phonons. In such a process there will be a conservation of energy and of wave vector to within a primitive reciprocal lattice vector.

In the following paragraphs of this section a brief sketch of the major points of infrared absorption is given. The proof of many of these statements is discussed in the quantum mechanical treatment of infrared processes in the next section.

Consider a single-phonon process in which a photon of frequency ν is annihilated and a single phonon is created. A photon in the infrared range has a wave vector $\underline{q} \sim 10^3$ cm^{-1}, while a phonon wave vector at the boundary of the Brillouin Zone has $\underline{K} \sim 10^8$ cm^{-1}. Thus, the wave vector of the phonon created is so small relative to the dimensions of the Brillouin Zone that it may be taken as zero. The conservation of energy condition then indicates that only optical phonons may be activated. In addition, since the polarization vectors of the radiation field are transverse to its direction of propagation $\underline{q}/|\underline{q}|$, it follows that only those modes at $\underline{K} \sim 0$ whose polarization vectors have a transverse component may be activated. The one-phonon process involves the direct coupling of a vibrational mode to the radiation field, and so can occur only in ionic crystals in which the constituents of the unit cell have an actual net charge. Absorption due to one-phonon processes is called Reststrahl absorption. The absorption spectrum shows very strong peaks corresponding to the "transverse" optical modes frequencies at $\underline{K}\sim 0$.

In a cubic ionic solid those $\underline{K} = 0$ optical modes which have eigenvectors transforming as the components of $\underline{r}$ (as T_{15} of O_h) and which are expected to be triply degenerate, actually split into two components. These optical modes may be classified as longitudinal (LO) and transverse (TO) only for $\underline{K} \neq 0$ and in particular directions of $\underline{K}$, but in the limit $\underline{K} \to 0$ the LO frequency is greater than the degenerate TO frequency. This is due to the fact that ions vibrating in the LO mode feel an additional restoring force from an electric field caused by the polarization of charge. A relation of the form

$$\omega^2_{LO}(\underline{K}\sim 0) = (\varepsilon_s/\varepsilon_\infty)\,\omega^2_{TO}(\underline{K}\sim 0) \quad \text{LST Relation}, \qquad (1)$$

where ε_s and ε_∞ are the static and high frequency dielectric constants, respectively, was first derived by Lyddane, Sachs, and Teller for a cubic diatomic crystal(1). This result has been generalized by Cochran and Cowley(2) and by Cochran(3) to the cases of cubic crystals with more complicated unit cells and of certain non-cubic crystals, respectively. In non-polarizable crystals the motions of atoms are not affected since the atoms have no net charge. Thus, these modes of vibration at $\underline{K}\sim 0$ remain triply degenerate, that is, $\omega_{LO}(\underline{K}\sim 0) = \omega_{TO}(\underline{K}\sim 0)$.

Since the radiation field couples only with TO modes in single-phonon processes, infrared absorption measurements provide no direct information about LO modes at $\underline{K}\sim 0$. For diatomic crystals in which the static and high frequency dielectric constants are accurately known, the LST relation can be utilized to find $\omega_{LO}(\underline{K}\sim 0)$. In more complex crystals the LO mode frequencies are commonly calculated from data of reflection measurements(4).

Thus far, only single-phonon processes have been considered. However, homopolar crystals display a weak absorption in the infrared region and ionic crystals display weak sidebands in addition to Reststrahl absorption. Such features arise as a result of multi-phonon processes. Consider a process in which a photon is destroyed and two phonons are created or one phonon is created and one destroyed. One proposed mechanism for the fluctuating dipole is that one phonon induces a charge on the atoms of the unit cell while the other produces a displacement of the induced charges. If two phonons are created, wave vector conservation requires that $\underline{K}_1 + \underline{K}_2 + \underline{q} = \underline{\varkappa}$, where $\underline{\varkappa}$ is a primitive reciprocal lattice vector. Setting $\underline{q} = 0$ and taking $\underline{\varkappa} = 0$ gives two phonons of opposite wave vector, $\underline{K}_1 = -\underline{K}_2$. The matrix element for such a transition is proportional to

$$\sqrt{\left[n(\underline{K}j) + 1\right] \left[n(-\underline{K}j') + 1\right]} \;, \qquad (2)$$

where $n(\underline{K}j)$ is the occupation number for mode $(\underline{K}j)$, and where the modes $(\underline{K}j)$ and $(-\underline{K}j')$ are no longer restricted to optical branches. At low temperatures the matrix element is practically temperature-independent. The resultant absorption bands are called summation bands. In the case of one phonon created and the other annihilated, the transition matrix element is proportional to

$$\sqrt{\left[n(\underline{K}j) + 1\right] \left[n(\ \underline{K}j')\right]} \;, \qquad (3)$$

so that the transition is strongly temperature-dependent and is very unlikely at low temperature. Again the modes $(\underline{K}j)$ and $(\ \underline{K}j')$ are not restricted to optical branches. The absorption bands for this type of transition are called difference bands.

For both summation and difference bands the details of the absorption spectrum are essentially determined by the density of the combined states. While two phonon processes should give rise to continuous absorption, absorption peaks can occur as a result of singularities in the combined state frequency distribution. Analysis of the absorption spectrum can be carried out with the use of critical point theory (5).

Finally, the third order anharmonic term in the expansion of the potential energy also can bring about two phonon processes. In this case the radiation field couples with a transverse optical mode which, in turn, couples to two other phonons via the third order anharmonic term. There is the same wave vector conservation requirement as for the two phonon processes discussed previously. The transition matrix element will have the same dependence on the occupation numbers of the phonon modes $(\underline{K}j)$ and $(\underline{K}'j')$ as given above.

This discussion could be extended to higher order phonon processes in the same manner as above, with the requirement of energy and wave vector conservation. However, as the number of phonons involved in the process increases, the transition probability for the process decreases, selection rules become less restrictive, and difficulty of analysis increases. Therefore, only one and two phonon processes are considered below.

II.B. Quantum Mechanical Treatment of Infrared Processes

In this section a quantum mechanical approach to lattice absorption due to Lax and Burstein(6,7) is briefly outlined. In the Born-Oppenheimer approximation(8) the unperturbed crystal wavefunction $\Psi_{in}(\underline{r},\underline{x})$ to lowest order is written as

$$\Psi_{in}(\underline{r},\underline{x}) = \varphi_i(\underline{r},\underline{x})\,\chi_{in}(\underline{x}), \tag{4}$$

where φ and χ are the respective electronic and vibrational wavefunctions of the crystal, with i denoting an electronic state and n a vibrational state. Here $\underline{r}$ denotes the collective electronic coordinates and $\underline{x}$ the collective nuclear coordinates. The nuclear coordinates $\underline{x}$ enter $\varphi_i(\underline{r},\underline{x})$ parametrically, and a non-degenerate electronic state is assumed. To calculate lattice absorption, the radiation field will be treated classically and taken to consist of a single frequency ν, with wave vector $\underline{q}$ and polarization $\hat{\underline{\pi}}$. The relevant interaction term for the radiation field and crystal lattice is $H' \propto \underline{A}\cdot\underline{p}$, where the radiation field couples to all charged particles. For the case of valence electrons, it follows from the form of the perturbation H' that the electronic wavefunctions $\varphi_i(\underline{r},\underline{x})$ may be replaced by Hartree products of localized orbitals in the calculation of relevant matrix elements. Thus, each electron can be associated with a particular nucleus. An approximate term for the the total interaction is given by H':

$$H' \propto \hat{\underline{\pi}}\cdot \sum_{ef}\sum_{J} e^{i\underline{q}\cdot\underline{R}(ef)} \left\{-e\underline{r}_J(ef) + Z(f)\left[\underline{R}(ef) + \underline{u}(ef)\right]\right\}, \tag{5}$$

where $\underline{R}$(ef) is the equilibrium position vector and $\underline{u}$(ef) the displacement vector for the nucleus at site (ef), where $\underline{r}_J$(ef) is the position vector for the Jth electron at nucleus (ef), and where Z(f) is the charge of nucleus f in the unit cell. There is a sum over all nuclei and electrons of the crystal. The above expression is obtained by replacing the instantaneous positions of the nuclei and the electrons in the exponential with the equilibrium positions of the nuclei and by then applying the dipole approximation.

For a purely vibrational transition, with the initial and final electronic state the same and labelled by i, the matrix element for a transition from vibrational state m to state n is proportional to

$$\hat{\underline{\pi}}\cdot\langle n|\,\underline{M}_{ii}\,|m\rangle, \quad \text{where} \tag{6}$$

$$\underline{M}_{ii}(\underline{x},\underline{q}) = \sum_{Jef} e^{i\underline{q}\cdot\underline{R}(ef)} \int \varphi_i^*(\underline{r},\underline{x}) \left\{-e\underline{r}_J(ef) + Z(f)\left[\underline{R}(ef) + \underline{u}(ef)\right]\right\} \varphi_i(\underline{r},\underline{x})\, d\underline{r}. \tag{7}$$

This equation may be put into a more suitable form for evaluation by re-expressing it in terms of electronic wavefunctions corresponding to the equilibrium configuration of nuclei. Since core electrons tend to move rigidly with their nucleus and so have wavefunctions independent of $\underline{x}$, they are completely taken into account by replacing the nuclear charge $Z(f)$ with $Z^{ion}(f)$, the charge of the ion consisting of the nucleus and core electrons. Those electrons whose wavefunctions have an $\underline{x}$ dependence are called "deformable." Making a change of variable from $\underline{r}_J(ef)$ to $\rho_J(ef) \equiv \underline{r}_J(ef) - [\underline{R}(ef) + \underline{u}(ef)]$, it follows that $\varphi_i(\underline{r},\underline{x}) \rightarrow \gamma(\rho,\underline{x})$, a many-body wavefunction for all deformable electrons of the crystal. Returning to equation (7), this change of variable and a subsequent expansion of $\gamma(\rho,\underline{x})$ in ionic displacements $\underline{u}(ef)$ gives

$$\underline{M}_{ii}(\underline{x},\underline{q}) = \sum_{J'ef} e^{i\underline{q}\cdot\underline{R}(ef)} \Big[Z^{ion}(f) \big[\underline{R}(ef) + \underline{u}(ef) \big] + \int \gamma^*(\rho,\underline{x}) \, (-e\rho_{J'}(ef)) \, \gamma(\rho,\underline{x}) \, d\rho \Big], \quad (8)$$

$$= \sum_{J'ef} e^{i\underline{q}\cdot\underline{R}(ef)} \Big\{ \Big[Z^{ion}(f)\, \underline{u}(ef) + \sum_{EF} \underline{\underline{M}}^{(1)}(J'ef;EF)\cdot\underline{u}(EF) \Big] + \tfrac{1}{2}\sum_{\substack{EF\\E'F'}} \underline{\underline{M}}^{(2)}(J'ef;EF,E'F'):\underline{u}(EF)\,\underline{u}(E'F') \Big\} + \ldots, \quad (9)$$

where the summation on J' is over deformable electrons.

From this expression it can be seen that one-phonon processes are due to a sum of contributions which includes electronic charge deformation, while two-phonon processes are due exclusively to electronic charge deformation. For homopolar crystals it can be shown that the symmetry properties of $\underline{M}_{ii}(\underline{x},\underline{q})$ forbid the occurrence of single-phonon processes.

Equation (9) in any case may be rewritten to second order in displacements in the form

$$\underline{M}(\underline{x},\underline{q}) = \sum_{\substack{ef\\e'f'}} e^{i\underline{q}\cdot\underline{R}(ef)} \Big[\underline{\underline{m}}(ef)\cdot\underline{u}(ef) + \underline{\underline{m}}(ef;e'f'):\underline{u}(ef)\underline{u}(e'f') \Big]. \quad (10)$$

The term $\underline{\underline{m}}(ef)$ is interpreted as the effective charge on the site (ef) with the lattice at equilibrium and the term $\underline{\underline{m}}(ef,e'f')\cdot\underline{u}(e'f')$ is interpreted as the charge induced at (ef) due to the displacement $\underline{u}(e'f')$ of the ion at (e'f').

Since the crystal is invariant under translations, it follows that $\underline{\underline{m}}(ef)$ should be independent of e, and that $\underline{\underline{m}}(ef;e'f')$ should depend on only (e-e')(6). Furthermore, it is expected that only neighboring atoms will act to induce effective charges on each other. Since the wavelength of infrared light is much greater than the dimensions of the unit cell, it is a good approximation to use $\underline{q} = 0$ in replacing $\underline{\underline{m}}(f)$ and $\underline{\underline{m}}(e-e',f;0f')$ by their infinite wavelength values, $\underline{\underline{m}}^{(o)}(f)$ and $\underline{\underline{m}}^{(o)}(e-e',f;0f')$. Using $e^{i\underline{q}\cdot\underline{R}(f)} \approx 1$ and expressing the displacements $\underline{u}(ef)$ in terms of phonon annihilation and creation operators, the relevant term for absorption of a photon and creation of a single phonon is

$$\underline{M}^{(1)}(\underline{x},\underline{q}) = \sum_{ef} e^{i\underline{q}\cdot\underline{R}(e)}\ \underline{\underline{m}}^{(o)}(f)\cdot \sum_{\underline{K}j} (NM_f)^{-\frac{1}{2}}\ \underline{e}(f|\underline{K}j)\ e^{-i\underline{K}\cdot\underline{R}(e)} \left(\frac{\hbar}{2\omega_j(\underline{K})}\right)^{\frac{1}{2}} b_{\underline{K}}^{j+}$$

$$= \sum_j \sqrt{\frac{N\hbar}{2\omega_j(\underline{q})}}\ b_{\underline{q}}^{j+} \sum_f \underline{\underline{m}}^{(o)}(f)\cdot\underline{e}(f|\underline{q}j)\Big/ M_f^{\frac{1}{2}}\ , \quad (11)$$

where $\underline{e}(f|\underline{q}j)$ is the polarization vector at site f for mode ($\underline{q}$j), where $b_{\underline{q}}^{j+}$ is the phonon creation operator for mode ($\underline{q}$j), and where wave vector conservation is required: $\underline{K} = \underline{q}$. Furthermore, energy conservation requires that only optical modes be activated for $\underline{q}\approx 0$. In addition, if the major contribution to $\underline{\underline{m}}^{(o)}(f)$ comes from ionic motion, then $\underline{\underline{m}}^{(o)}(f)$ is, for the most part, a scalar. Since the polarization vectors of light are transverse to its direction, only those branches which have polarization vectors with components in the transverse direction can couple with the radiation field. If the vibrational polarization vectors for wave vector $\underline{K}\approx 0$ are approximated as being purely longitudinal or purely transverse, then only the $\underline{K}\approx 0$ transverse optical modes can be activated.

The above approach indicates that a set of absorption lines should be observed in infrared measurements. The lines should appear at the frequencies of the transverse optical modes for $\underline{K}=0$. The absorption lines will actually be broadened by anharmonic effects.

Similar considerations give for the second order term in displacements

$$\underline{M}^{(2)}(\underline{x},\underline{q}) = \sum_{\substack{\underline{K}_1 j_1 \\ \underline{K}_2 j_2}} Q^{j_1}_{\underline{K}_1} Q^{j_2}_{\underline{K}_2} \sum_{ff'} \underline{e}(f|\underline{K}_1 j_1)\, \underline{e}(f'|\underline{K}_2 j_2)$$

$$\times\ (M_f M_{f'})^{-\frac{1}{2}} : \sum_E \underline{\underline{m}}^{(o)}(0f;Ef')\, e^{i\underline{K}_2\cdot\underline{R}(E)} \quad , \quad (12)$$

along with the wave vector conservation rule

$$\underline{K}_1 + \underline{K}_2 \pm \underline{q} = \underline{\mathcal{K}}, \quad (13)$$

where $\underline{\mathcal{K}}$ is a primitive reciprocal lattice vector. $Q^j_{\underline{K}}$ can be expressed in terms of annihilation and creation operators so that two phonons can be created or one phonon created and another destroyed. It is clear that any type of mode may participate in two phonon processes, so that one would expect a continuous absorption. The exact structure of this absorption will depend on the frequency distribution for lattice modes. An analysis of two phonon absorption can be performed by considering critical points in the sum of frequencies spectrum. Such an analysis was carried out by Johnson for diamond and zinc-blende crystals(5).

The two phonon process may also come about by a second order process involving the third order anharmonic term in the expansion of V(...$\underline{u}$(ef) ....). In this mechanism the radiation field interacts with a $\underline{K} \sim \underline{q}$ "transverse" optical mode which is, in turn, coupled to two other modes via the anharmonic interaction. The result is that two phonons are created or one phonon is created and one destroyed. From previous considerations, the phonon activated in the intermediate state must be "transverse" optical. It can be shown(8) that there is wave vector conservation in an anharmonic interaction so that there is wave vector conservation from initial to intermediate and from intermediate to final states in the matrix element. Every transverse optical phonon at K = q can serve as a virtual phonon in the processes, regardless of energy. The anharmonic two-phonon process cannot occur in homopolar crystals, of course. For polar crystals it is believed that the anharmonic process predominates over the two-phonon mechanism discussed previously (7).

III. RAMAN SCATTERING

III.A. Characteristics of Raman Scattering in Crystals

Further information about vibrational modes of crystals may be obtained from the inelastic scattering of light by the crystals. A shift in the frequency of the scattered from that of the incident light may correspond to the creation or destruction of lattice phonons. The light source used in such scattering measurements has a frequency somewhat greater than that of the lattice modes, so as to avoid lattice absorption or emission, but a frequency much less than is required for electronic excitation. As in the case for infrared absorption the wave vector of the light will be so much smaller than the dimensions of the Brillouin Zone that it may be approximated by $\underline{q} \sim 0$. In the scattering process there will be a conservation of energy and of wave vector to within a primitive reciprocal lattice vector, so that in a process involving a single phonon the shift in frequency for incident and scattered light gives directly a vibrational mode frequency very near the Brillouin Zone center, $\underline{K} \sim 0$. Scattering processes involving only optical phonons are termed first order Raman Scattering, while those involving only acoustic phonons are called Brillouin Scattering. Higher order processes may involve a combination of optical and acoustic phonons and are called higher order Raman Scattering. Unless otherwise specified, the term Raman Scattering is used here to refer to the first order process.

Of primary interest here are the optical modes near the center of the zone, so first order Raman Scattering is considered. For crystals of reasonably high symmetry the optical dispersion curves should be quite flat about the Brillouin Zone center, and so approximate values for a large number of mode frequencies may be obtained. It will be seen that not all optical modes at $\underline{K} \sim 0$ are active in Raman Scattering processes. The selection rules governing these processes differ from those for infrared absorption. Thus, infrared absorption and Raman Scattering measurements independently supply information about lattice optical modes.

Now consider the physical mechanism through which the scattering of light occurs. The incident light may be thought of as a radiation field, with the electric field given by $\underline{\mathcal{E}} = -\partial \underline{A}/\partial t$. The electric field acts on the charged particles of the crystal, but only the electrons are able to follow the rapidly fluctuating field. As a result an electric dipole moment is induced from the motion of the electrons; this dipole may absorb a photon of the radiation field having the same frequency. The electrons of the crystal also experience a field from the oscillating nuclei which acts to modulate the frequency of the electric dipole moment. Thus, photons emitted by the oscillating dipole may have a shift in frequency due to this modulation. The difference in this mechanism from lattice absorption should be emphasized. For infrared absorption a resonance condition exists between the fluctuating dipole moment originating with lattice vibrations and the radiation field; for light scattering the radiation field creates a dipole moment which absorbs and re-emits photons.

For a first order scattering process involving a phonon of frequency ω and wave vector $\underline{K}$, energy and wave vector conservation imply

$$\nu_i = \nu_s \pm \omega(\underline{K}) \quad \text{and} \quad \underline{q}_i = \underline{q}_s \pm \underline{K} , \tag{14}$$

where + and - refer to phonon creation and annihilation, respectively, and where the subscripts i and s denote incident and scattered photon values, respectively. Since $|\underline{q}_i| \sim |\underline{q}_s|$, one makes the approximation

$$|\underline{K}| \sim 2 |\underline{q}_i| \sin\Psi , \tag{15}$$

where (2Ψ) is the angle between the directions $\hat{\underline{q}}$ of incident and scattered photons. Thus, the frequency shift for Brillouin Scattering is expected to be highly dependent on angle Ψ, while that for Raman Scattering to be insensitive to changes in Ψ. Raman Scattering measurements are commonly performed with the angle of scattering (2Ψ) at $\pi/2$ (9).

III.B. Theoretical Treatment of Raman Scattering

The theory of Raman Scattering can be approached from a purely quantum mechanical or a semi-classical point of view. Under certain specific approximations, it is found that they give roughly equivalent results(9). In practice, one seldom has sufficient information with regard to electronic states of the crystal to make a calculation of the electronic polarizability tensor associated with Raman Scattering. Of interest here is a qualitative description of Raman Scattering processes along with considerations of the symmetry properties of the polarizability tensor and resultant selection rules for the allowed vibrational transitions.

In the semi-classical description of Raman Scattering the radiation field is treated classically and the lattice vibrations quantum mechanically(8). The approach, due to Born and Bradburn(10), consists in representing the interaction of light and lattice vibrations as a time-dependent electric field first inducing and then interacting with the dipole moment of the crystal. The perturbation H' is taken to be of the $\mathcal{E} \cdot \underline{r}$ rather than $\underline{A} \cdot \underline{p}$ form, and the radiation field is represented by a two-component electric field to account for photon creation and annihilation

$$H' = -\underline{M} \cdot \underline{\mathcal{E}}(t), \tag{16}$$

$$\underline{\mathcal{E}}(t) = \underline{\mathcal{E}}^{+} e^{i\nu t} + \underline{\mathcal{E}}^{-} e^{-i\nu t}, \text{ with } [\underline{\mathcal{E}}^{+}]^{*} = \underline{\mathcal{E}}^{-}, \tag{17}$$

$$\underline{M} = \sum_{Jef} \left\{ Z^{ion}(f) \left[\underline{R}(ef) + \underline{u}(ef) \right] - e\underline{r}_J(ef) \right\}. \tag{18}$$

Although assumption of a spatially independent electric field is not really justified for interaction with a crystal and eliminates wave vector conservation in the equations, its net result is the omission of factors $e^{\pm i\underline{q}\cdot\underline{R}(ef)}$ in H' which do not affect the qualitative results. The perturbed crystal wavefunctions are found in the usual way from time-dependent perturbation theory and the convention is adopted of labelling perturbed states by capitalized letters ($|L\rangle$) and unperturbed states by small letters ($|1\rangle$). The time-dependent moment, denoted by $\underline{m}(t)$, which has been induced when the crystal is in state $|L\rangle$ may be calculated from the term $\langle L|\underline{M}|L\rangle$. This moment may absorb and re-emit light of the same frequency, corresponding to Rayleigh scattering.

Consider the case of a light scattering event in which the crystal makes a transition from state $|L\rangle$ to state $|G\rangle$. To calculate the induced moment one examines the expression $\langle L|\underline{M}|G\rangle$ and $\langle G|\underline{M}|L\rangle$ and identifies terms appropriate to the $L \rightarrow G$ transition by their time factor, namely, $\exp[i(\omega_{1g} + \nu)t]$. This time factor corresponds to emission of light with frequency $(\omega_{1g} + \nu)$, where $(\hbar\omega_{1g})$ is the energy difference between the unperturbed crystal states $|1\rangle$ and $|g\rangle$. Omitting the details of the calculation(8), the result is

$$m_{\mu}(t) = \sum_{\sigma} \left\{ \alpha^{1g}_{\mu\sigma}(\nu)^{*} \mathcal{E}^{-}_{\sigma} e^{-i(\nu+\omega_{1g})t} + \alpha^{1g}_{\mu\sigma}(\nu) \mathcal{E}^{+}_{\sigma} e^{+i(\nu+\omega_{1g})t} \right\}, \tag{19}$$

$$\text{where } \alpha^{1g}_{\mu\sigma}(\nu) = \hbar^{-1} \sum_{p} \left\{ \frac{\langle 1|M_{\mu}|p\rangle \langle p|M_{\sigma}|g\rangle}{\omega_{pg} + \nu} + \frac{\langle 1|M_{\sigma}|p\rangle \langle p|M_{\mu}|g\rangle}{\omega_{p1} - \nu} \right\}. \tag{20}$$

$\alpha^{1g}_{\mu\sigma}(\nu)$ is called the transition polarizability tensor from state 1 to g.

From the expression for the oscillating dipole moment, one can calculate the flux of energy radiated by the moment by means of classical electromagnetic theory(11). For Rayleigh scattering in which no energy transfer occurs between crystal and field, the initial and final states are the same. For the Raman scattering under consideration, the initial and final states of the crystal differ only in their vibrational portion.

If the validity of the Born-Oppenheimer approximation is assumed, then the total crystal wavefunction $\Psi_{in}(\underline{x},\underline{r})$ may be denoted as $|in\rangle$. Generally, one wishes to obtain information about the vibrational modes associated with the ground electronic state. Thus, the transition polarizability tensor for this case is written as $\alpha_{\mu\sigma}^{nn'}(\nu)$, where n and n' are the final and initial vibrational states, respectively. The polarizability tensor may be separated into ionic and electronic parts as follows (8):

$$\alpha_{\mu\sigma}^{nn'}(\nu) \simeq \sum_{n''} \left\{ \frac{\langle On|M_\mu|On''\rangle \langle On''|M_\sigma|On'\rangle}{E(On'') - E(On') + \hbar\nu} + \frac{\langle On|M_\sigma|On''\rangle \langle On''|M_\mu|On'\rangle}{E(On'') - E(On) - \hbar\nu} \right\}$$
$$+ \sum_{b\neq 0} \sum_{n''} \left\{ \frac{\langle On|M_\mu|bn''\rangle \langle bn''|M_\sigma|On'\rangle}{E_b - E_o + \hbar\nu} + \frac{\langle On|M_\sigma|bn''\rangle \langle bn''|M_\mu|On'\rangle}{E_b - E_o - \hbar\nu} \right\}, \quad (21)$$

where vibrational energy changes are neglected in the second bracket, the electronic part, and where E_b is the energy of the unperturbed state $|b\ 0\rangle$. For scattering measurements where $\hbar\nu \gg \{E(On'') - E(On)\}$, the electronic part is expected to predominate(8). Hereafter, the ionic part is neglected and the tensor α denotes the electronic polarizability tensor.

In principle one could explicitly calculate the polarizability tensor from equation (21). However, lack of knowledge of wavefunctions of the excited electronic states makes such a calculation virtually impossible. In a qualitative treatment, it is customary to express $\alpha_{\mu\sigma}^{nn'}(\nu)$ as a vibrational matrix element of the form $\langle n|\beta_{\mu\sigma}(\nu,\underline{x})|n'\rangle$(9). The tensor β may be expanded in a Taylor series in normal coordinates, with the first term in the series corresponding to Rayleigh scattering, the second to one-phonon Raman scattering, the third to two-phonon Raman scattering, and so forth. Phonon creation or annihilation also introduces a factor of $\sqrt{n(\underline{K}j)+1}$ or $\sqrt{n(\underline{K}j)}$ into the polarizability tensor, and thereby a temperature dependence for scattering intensity. If the frequency of the light source satisfies $\nu \ll \hbar^{-1}(E_b - E_o)$ for $b \neq 0$, then the tensors α and β are symmetric in their lower indices.

From classical radiation theory the radiation emitted from an oscillating dipole may be calculated from the Poynting vector. Applying this method to the dipole $\mathfrak{m}(t)$ oscillating with frequency $[\nu + \hbar^{-1}(E(0n') - E(0n))]$, the Raman scattering intensity per unit time into $\hat{\underline{n}}\, d\Omega$ associated with the transition $n \rightarrow n'$ is proportional to (11):

$$\left\{\nu + \hbar^{-1}[E(0n')-E(0n)]\right\}^4 \sum_i \sum_{\substack{\beta\gamma \\ \mu\sigma}} n^i_\beta \quad n^i_\gamma \left[\alpha^{nn'}_{\beta\mu}(\nu)^* \alpha^{nn'}_{\gamma\sigma}(\nu)\right] \mathcal{E}^-_\mu \mathcal{E}^+_\sigma , \tag{22}$$

where $\hat{\underline{n}}^1$ and $\hat{\underline{n}}^2$ are unit vectors forming an orthonormal triad with $\hat{\underline{n}}$. If the incident light is polarized in the ρ direction and one observes only that scattered polarized light in the $\mathcal{E}$ direction, then

$$I_\mathcal{E} \propto \left\{\nu + \hbar^{-1}\left[E(0n') - E(0n)\right]\right\}^4 I_\rho \left|\alpha^{nn'}_{\mathcal{E}\rho}(\nu)\right|^2 , \tag{23}$$

where $I_\mathcal{E}$ and I_ρ are the intensities per unit solid angle of the scattered and incident light, respectively.

As mentioned previously, one of the difficulties with the above method is that the unperturbed electronic wavefunctions are not easily found, since they correspond to non-equilibrium configurations of the nuclear coordinates. In some purely quantum mechanical calculations of Raman scattering, this difficulty has been avoided by referring electronic wavefunctions to the equilibrium nuclear configuration and including an electron-phonon interaction explicity(9). The predominant mechanism is thought to involve the electrons of the crystal as an intermediary between the radiation field and lattice modes, absorbing and emitting a photon, and creating or destroying phonons. First order Raman scattering is calculated from third order, time-dependent perturbation theory, with the perturbation term H' given by

$$H' = H_{el\text{-}rad} + H_{el\text{-}vib}. \tag{24}$$

One obtains explicit wave vector conservation in the scattering event. Applying this technique to certain semiconductors, Loudon(12) has calculated a "Raman tensor" which plays an analogous role to the polarizability tensor in determining scattering intensity. In specific approximations a proportionality exists between the two tensors.

IV. NEUTRON SCATTERING

IV.A. Characteristics of Neutron Scattering From Crystals

Neutron scattering experiments are of great importance in obtaining information about lattice dynamics. The usefulness of the method is due to the fact that experimentally significant changes in both the wave vector and the energy of a thermal neutron ($\sim$.025 ev) may occur in a single scattering event. There is no restriction on the type (acoustic or optical) or wave vector of lattice modes which may interact with the incoming neutrons. In the typical scattering experiment a beam of mono-energetic neutrons is incident on a crystal and neutrons are scattered both elastically and inelastically. From a semi-classical viewpoint in which a single incident neutron is considered as a wave, any change in the neutron's wavelength upon scattering can be thought of as a Doppler shift resulting from the motion of the scatterer, that is, from the vibrational motion of nuclei of the crystal.

In the scattering of thermal neutrons each nucleus of the crystal is considered to interact with the incident neutron wave through short-range nuclear forces. As a result of the presence of elemental isotopes and different orientations of nuclear spin for members of the crystal, the _scattering length_ of a particular site in the unit cell may vary significantly from cell to cell.

In the calculation of ordinary and differential cross sections one must evaluate summations over all crystal nuclei of terms containing scattering lengths and products of scattering lengths. Spin and isotope variations from cell to cell destroy the translational invariance and so, assuming a random distribution of these variations, one takes average values of the terms. Hereafter, such variations are denoted as disorders, but are not to be confused with actual defects in the crystal structure. One may conveniently divide the total scattering into two parts, called _coherent_ and _incoherent_ scattering. Consider an imaginary crystal in which the actual value of a site's scattering length is replaced by the average value for the site. The imaginary crystal has translational invariance. The scattering from the imaginary crystal involves the interference of neutron waves scattered by all nuclei and is called _coherent scattering_. The remainder of the total scattering from the actual crystal involves neutron waves scattered independently from the nuclei and is termed _incoherent scattering_. In every neutron scattering experiment both types of scattering are observed, with the relative importance of each type dependent on the crystal under study.

While in any scattering event there is energy conservation, coherent scattering also requires a conservation of wave vector. The combination of the two conservation rules makes coherent inelastic scattering invaluable in the determination of dispersion curves. If the incoming and scattered neutron beams are represented as plane waves with wave vectors $\underline{q_i}$ and $\underline{q_s}$, respectively, then for scattering events involving a single phonon the conservation of energy and wave vector results in the following equations:

$$\underline{Q} \equiv \underline{q_i} - \underline{q_s} = \pm \underline{K} + \underline{\kappa}, \tag{25}$$

$$\text{and } \Delta E = \frac{\hbar^2}{2m} \left(q_i^2 - q_s^2 \right) = \pm \hbar\omega_j(\underline{Q}) = \pm\hbar\omega_j(\underline{K}), \tag{26}$$

where $\hbar\underline{Q}$ is the quasi-momentum and ΔE the energy transferred to the crystal, where the periodicity of the eigenfrequencies in reciprocal space has been used, and where the plus and minus signs refer to phonon creation and annihilation, respectively. On account of the finite number of branches j, the conservation equations can be satisfied simultaneously only for certain values of $\underline{q_i}$ and $\underline{q_s}$.

In the typical scattering experiment the energy of the incident neutron beam may be pre-selected and the intensity of scattering in a particular energy interval and direction is found by a chopper or a crystal diffraction technique. A peak in intensity of coherent scattering at some energy and direction corresponds to the two conservation equations being satisfied. A measurement immediately gives $\underline{q_s}$, and the conservation equations then give the frequency of some branch j at the wave vector $\underline{K}$ of the Brillouin Zone. By varying the incident beam energy the eigenfrequencies at every point of the Zone can be determined, in principle. It is found that the scattering intensity also depends on the polarization vectors.

An additional comment is appropriate at this point. Only the single phonon processes have been discussed; the reason is that multi-phonon processes result in a weak and relatively constant scattering intensity, experimentally resembling background, and so are of little importance.

IV.B. Theory of Neutron Scattering

In this section a simplified account of neutron scattering by crystals is outlined(13). Among the assumptions are that neutron absorption by the scatterer can be neglected and that each neutron undergoes a single scattering event. For simplicity spin disorder is also neglected. The incident beam is represented as a plane wave with wave vector $\underline{q}_i$. The target consists of an aggregate of nuclei dynamically coupled through the lattice modes. The first problem is obtaining an adequate representation of the neutron-nucleus interaction. The usual approach is to represent the two-particle interaction by the Fermi pseudo-potential $V(\underline{r})$ in the Born approximation for calculating ordinary and differential cross sections(13), with

$$V(\underline{r}) = \frac{2\pi\hbar^2}{m} \; a \; \delta(\underline{r} - \underline{R}_{nuc}) , \qquad (27)$$

where m is the mass of the neutron, a is the scattering length of the nucleus, $\underline{R}_{nuc}$ is the position of the nucleus, and a is chosen so as to give the correct isotopic scattering cross section.

The interaction of a monochromatic neutron beam with a crystal may be written simply as a sum of Fermi pseudo-potentials over all nuclei of the crystal

$$V(\underline{r}) = \sum_{ef} \frac{2\pi\hbar^2}{m} \; a(f) \; \delta(\underline{r} - \underline{R}(ef) - \underline{u}(ef) \;), \qquad (28)$$

where a(f) is the scattering length of the $f^{\underline{th}}$ atom in the unit cell. The appearance of displacement $\underline{u}(ef)$ shows the time-dependence of the interaction. In the Born approximation, the differential cross section for a scattering event with incident and scattered neutron wave vectors $\underline{q}_i$ and $\underline{q}_s$, respectively, and initial and final lattice vibrational states n and n', respectively, is given by (14)

$$\frac{d\sigma}{d\Omega} = \left(\frac{m}{2\pi\hbar^2}\right)^2 \frac{q_s}{q_i} \left| \left\langle n'; \; \underline{q}_s \middle| V(\underline{r}) \middle| \underline{q}_i; \; n \right\rangle \right|^2 , \qquad (29)$$

where $|\underline{q}; n\rangle \doteq e^{i\underline{q}\cdot\underline{r}}|n\rangle$, a product wavefunction. The explicit form of the matrix element in the above equation is

$$\frac{2\pi\hbar^2}{m} \sum_{ef} a(f) \; e^{i\underline{Q}\cdot\underline{R}(ef)} \left\langle n' \middle| e^{i\underline{Q}\cdot\underline{u}(ef)} \middle| \; n \right\rangle , \qquad (30)$$

where $\underline{Q} \equiv \underline{q}_i - \underline{q}_s$. If the displacements $\underline{u}(ef)$ are expressed in terms of the phonon creation and annihilation operators, and the vibrational wavefunction is written in the occupation number representation, then the matrix element becomes

$$\frac{2\pi\hbar^2}{m} \sum_{ef} a(f) \; e^{i\underline{Q}\cdot\underline{R}(ef)} \prod_{\underline{K}j} \left\langle n(\underline{K}j)' \middle| e^{i(\alpha^{ef}_{\underline{K}j} b^j_{\underline{K}} + \chi^{ef*}_{\underline{K}j} b^{j+}_{\underline{K}})} \middle| n(\underline{K}j) \right\rangle , \qquad (31)$$

where

$$\chi_{\underline{K}j}^{ef} = \left(\frac{\hbar}{2N\, M_f\, \omega_j(\underline{K})} \right)^{\frac{1}{2}} \underline{e}(f|\underline{K}j).\underline{Q}\; e^{-i\underline{K}.\underline{R}(e)} \tag{32}$$

Since the actual initial vibrational state of the crystal is not known, a calculation of the differential cross section requires that a thermal average of the square of the absolute value of the above matrix element be taken.

The primary source of information about vibrational modes is the measurement of inelastic neutron scattering in which a single phonon is created or destroyed. Of importance for this case is the double differential cross section. Equation(29) is transformed into the double differential cross section by explicitly displaying energy conservation and summing over the possible final states

$$\left(\frac{d^2\sigma}{d\Omega\, d\epsilon}\right)_1 = \frac{q_s}{q_i} \sum_{\{n(\underline{K}j)\}'} \overline{\left| \left\langle \{n(\underline{K}j)\}'; \underline{q}_s \left| V(\underline{r}) \right| \underline{q}_i; \{n(\underline{K}j)\} \right\rangle \right|^2}$$

$$\times\; \delta\left(\frac{\hbar^2 q_i^2}{2m} - \frac{\hbar^2 q_s^2}{2m} \mp \hbar\omega_j(\underline{K}) \right), \tag{33}$$

where the "1" denotes single phonon inelastic scattering, where the symbol $\{n(\underline{K}j)\}$ denotes the set of occupation numbers for all modes, where the sum is over all final states in which a single phonon has been created or destroyed, where the bar over the square of the absolute value of the matrix element indicates a thermal average is to be taken, and where the - and + signs refer to phonon creation and annihilation, respectively. The details of the rather cumbersome calculation of the thermal average are omitted here, but may be found in reference (13). The result obtained for the double differential cross section in the case of annihilation of a phonon is

$$\left(\frac{d^2\sigma}{d\Omega\, d\epsilon}\right)_1 = \frac{q_s}{q_i} \sum_{\underline{K}j} \frac{\hbar}{2N\,\omega_j(\underline{K})\,\left(e^{+\hbar\omega_j(\underline{K})/kT} - 1\right)}$$

$$\times \sum_{e,e'} e^{i(\underline{Q}+\underline{K}).\left[\underline{R}(e) - \underline{R}(e')\right]}$$

$$\times \left| \sum_f \frac{a(f)}{M_f^{\frac{1}{2}}}\, e^{-W(f)}\, e^{i\underline{Q}.\underline{R}(f)}\; \underline{Q}.\underline{e}(f|\underline{K}j) \right|^2 \times$$

$$\times \, \delta\left(\frac{\hbar^2 q_i^2}{2m} - \frac{\hbar^2 q_s^2}{2m} + \hbar\,\omega_j(\underline{K}) \right), \qquad (34)$$

where

$$W(f) = \frac{\hbar}{4\,N\,M_f} \sum_{\underline{K}j} \frac{|\underline{Q}.\underline{e}(f|\underline{K}j)|^2}{\omega_j(\underline{K})} \coth\left(\frac{\hbar\omega_j(\underline{K})}{2kT} \right). \qquad (35)$$

The term W(f) is called the Debye-Waller factor for nucleus f.

Finally, there is an additional complicating feature which occurs, namely, the presence of isotopic disorders which destroy the translational invariance of the crystal. It is generally assumed that the isotopes are randomly distributed throughout the crystal, an average is taken over the scattering lengths for each site of the unit cell, and mass differences are neglected (12,15). When the isotopic averaging calculation is performed, the double differential cross section separates into two parts. One part depends on the mean square deviation of the scattering lengths and is called the incoherent one-phonon double differential cross section. This part vanishes for the case of no isotopic disorder. The second part depends only on the average value of the scattering length and is called the coherent one-phonon double differential cross section. Replacing scattering length a(f) by its average value $\overline{a(f)}$ and evaluating the summation over e,e' in equation (34), the coherent term is found to be

$$\left(\frac{d^2\sigma}{d\Omega d\epsilon}\right)_1^{coh} = \frac{q_s}{q_i} \frac{(2\pi)^3}{\Omega_a} \sum_{\underline{K}j} \sum_{\underline{\kappa}} \delta(\underline{Q} \mp \underline{K} - \underline{\kappa}) \frac{\hbar\left(\overline{n(\underline{K}j)} + \frac{1}{2} \pm \frac{1}{2}\right)}{2\,\omega_j(\underline{K})}$$

$$\times \left| \sum_f \overline{a(f)}\, e^{i\underline{Q}.\underline{R}(f)} \frac{\underline{Q}.\underline{e}(f|\underline{K}j)}{M_f^{\frac{1}{2}}} e^{-W(f)} \right|^2$$

$$\times \, \delta\left(\frac{\hbar^2 q_i^2}{2m} - \frac{\hbar^2 q_s^2}{2m} \mp \hbar\,\omega_j(\underline{K}) \right), \qquad (36)$$

where Ω_a is the volume of a unit cell, the sum over $\underline{\kappa}$ is over all primitive reciprocal lattice vectors, $\overline{n(\underline{K}j)}$ is the average value of the occupation number for mode ($\underline{K}$j), and the upper sign corresponds to phonon creation and the lower sign to phonon annihilation. Coherent scattering arises from interference between waves scattered by all the nuclei. The term enclosed by absolute value signs in the above expression is called the inelastic structure factor. In coherent scattering there are requirements of conservation of both energy and wave vector, while in incoherent scattering there is only energy conservation.

Since coherent scattering obeys an additional condition, its peaks in scattering intensity tend to be more well defined than for the case of incoherent scattering. Usually incoherent scattering results in a fairly constant counting rate, although occasionally it produces an intensity peak resembling that of coherent scattering. The relative amounts of incoherent and coherent scattering determine the suitability of a material for examination by neutron scattering measurements. If one wishes to measure dispersion curves, coherent scattering must be strong relative to background (which here includes incoherent and multiphonon scattering).

As discussed in the previous section, the conservation conditions on energy and wave vector for one-phonon coherent scattering allow a determination of the values of the dispersion curves at any point of the Brillouin Zone. In practice, one usually performs measurements for wave vectors in planes, along lines, or at points of higher symmetry in the Brillouin Zone. Thus, the change in wave vector $\underline{Q}$ is chosen so that its value in the Zone, that is, wave vector $\underline{K}$, possesses this higher symmetry. The activation of mode $(\underline{K}j)$ may be accompanied by any change in the neutron wave vector $\underline{Q} = \underline{K} + \underline{\varkappa}$, where $\underline{\varkappa}$ is a primitive reciprocal lattice vector, subject to energy conservation. To find the mode frequencies at $\underline{K}$, the experimenter may choose any of these $\underline{Q}$ values and, holding $\underline{Q}$ constant, vary $\underline{q}_i$ and $\underline{q}_s$.

Aside from the factor (q_s/q_i), the double differential cross section for coherent scattering depends only upon the change in wave vector, $\underline{Q}$, and in energy, $\Delta E = \pm\,\omega_j(\underline{K})$, of the scattered neutron. On account of the product $\underline{Q}.\underline{e}(f|\underline{K}j)$ there is also a direct dependence on the polarization vectors of the modes involved in the scattering process. Through measurements of absolute scattering intensities it is possible to use the various values or, more importantly, various directions of $\underline{Q} = \underline{K} + \underline{\varkappa}$ as a probe to determine polarization vectors at points of higher symmetry $\underline{K}$. Polarization vectors obey orthogonality relations; in addition, information about their degeneracies and orientations is obtained from group theory. When these theoretical considerations are combined with intensity measurements, one has, in principle, sufficient data to calculate polarization vectors.

While a complete measurement of dispersion curves throughout the Brillouin Zone is out of the question, one can measure the curves at points and along lines of high symmetry in the Zone and argue to the general shape of the curves from symmetry considerations (compatibility relations, intersection of curves with planes of symmetry, and so forth). If the frequency distribution is desired, however, such generality is unsatisfactory, and calculation must be performed for the entire Zone by means of some model(15). Two approaches are commonly employed: (1) calculation by a fundamental model utilizing adjustable parameters which are subsequently fitted to known values of eigenfrequencies and polarization vectors, (2) calculation by a phenomenological model utilizing interatomic force constants which have been determined from known eigenfrequencies and polarization vectors. The second approach is generally more reliable. Each element of the dynamical matrix $D(\underline{K})$ is a linear combination of interatomic force constants. A knowledge of eigenfrequencies and polarization vectors at some point $\underline{K}$ allows one to invert the dynamical equation and obtain linear equations relating elements of $D(\underline{K})$ and, therefore, relating the interatomic force constants. The equations are simplified at points of higher symmetry on account of the symmetry of $D(\underline{K})$. The ultimate complexity of the problem will depend on the range of interatomic potential assumed. Both approaches absolutely require the correct assignment of branches wherever measurements are performed.

V. VIBRONIC SPECTRA

V.A. Magnetic Ions in Host Lattices

When an optically active ion is introduced into a crystal, the interaction of the ion with other constituents of the crystal results in a modification of its electronic wavefunctions and energy levels. This modification depends on the nature and strength of the interaction. Consideration of vibronic spectra will be limited to the spectra of magnetic ions in ionic host lattices. It will generally be assumed that the concentration of the magnetic ions is sufficiently dilute that they may be treated as non-interacting impurities. Finally, for the most part the specifics of vibronic transitions associated with fluorescence will be dealt with here. The extension of the theory to vibronic transitions associated with absorption should become quite clear.

The term "magnetic ions" refers to a class of ions which, upon introduction into an ionic crystal, display certain magnetic and optical properties as a result of partially filled electronic shells. The class of magnetic ions is comprised of the transition metal, the rare earth (lanthanide), and the actinide ions. It is characteristic of the transition metal ions to have at least one ionic state in which a partially filled d-shell is the outer electronic shell for the ground state configuration. Transition metal ions are further classified by series according to whether the outer d-shell is the 3d (first series), 4d (second series), or 5d (third series) shell. The rare earth ions have in the ground state configuration at least one ionic state with a partially filled 4f shell which is interior to the filled 5s and 5p shells. Finally, the actinide ions have in the ground state configuration at least one ionic state in which the 5f shell is partially filled and is the outermost shell.

There are several approaches that may be used in calculating the energy levels of a magnetic ion impurity in an ionic solid. A simple approach which often gives satisfactory results is the crystal field approximation (16-18). The interaction of the electrons of the impurity ion with the rest of the crystal is represented as being purely electrostatic in origin. The impurity ion is subjected to an external electric field which has as its sources the other electrons and nuclei of the crystal. A major assumption is that the electronic wavefunctions of the impurity ion are quite localized, so that the amount of overlap with wavefunctions of the neighboring ions (ligands) is not significant. The strength of the Coulomb interaction depends on the electronic configuration of the magnetic ion. Since their optically active electrons lie in the outermost shell, the interaction is expected to be relatively large for transition metal ions. The interaction for rare earth ions should be much smaller on account of the shielding effect of the 5s and 5p shells. A comparison of spectra for the free ion and the ion imbedded in a host lattice shows that this is the case(16). Relatively minor shifts occur for rare earth ions, while the changes in energy for transition metal ions are considerable. As discussed in earlier articles, there are several shcemes which may be used in the application of crystal field theory (17,18). The appropriate scheme to be used depends on the relative magnitudes of three terms in the electronic Hamiltonian of the impurity ion: (1) crystal field interaction, (2) electron-electron interaction (Coulombic), (3) spin-orbit interaction. It may happen that crystal field proves inadequate on account of wavefunction overlap. In this case, a molecular orbital approach can be utilized (17,18). In the subsequent development of a simplified model for vibronic transitions, the validity of the crystal field scheme is assumed.

V.B. Characteristics of Vibronic Spectra

In measurements of the absorption and fluorescence spectra of the magnetic ion impurities, certain features are seen which cannot be explained in terms of purely electronic transitions between levels. For instance, it may be found that an ion in a site having inversion symmetry has undergone an intra-configurational transition of electric dipole character. Furthermore, the spectra often exhibit peaks in the electronic region which are temperature dependent and which are inconsistent with energy level schemes. The additional factor that must be taken into account is the dynamic character of the crystal field; its sources are undergoing vibrational motion. The dynamic portion of the crystal field allows transitions in which both photons and phonons are created and destroyed simultaneously. Such transitions are known as vibrational-electronic or vibronic transitions. From a physical point of view vibronic transitions occur in the following way: a radiation field is applied upon the crystal and it induces an electronic moment at the impurity ion. The time-dependent part of the crystal field modulates this moment with frequencies of vibrational modes, so that the photon absorbed or emitted in the transition has an energy which differs from the energy gap between participating energy levels. Energy is conserved through the simultaneous creation or annihilation of phonons.

In fluorescence experiments a radiation field generally is used to pump energy into the broad absorption band regions of the impurity ion. Ions excited into these bands quickly fall into metastable levels by very fast radiationless processes. A transition from a metastable to ground level may proceed by purely radiative, vibronic, or radiationless means. At low temperatures the typical fluorescence spectrum shows narrow lines for the purely radiative transitions from metastable levels. Each line is accompanied by a sideband structure, more prominent on the low energy side and extending up to hundreds of cm^{-1}. The sidebands are temperature-dependent and are due to vibronic transitions. In Figure 1 fluorescence spectra of dilute V^{2+} ions in MgO are shown in the 7000 - 10,000 Å region for various temperatures. As the temperature increases the vibronic sidebands increase in intensity relative to the electronic (no-phonon) line and become broad and featureless.

Before turning to the theory of vibronic transitions, consideration should be given to the effect of introducing an impurity ion into an ionic crystal. The mass of the impurity will almost certainly be different from that of the ion it replaces. In addition, there generally will be a different set of interatomic force constants between the impurity and its neighbors. It follows that there will be a perturbation of the lattice modes which may significantly affect the motion of the crystal constituents in the vicinity of the impurity ion(19). It may also happen that the impurity ion and the ion it replaces have different ionic charges, possibly resulting in a vacancy in the lattice for charge compensation. A knowledge of the position of the vacancy relative to the impurity is necessary for any sort of symmetry considerations.

The perturbation of lattice modes must be considered whenever an interpretation of vibronic data is attempted. The nature and

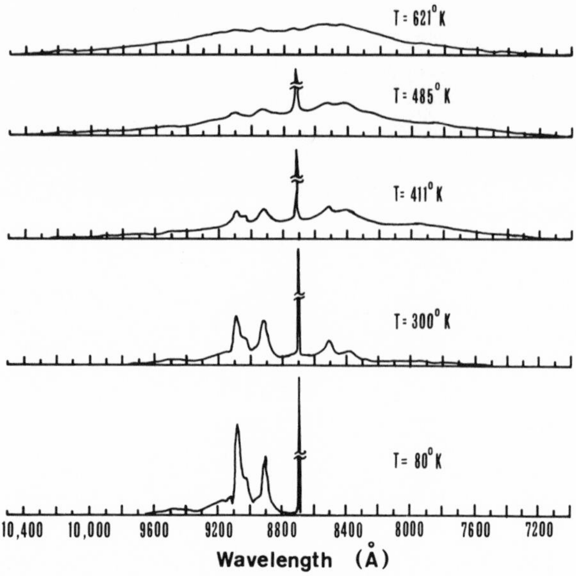

Figure 1 Temperature Dependence of the $^2E_g \rightarrow {}^4A_{2g}$ Emission Spectrum of $MgO:V^{2+}$ (DiBartolo) (18,43).

range of the interaction between the impurity ion and its surroundings determine whether the vibronic sidebands reflect local modes, lattice modes, or both. In view of the localization and shielding of the optically active electrons of rare earth ions, one might expect a Coulombic, long range interaction and a vibronic spectrum reflecting lattice modes. On the other hand, the electronic configuration and possible wavefunction overlap of transition metal ions suggest an interaction having a relevant short range character. In this case the perturbed local modes may not only be important, but may even predominate in determining the vibronic sideband structure.

Group theoretical analyses of vibronic transitions have followed either of two extreme assumptions, namely, that the sidebands reflect (1) only the modes of a molecular complex consisting of the impurity ion and certain of its neighbors or, (2) only lattice modes. The initial calculation of vibronic selection rules by Satten (20,21) was based on the first assumption; if the vibrational frequencies of the molecular complex were already known from other types of measurements on molecules, then a correlation of these frequencies with peaks in the vibronic spectrum might be possible. The molecular complex model has been applied only to a very limited number of materials and, in general, has proved inadequate to explain the complexities of vibronic sidebands.

In the initial calculation of vibronic selection rules using lattice modes, an attempt was made to interpret the vibronic spectra of Pr^{3+} in $LaCl_3$ and $LaBr_3$ by means of only the $\underline{K} = 0$ optical modes(22). The frequencies of many of the $\underline{K} = 0$ optical modes were available from infrared absorption and Raman scattering measurements and were compared with peaks in the vibronic spectrum. It was found that the $\underline{K} = 0$ modes were insufficient to account for the total vibronic structure; a subsequent analysis of the $LaCl_3$: Pr^{3+} system by Satten(23)included all lattice modes, with special attention for those modes with wave vectors of higher symmetry in the Brillouin Zone. The lattice mode model has been applied to impurities (especially rare earths) in various host lattices and the analysis of the vibronic sidebands has generally been at least reasonable.

If the assumption is made that the impurity ion interacts only with the lattice modes of the perfect crystal, then selection rules for vibronic transitions can be formulated. However, the interpretation of observed vibronic peaks from these rules is not obvious, since so large a number of modes may participate. This is due to the fact that the impurity ion destroys the translational symmetry of the crystal, so that there is no conservation law for the wave vector. The contribution to the vibronic sidebands from a single mode will be quite small. Peaks in the sidebands result from the cumulative effect of a large number of modes in a narrow frequency range. The vibronic intensity thus depends on the frequency distribution of lattice modes, so that at first sight one might attempt to identify the frequency distribution with the sideband structure. However, the matrix elements for vibronic transitions are themselves functions of frequency and are subject to selection rules. Apart from such symmetry considerations, in the simplest model the vibronic spectrum reflects the product of two frequency-dependent functions, one of these being the vibrational frequency distribution.

V.C. Theory of Vibronic Transitions

In this section a model for vibronic transitions is presented. Depsite the large number of simplifications used in the derivation, the expressions obtained are expected to be at least qualitatively valid for many impurity ion-host lattice systems and are useful in formulating analyses based on selection rules. Among the assumptions are the validity of the Born-Oppenheimer approximation for the impurity ion, the validity of the crystal field approximation, and the validity of neglecting local mode effects.

The equilibrium position of the nucleus of the impurity ion is chosen as the origin of the coordinate system and the position vector of the Jth optically active electron of this ion is denoted by $\underline{r}_J$. Let N be the total number of ions in the crystal, n the total number of optically active electrons in the impurity ion, and (ef)=(11) label the impurity ion. Knowledge of the electronic states of the free ion is assumed, and the perturbation of these states when the free ion is introduced into the host is to be explicitly considered. The interaction of the optically active electrons with the other constituents of the crystal is a function of the instantaneous positions of the electrons and ions, according to the crystal field approximation: $V = V(..,\underline{R}(ef) + \underline{u}(ef),..,\underline{r}_J)$, where (ef) runs over all values except (11) and J takes all values from 1 to n. The instantaneous position of electron J can be expressed in terms of the displacement of the impurity nucleus and a radial vector $\underline{\rho}_J$, so that $\underline{r}_J = \underline{\rho}_J + \underline{u}(11)$. The optically active electrons and the ion constituents of the crystal are considered as point charges, and their interaction is a function of the distance between them. A Taylor expansion may be made for the interaction V in terms of the equilibrium positions of all ions in the crystal, including the impurity. The method of obtaining the expansion is as follows: first, an expansion is made in the displacements $\underline{u}(ef)$, where $(ef) \neq (11)$. The displacement $\underline{u}(11)$ appears in the interaction of electron J with each ion $(ef) \neq (11)$ and so does not play a role equivalent to other displacements. The coefficients in the initial expansion in $\underline{u}(ef) \neq \underline{u}(11)$ then are expanded in $\underline{u}(11)$ so as to give the result

$$\begin{aligned} V = V_o &+ \sum_{\alpha ef}{}' u_\alpha(ef) \left.\frac{\partial V}{\partial u_\alpha(ef)}\right|_o + \sum_{\alpha J} u_\alpha(11) \left.\frac{\partial V}{\partial \rho_{J\alpha}}\right|_o \\ &+ \frac{1}{2} \sum_{\alpha ef}{}' \sum_{\beta e'f'}{}' u_\alpha(ef)\, u_\beta(e'f') \left.\frac{\partial^2 V}{\partial u_\alpha(ef)\, \partial u_\beta(e'f')}\right|_o \\ &+ \frac{1}{2} \sum_{\alpha J} \sum_{\beta J'} u_\alpha(11)\, u_\beta(11) \left.\frac{\partial^2 V}{\partial \rho_{J\alpha}\, \partial \rho_{J'\beta}}\right|_o \\ &+ \sum_{\alpha ef}{}' \sum_{\beta J} u_\alpha(ef)\, u_\beta(11) \left.\frac{\partial^2 V}{\partial u_\alpha(ef)\, \partial \rho_{J\beta}}\right|_o + \ldots, \end{aligned} \tag{37}$$

where the prime on the summation symbol indicates that the term (ef) = (11) is omitted, where the partial derivatives are evaluated at the equilibrium positions of all nuclei, and where the partial derivatives with respect to $\underline{u}(11)$ have been replaced by those with respect to the ξ_J's. Each of the terms above is dependent on the instantaneous positions of the electrons ξ_J. The term V_o is the interaction with the static crystal field previously referred to. The dynamic part of the interaction potential due to the vibrational motion of the ions of the crystal is given by

$$V - V_o = \left\{ \sum_{ef} \underline{u}(ef) \cdot \underline{\nabla}_{ef} + 1/2 \left[\sum_{ef} \underline{u}(ef) \cdot \underline{\nabla}_{ef} \right]^2 + \ldots \right\} V \Big|_o \, , \tag{38}$$

where $\underline{\nabla}_{11}$ is understood to mean differentiation with respect to coordinates of the n electrons. The term $(V - V_o)$ is the vibrational-electronic interaction term and is hereafter denoted by H_{el-vib}.

If the simplifying assumption is made that the substitution of an isolated impurity into a crystal essentially leaves the vibrational spectrum unchanged, then the displacements of ions can be expressed in terms of normal coordinates for a perfect crystal

$$\begin{aligned} H_{el-vib} = & \sum_{\underline{K}J} \left(\sum_{\alpha ef} (NM_f)^{-1/2} e^{-i\underline{K}\cdot\underline{R}(e)} e_\alpha(f|\underline{K}J) \, (\underline{\nabla}_{ef})_\alpha V \Big|_o \right) \cdot Q_{\underline{K}}^J \\ & + 1/2 \sum_{\underline{K}J} \sum_{\underline{K}'J'} \left[\sum_{\alpha ef} \sum_{\beta e'f'} (N^2 M_f M_{f'})^{-1/2} e^{-i\underline{K}\cdot\underline{R}(e)} e^{-i\underline{K}'\cdot\underline{R}(e')} \right. \\ & \left. \times \, e_\alpha(f|\underline{K}J) \, e_\beta(f'|\underline{K}'J') \, (\underline{\nabla}_{ef}) \, (\underline{\nabla}_{e'f'}) \, V \Big|_o \right] \cdot Q_{\underline{K}}^J Q_{\underline{K}'}^{J'} + \ldots \\ \equiv & \sum_{\underline{K}J} f_{\underline{K}}^J Q_{\underline{K}}^J + \sum_{\underline{K}J} \sum_{\underline{K}'J'} g_{\underline{K}\underline{K}'}^{JJ'} Q_{\underline{K}}^J Q_{\underline{K}'}^{J'} + \ldots . \end{aligned} \tag{39}$$

The terms $f_{\underline{K}}^J$ and $g_{\underline{K}\underline{K}'}^{JJ'}$ are functions only of the electronic coordinates of the impurity ion at equilibrium, while the normal coordinates are functions of the ionic displacements alone.

From a quantum mechanical viewpoint, the system consists of the free ion, the lattice vibrations, and the total interaction potential V so that

$$\begin{aligned} H &= H_{free\ ion} + H_{vib} + V \\ &= (H_{free\ ion} + V_o) + H_{vib} + H_{el-vib} \end{aligned} \tag{40}$$

The eigenfunctions of $(H_{free\ ion} + V_o)$ are assumed known from a crystal field calculation and are denoted by $|\psi^{el}\rangle$, and H_{el-vib} is treated as a time dependent perturbation. At this point the vibronic perturbed wavefunctions and energies can be calculated. However, the major interest is in fluorescence and absorption, wherein a radiation field couples with the optically active electrons of the impurity. It is more suitable to consider the vibronic and radiation interactions simultaneously.

Consider the quantized radiation field in the occupation number representation interacting with the electrons of the impurity through $(\underline{A}\cdot\underline{p})$. For frequencies in the visible light region, the spatial variation of the vector potential $\underline{A}$ with displacement of the impurity ion may be neglected. In addition, electronic motion is much faster than nuclear motion, so one can set $\dot{\underline{r}}_J \approx \dot{\rho}_J$. The sum $(H_{el-rad} + H_{el-vib})$ is the perturbation term, and the unperturbed wavefunctions are in the form of products $|\psi^{el}\rangle|\{n_{q\gamma}\}\rangle\,|\{n(\underline{K}j)\}\rangle$, where $n_{q\gamma}$ is the radiation occupation number for photons of wave vector $\bar{q}$ and polarization γ. Term H_{el-rad} by itself results in purely radiative electronic transitions, H_{el-vib} in radiationless electronic transitions. Vibronic transitions come about in second order perturbation theory.

The matrix element involved in a vibronic transition with the simultaneous creation or annihilation of a photon of frequency $\nu_{\underline{q}}$ and polarization $\hat{\underline{\pi}}^{\gamma}_{\underline{q}}$ and the creation or annihilation of a phonon of frequency $\omega_J(\underline{K})$ is given by:

$$M(\underline{K}J;\underline{q}\gamma) = \sum_j \frac{\langle \psi_f^{el};\ ..\ n_{q\gamma} \pm 1;\ ...\ n(\underline{K}J) \pm 1 ..| H_{el-rad} + H_{el-vib} | \psi_j \rangle}{E_i - E_j}$$

$$\times \langle \psi_j | H_{el-rad} + H_{el-vib} | \psi_i^{el};\ ...\ n_{\underline{q}\gamma}\ ...;\ ...\ n(\underline{K}J)\ ...\rangle \qquad (41)$$

where the summation is overall intermediate states. Using equation (39) and the expressions

$$Q^J_{\underline{K}} = \left(\frac{\hbar}{2\omega_J(\underline{K})}\right)^{1/2} (b^J_{\underline{K}} + b^{J+}_{-\underline{K}}), \qquad (42)$$

$$\underline{A}(\underline{r}) = \sum_{\underline{q}\gamma} \left(\frac{hc^2}{\nu_{\underline{q}} V}\right)^{1/2} \hat{\underline{\pi}}^{\gamma}_{-\underline{q}} \cdot (a^{\gamma}_{\underline{q}}\, e^{i\underline{q}\cdot r} + a^{\gamma +}_{\underline{q}}\, e^{-i\underline{q}\cdot\underline{r}}), \qquad (43)$$

and considering the concrete case of a fluorescent type of process in which a phonon of mode $(\underline{K}J)$ is created and the electronic transition is from metastable state i to ground state f, one finds

$$M(\underline{K}J;\underline{q}\gamma) = \frac{e}{m}\left(\frac{\hbar}{2\omega_J(\underline{K})}\right)^{1/2}\sqrt{n(\underline{K}J)+1}\sqrt{n_{\underline{q}\gamma}+1}\left(\frac{h}{\nu_{\underline{q}}V}\right)^{1/2}$$

$$\times\sum_j\left\{\frac{\langle\psi_f^{el}|\sum e^{i\underline{q}\cdot\underline{r}}\ \underline{p}\cdot\hat{\underline{\pi}}_{\underline{q}}^{\gamma}|\psi_j^{el}\rangle\langle\psi_j^{el}|f_{\underline{K}}^J|\psi_i^{el}\rangle}{E_i^{el}-(E_j^{el}+\hbar\omega_J(\underline{K}))}\right.$$

$$\left.+\frac{\langle\psi_f^{el}|f_{\underline{K}}^J|\psi_j^{el}\rangle\langle\psi_j^{el}|\sum e^{i\underline{q}\cdot\underline{r}}\ \underline{p}\cdot\hat{\underline{\pi}}_{\underline{q}}^{\gamma}|\psi_i^{el}\rangle}{E_i^{el}-(E_j^{el}+\hbar\nu_{\underline{q}})}\right\} \quad (44)$$

The matrix element clearly depends on the frequency of the phonon activated.

It may happen that the major contribution to the matrix element comes from either that term in the summation for which $|\psi_j^{el}\rangle=|\psi_i^{el}\rangle$, or from those terms in the summation for intermediate states close in energy to the initial state. In this case, the situation that $\nu_{\underline{q}} \gg \omega_J(\underline{K})$ suggests that the terms in the parentheses with phonon freqüencies in their denominator will predominate. Another possible occurrence is that the vibronic sidebands show essentially the same polarization properties in emission as does the no-phonon line. For the first term in the parentheses, this implies that $|\psi_j^{el}\rangle=|\psi_i^{el}\rangle$, while for the second term that $|\psi_j^{el}\rangle=|\psi_f^{el}\rangle$. Of course, the matrix element has a different dependence on the phonon frequency for each of these cases. In the analysis of any vibronic system, one must consider such factors as the energy levels of the magnetic ion, the electronic wavefunctions (including parity), the polarization characteristics of the emission, and so forth.

The matrix element for a process involving the creation of a photon and absorption of a phonon is found in the same manner, but among other differences has a factor of $\sqrt{n(KJ)}$ rather than $\sqrt{n(\underline{K}J)+1}$. This matrix element corresponds to emission in the sideband on the high energy side of the no-phonon line. At very low temperature ($T\approx0$), the phonon occupation number vanishes and the high energy vibronic sideband disappears.

When a time-dependent perturbation is applied, the probability per unit time for a system to make a transition from state i to state f is given by the Golden Rule

$$P_{i\to f} = \frac{2\pi}{\hbar}\left|M_{i\to f}\right|^2 \rho(E_f), \quad (45)$$

where $M_{i\to f}$ is the matrix element of the perturbation and $\rho(E_f)$ is the density of final states. If all vibronic transitions proceed from the same initial to the same final electronic state of the impurity and involve the creation of a single phonon, then the vibronic sidebands result from a sum over all vibrational modes and show an intensity dependent on the frequency distribution of lattice modes, subject to appropriate selection rules.

When a magnetic ion is placed in an ionic crystal, its electronic levels are not only shifted, but also are broadened by various mechanisms, with the predominant mechanism generally being the Raman scattering of phonons (18). In applying the Golden Rule this broadening of a level can be taken into account by an energy level distribution function $\rho(E^{el})$, defined over the width ΔE^{el} of the level. The intensity of fluorescent emission with frequency ν and polarization γ resulting from vibronic transitions between levels i and f has the following proportionality:

$$I_\gamma(\nu_q) \propto \sum_{\underline{K}J} \int_{\Delta E_i^{el}} dE_i^{el} \int_{\Delta E_f^{el}} dE_f^{el}\ \rho(E_i^{el})\,\rho(E_f^{el})$$

$$\times \left| M(\underline{K}J;\underline{q}\gamma) \right|^2 D(\nu_q) \times \delta(E_i^{el}-E_f^{el}-\hbar\omega_J(\underline{K})-\hbar\nu_q) \quad , \quad (46)$$

where $D(\nu_q)$ is the density of photon states and has the dependence $D(\nu_q) \propto \nu_q^2$. Over the usual frequency range of vibronic sidebands this function is relatively constant and is hereafter neglected. If the fluorescent system under consideration displays a narrow linewidth for the purely electronic transition i $\rightarrow$ f (assuming this transition to be allowed), then the width in energy of these levels will be narrow relative to structure in the vibrational frequency distribution and may be accurately approximated as a δ-function in energy. This approximation will generally be valid for magnetic ions. With these simplications the above equation reduces to

$$I_\gamma(\nu_q) \propto \sum_{\underline{K}J} \left| M(\underline{K}J;\underline{q}\gamma) \right|^2 \delta(E_i^{el}-E_f^{el}-\hbar\nu_q-\hbar\omega_J(\underline{K})). \quad (47)$$

The δ-function allows only those modes with frequency

$$\Omega \equiv \hbar^{-1}(E_i^{el}-E_f^{el}) - \nu_q \quad (48)$$

to contribute to the intensity $I_\gamma(\nu)$, while the selection rules further limit even these modes. It proves convenient at this point to speak of intensity of emission at frequencies measured relative to the purely electronic transition frequency $(E_i^{el}-E_f^{el})$, so that $I_\gamma(\nu_q) \rightarrow I_\gamma(\Omega)$. If the individuality of modes is no longer specified and the matrix element is simply written as a function of mode frequency such that $M(\underline{K}J;\underline{q}\gamma) \rightarrow M(\omega;\underline{q}\gamma)$, then the sum in equation (47) may be written as an integral involving the vibrational frequency distribution

$$I_{\gamma}(\Omega) \propto \int d\omega \left| M(\omega;\underline{q}\gamma) \right|^2 g(\omega) \; \delta(\omega-\Omega) \tag{49}$$

$$\propto \left| M(\Omega\,;\underline{q}\gamma) \right|^2 g(\Omega). \tag{50}$$

Thus, the emission intensity in the sideband region is found to be proportional to the product of two functions of vibrational frequency; one of these is the vibrational frequency distribution. Under the best circumstances, the frequency dependence of the matrix elements alters the frequency distribution only slightly, so that maxima in the sidebands correspond to maxima in the frequency distribution.

This result was obtained by suppressing the role of selection rules. When selection rules are taken into account, they modify the frequency distribution by eliminating contributions from many vibrational modes. Their effect is particularly apparent when a critical point in the Brillouin Zone corresponding to a region of high density in the distribution is not allowed to participate in vibronic transitions.

V.D. Vibronic Selection Rules From Group-Theoretical Analysis

Under the simplifying assumptions made in the previous section, the matrix element for a single phonon vibronic transition is proportional to

$$\left\langle \Psi_f^{el} \middle| e^{i\underline{q}\cdot\underline{r}}\,\underline{p}\cdot\hat{\underline{\pi}}_q^{\gamma} \middle| \Psi_j^{el} \right\rangle \left\langle \Psi_j^{el} \middle| f_{\underline{K}}^{J} \middle| \Psi_i^{el} \right\rangle . \tag{51}$$

From this expression, selection rules can be derived from group theory. However, it is necessary to digress briefly at this point and consider the symmetry properties of lattice vibrations. Since this topic is somewhat complex, only the general results can be given here. For those interested, an elegant and detailed account of crystal symmetry and lattice vibrations can be found in an article by Maradudin and Vosko(24).

In the article just cited, it is shown that for a given $\underline{K}$-vector, the 3r-component eigenvectors of the dynamical matrix transform under certain symmetry operations according to irreducible representations of the group of the $\underline{K}$-vector, denoted as $\mathcal{K}$. Using the notation of Koster(25), the group $\mathcal{K}$ consists of all those elements $\{\beta|\underline{b}\}$ of the space group G which have the property that $\beta\underline{K} = \underline{K} + \mathcal{K}$, that is, which have rotational operations β which leave the $\underline{K}$-vector invariant to within a reciprocal lattice vector.

In a perfect crystal, each mode ($\underline{K}j$) is labelled by a particular irreducible representation of the group of its $\underline{K}$-vector. For a given crystal, those irreducible representations of group $\mathcal{K}$ actually characterizing modes are explicitly determined by the construction and subsequent reduction of a set of matrices commuting with the dynamical matrix. Those normal coordinates for $\underline{K}$-vectors in the star of $\underline{K}$, denoted by $\{\underline{K}\}$, which are degenerate as a requirement of symmetry (not accidentally degenerate) then form bases for irreducible representations of the space group G. The irreducible representations of G which occur in this way are completely specified, however, by the irreducible representations of $\mathcal{K}$ describing modes ($\underline{K}j$).

The translational invariance of the crystal is destroyed by the presence of an impurity ion, resulting in a crystal symmetry described by a point group A with origin at the impurity ion. Before the substitution of an impurity, the normal coordinates form bases for irreducible representations of the space group G. The presence of the impurity should partially lift the degeneracy of the normal coordinates so that linear combinations of normal coordinates form bases for representations of point group A. However, in the approximation of neglecting the perturbation of lattice modes, the modes continue to be characterized by space group representations. The formulation of selection rules is obtained by projecting the space group representations onto the point group representations, that is, reducing the space group representations in terms of irreducible representations of A.

The interaction potential V is invariant under the group A and so, if V is expanded in a power series, each non-vanishing term of the series must be invariant under A. Referring to (39), since $Q^J_{\underline{K}}$ transforms according to an irreducible representation of A, say $\Gamma(Q^J_{\underline{K}})$, then $f^J_{\underline{K}}$ must transform according to the representation $\Gamma(Q^J_{\underline{K}})^*$. Similarly, $g^{JJ'}_{\underline{KK}'}$ must transform according to

$$(\ \Gamma(Q^J_{\underline{K}})^* \times \Gamma(Q^{J'}_{\underline{K}'})^*\ ).$$ The electronic wavefunctions are also

characterized by irreducible representations of the site symmetry group A. The term $e^{i\underline{q}\cdot\underline{r}}\hat{\pi}_{\underline{q}}\cdot\underline{p}$ may be expanded in the usual way to give an electric dipole operator transforming as a polar vector, a magnetic dipole operator as an axial vector, and so forth. The components of each operator transform accoding to irreducible representations of the group A and, in some cases, components associated with particular transitions may be identified by polarization studies of the vibronic spectra.

The matrix element for a vibronic transition is characterized by the initial and final electronic state, a particular radiation operator, and a particular vibrational mode. Since each is characterized by an irreducible representation of group A, group theory says that a particular transition is allowed only if

$$\Gamma(\psi_f^{el}) \in \left[\Gamma(\text{Rad Op}) \times \Gamma(\text{Active Vib Mode}) \times \Gamma(\psi_i^{el})\right]. \quad (52)$$

In a previous article, Orbach shows that critical points are responsible for certain features in the frequency distribution. In the model under consideration, critical points are expected to play a similar role for vibronic sideband features. Vibronic selection rules prove to be most useful in considering the contribution of the various critical points. Critical points of dispersion curves are identified from the periodicity and point symmetry of the eigenfrequencies in the Brillouin Zone. The technique is basically the same as that for electronic energy bands in crystals, such as is found in Cornwell(26),for instance. Specification of the type (index) of a critical point generally requires additional information, such as provided by infrared absorption, Raman scattering, or thermal neutron scattering measurements. Such information may show the existence of other critical points not required by symmetry.

V.E. Vibronic Theory: Alternate Approaches and Numerical Calculations

In Section V.C., a formalism for vibronic transitions has been presented which is particularly well suited for analysis by group theoretical techniques. In recent years, however, there have been a number of publications concerning the actual calculation of vibronic sidebands of magnetic ions in simple ionic host lattices and employing an alternate formalism. It seems worthwhile at this point to review some of these attempts, along with the assumptions made regarding the nature of the interaction of the ion impurity with its surroundings. The complexity of the calculations necessitates that any review of these approaches be general and qualitative.

Perhaps the earliest attempt at a quantitative reproduction of vibronic sideband structure was made by Timusk and Buchanan for the system KBr:Sm^{2+}(27).They assumed that the interaction H_{el-vib} was of an electrostatic and long range character and calculated the time-dependent electric field at the impurity site due to lattice vibrations of all other ions. A shell model was employed to describe vibrations of KBr and subsequent expressions were obtained for the sideband intensity associated with the $^5D_0 \rightarrow {}^7F_0$ intraconfigurational ($4f^6$) transition. The calculation was quite simple in that it neglected perturbation of modes and the modification of equilibrium positions of ions due to the presence

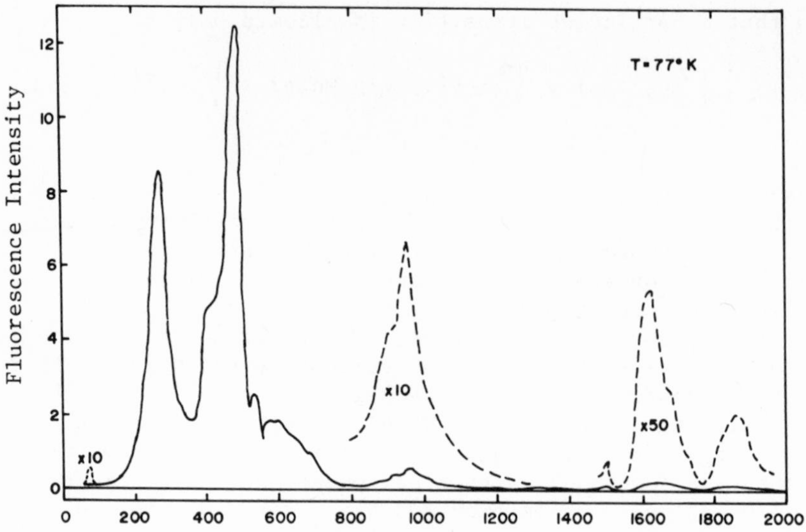

Figure 2 Vibronic Sideband (cm^{-1}) of $MgO:V^{2+}$ Relative to No-Phonon Line at 77°K (Weber and Schaufele) (46).

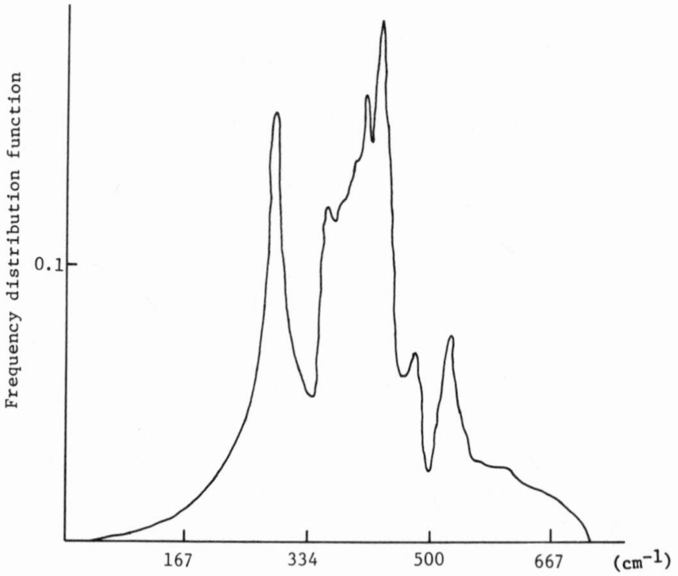

Figure 3 Frequency Distribution of MgO (Sangster et al.) (38).

of both the impurity and a charge compensating vacancy. There was, however, a reasonable reproduction of sideband structure and an identification of certain prominent vibronic features with critical points in the dispersion curves of KBr.

Following this publication, a number of articles appeared which proposed various schemes for calculating vibronic sidebands. To mention but a few of the subsequent calculations, vibronic sidebands due to Sm^{2+} in various alkali halide host crystals were considered by Buchanan and Woll(28), by Bron(29), and by Kühner and Wagner(30), and vibronic sidebands due to transition metal impurities in MgO were investigated by Sangster and McCombie ($MgO:V^{2+}$ and $MgO:Ni^{2+}$)(31), by Manson ($MgO:Ni^{2+}$)(32), and by Sangster ($MgO:Cr^{3+}$)(33). Generally, both the assumptions and the methods of calculation employed by these authors differ. However, with the exception of the initial article by Timusk and Buchanan, the approaches have the following similarities: (1) through some technique the electronic and vibrational systems are decoupled; (2) the electronic portion of the expression for vibronic intensity is not evaluated, but is treated as a single or set of adjustable parameters; (3) the vibrational system is treated in a Green's function formalism for ease of calculation of the perfect crystal lattice dynamics by shell models, as well as for ease in incorporating perturbations of lattice dynamics due to the presence of the impurity; and (4) changes in atomic force constants and mass at the impurity site, as well as charge compensating vacancies near this site are treated by a model involving a series of springs joining the impurity and its neighbors; the spring constants are generally treated as adjustable parameters also. These parameters are then evaluated for a best fit of calculations to experimental data. The value of the parameters not only determines predicted intensities, but also can shift the positions of prominent peaks due to critical points in the dispersion curves of the perfect crystal to different positions in the vibronic sideband. There appear to exist no "first principles" calculations of electronic matrix elements in the expressions of these authors, so that interpretations of vibronic sidebands are based on a number of adjustable parameters.

VI. EXAMPLE: ANALYSIS OF $^2E_g \rightarrow {}^4A_{2g}$ SIDEBAND OF $MgO:V^{2+}$

As an illustrative example of analysis of vibronic sidebands from symmetry, the sideband emission accompanying the $^2E_g \rightarrow {}^4A_{2g}$ transition of the system $MgO:V^{2+}$ is now considered. It must be admitted at the outset that the validity of the perfect lattice approach for this particular system is open to question in view of the expected amount of wavefunction overlap. Nonetheless, the system provides an excellent example, even if artificial, for the following reasons: (1) MgO has a simple structure, belongs to symmorphic space group O_h^5, and has been extensively studied by various techniques(34-38). There are shell models to calculate its dispersion curves and frequency distribution(36-39);(2) the low temperature vibronic sideband of $MgO:V^{2+}$ is relatively simple, is associated with a single electronic transition, and shows a remarkable qualitative similarity to the calculated frequency distribution of MgO (see Figures 2 and 3). The system has several disadvantages, in that there is no polarization of emission and that absorption studies cannot be accurately performed on account of an overlapping absorption region of the trivalent vanadium impurity ion V^{3+}. It might be noted that selection rules for the $MgO:V^{2+}$ system have been described previously by Loudon (40).

The calculation of selection rules requires the characterization of electronic wavefunctions of V^{2+} ($3d^3$) by irreducible representations (abbreviated as irr. reps) of the local site group O_h. In the strong field scheme the ground and first excited (metastable) states of the lowest energy crystal configuration (t^3) are characterized by the irr. reps $^4A_{2g}$ and ^2E_g, respectively(18). In the BSW notation(26) these irr. reps are denoted by $^4\Gamma_2$ and $^2\Gamma_{12}$, respectively. Broad bands above the ^2E_g level belong to the (t^2e) configuration and populate this level by fast radiationless processes. The metastable to ground state transition requires spin-orbit interaction. This interaction does not split either level, but does result in each state being characterized by the double-valued irr. rep $D_{3/2}$ ($=\Gamma_8^+$). It turns out that the vibronic selection rules governing active vibrational modes are unchanged by this interaction for the case of the $MgO:V^{2+}$ system.

A number of authors have published sideband spectra of the system $MgO:V^{2+}$(41-44). Figure 2 shows the vibronic spectrum found by Weber and Schaufele(46) (unpublished) and it is taken as the standard in this section. An early study of the Zeeman splitting of the no-phonon line led to identification of the purely electronic transition as magnetic dipole in character(45). Studies of the low frequency region of the emission sideband ($\omega \to 0$) have shown an ω^5 dependence of the intensity, indicating an electric dipole character for the vibronic sideband (44).

The MgO crystal has the rocksalt structure (fcc), with atoms at (000) and (a/2 00) in the unit cell. Physically, it is almost totally ionic and has a high coherent scattering cross section for thermal neutrons, a fact which has led to measurement of dispersion curves in many directions(37). Since its space group is symmorphic, the irr. reps of pure rotations of $\mathcal{K}$ are the irr. reps of ordinary point groups. The unit cell in reciprocal space is the body-centered cubic cell. The transformation properties of the polarization vectors are easily found by the procedure of reference 24. The periodicity and point symmetry of $\omega_j^2(\underline{K})$ specify that points Γ, X, L, W in the Brillouin Zone are critical points. In Table 1 is a compilation of critical points, together with irr. reps of $\mathcal{K}$ characterizing

Table 1. Critical Points for MgO.

Critical Point	Branch	Index	Irr Rep of χ	Irr Rep of G $\rightarrow$ Irr Reps O_h (Γ's)
Γ	TO	J=0 J=0	Γ_{15} (Degen.)	15
	LO	J=3	split	
X	TA	J=1 J=1	X_5' (Degen.)	15, 25
	LA	J=3	X_4'	15
	TO	J=1 J=3	X_5' (Degen.)	15, 25
	LO	J=2	X_4'	15
L	TA	J=1 J=1	L_3' (Degen.)	12', 15, 25
	TO	J=0 J=0	L_3 (Degen.)	12, 15', 25'
	LA	J=3	L_1	1, 25'
	LO	J=2	L_2'	2', 15
W		J=3	W_1 or W_2'	1, 12, 25 or 2, 12, 15
		Singular	W_3 (Degen.)	15', 25', 15, 25
		J=2	W_1 or W_2'	1, 12, 25 or 2, 12, 15
		Singular	W_3 (Degen.)	15', 25', 15, 25
Σ		J=1	Σ_4	2, 12, 15', 15, 25
		J=0	Σ_4	2, 12, 15', 15, 25
		J=3	Σ_1	1, 12, 25', 15, 25

Table 2. Features in the frequency distribution of MgO with associated critical points, versus similar features in vibronic sideband.

Freq. Dist. Position (cm^{-1})	Associated Critical Pts.	Sideband Pos. (cm^{-1})
~287	TA branches at X, L	~274
~344	TO branch at L	not apparent
~357	branch 1 at W, "TO" branch along Σ	~420
~395	TO branches at point Γ	not resolved
~436	LA branch at X, branch 4 at W	~487
~455	"LA" branch along Σ	~510
~497	branch 6 at W	~550
~609	LO branch at L	not apparent
~720	LO branch at Γ	~780

the branches at each critical point. Also given is a reduction of the corresponding space group rep into irr. reps of the point group O_h. From neutron scattering data and shell model calculations the indices and mode frequencies for the critical points are found, and other critical points along line Σ which are not required by symmetry are identified. Table 2 lists the approximate positions of some prominent features in the frequency distribution as calculated by Sangster et al.(38), together with the critical points which may be responsible. Also shown in this table are the positions of some apparently corresponding features in the vibronic sideband. It can be seen that the vibronic sideband can be thought of as consisting of two segments. The segment of sideband nearer the no-phonon line has its features shifted downward some 15 cm^{-1} relative to the frequency distribution, while the second segment has an upward relative shift of about 50 cm^{-1}. For a discussion of the reason for these shifts the article by Sangster and McCombie may be consulted(31).

Using equation (52), and a multiplication table for irr. reps of O_h it is readily seen that for the $MgO:V^{2+}$ system the restriction on participating modes by selection rules is equivalent to rules of parity: the rep $\Gamma(Q_K^J)$ of O_h must have odd parity (negative character for inversion operation) if it is to contribute to the sideband. As for lattice modes, when their associated space group irr. rep is projected onto irr. reps of O_h, the lattice modes can participate in a vibronic transition only if at least one odd parity rep of O_h results. A check of Table 1 shows that only two modes at critical point cannot contribute, L(LA) and L(TO). It is expected that the mode whose space group rep projects onto both even and odd irr. reps of O_h will have a reduced strength in comparing the vibronic spectrum with the frequency distribution.

The interpretation of the vibronic spectrum of $MgO:V^{2+}$ is rather lengthy, complicated, and only a portion of the analysis can be included here. It should be kept in mind that shell model calculations may subject to significant error. Critical points for modes of "totally odd" parity should produce sharper features than for those of "mixed" parity, and may result in greater relative intensities in the vibronic sideband.

Consider the major peak at $\sim$274 cm^{-1} in the vibronic spectrum, corresponding to a peak at $\sim$287 cm^{-1} in the frequency distribution and associated with critical points (J=1) at X(TA) and L(TA). The modes are "totally odd", and so the intensity of this peak is enhanced relative to most of the spectrum, when comparison is made with the frequency distribution.

Now consider the shoulder on the low energy side of the second major peak at 487 cm^{-1}. This corresponds to a similar region of sharper structure and greater strength in the frequency distribution, associated with critical points(J=0) at L(TO) and along $\Sigma(\Sigma_4)$. The L(TO) mode is forbidden by selection rules to contribute to the vibronic spectrum, and this is probably the reason for the gradual slope and reduced strength of the vibronic shoulder.

The second major peak in the vibronic spectrum lies at $\sim$487 cm^{-1}. It corresponds to peaks in the frequency distribution at $\sim$414 cm^{-1} and $\sim$436 cm^{-1}, but its slight asymmetry suggests a comparison with the latter peak is appropriate. The structure of the $\sim$436 cm^{-1} peak is associated with critical points(J=3) at X_4'(LA) and (J=2) at $W(W_1$ or $W_2')$. The totally odd parity character of X_4'(LA) enhances the very rapid descent of the vibronic intensity on the high(phonon) energy side of the peak.

This same sort of analysis could be performed for other vibronic features. The above outline was designed to demonstrate the general technique to be applied. It might be noted that the consistency of the shift of the vibronic sideband extends to the cutoff of the frequency distribution at $\sim$720 cm^{-1} and of the one-phonon sideband at $\sim$780 cm^{-1}. Again the sideband feature has an upward shift of $\sim$60 cm^{-1} relative to the frequency distribution.

REFERENCES

1. R. M. Lyddane, R. G. Sachs, E. Teller, Phys. Rev. 59, 673 (1941).
2. W. Cochran and R. A. Cowley, J. Phys. Chem. Solids 23, 447 (1962).
3. W. Cochran, Adv. Phys. 9, 387 (1960).
4. S. S. Mitra, in Optical Properties of Solids, S. Nudelman and S. S. Mitra (eds.), Plenum (1969).
5. F. A. Johnson, Prog. Semiconductors 9, 179 (1965).
6. M. Lax and E. Burstein, Phys. Rev. 97, 39 (1955).
7. E. Burstein, in Phonons and Phonon Interactions, T. A. Bak (ed), Benjamin, (1964).
8. M. Born and K. Huang, Dynamical Theory of Crystal Lattices, Clarendon Press, (1954).
9. R. Loudon, Adv. Phys. 13, 423 (1964).
10. M. Born and M. Bradburn, Proc. Roy. Soc. A, 188, 161 (1947).
11. W. Panofsky and M. Phillips, Classical Electricity and Magnetism, Addison-Wesley, (1962), p. 358.
12. R. Loudon, Proc. Roy. Soc. A 257, 218 (1963).
13. L. S. Kothari and K. S. Singwi in Vol. 8, Solid State Physics, F. Seitz and D. Turnbull (eds.), Academic Press (1959).
14. L. I. Schiff, Quantum Mechanics, McGraw-Hill, (1968), p. 332.
15. W. M. Lomer and G. G. Low, in Thermal Neutron Scattering, P. A. Egelstaff (ed.), Academic Press (1965).

16. J. L. Prather, Atomic Energy Levels in Crystals, NBS Monograph 19 (1961).
17. J. S. Griffith, The Theory of Transition Metal Ions, Cambridge (1961), Chapters 8 and 9.
18. B. DiBartolo, Optical Interactions in Solids, Wiley (1968).
19. A. A. Maradudin in Vol 18, 19 Solid State Physics, F. Seitz and D. Turnbull (eds.), Academic Press (1966).
20. R. A. Satten, J. Chem. Phys. 27, 286 (1957).
21. R. A. Satten, J. Chem. Phys 29, 658 (1958).
22. I. Richman, R. A. Satten, E. Y. Wong, J. Chem. Phys. 39, 1833 (1963).
23. R. A. Satten, J. Chem. Phys. 40, 1200 (1964).
24. A. A. Maradudin and S. H. Vosko, Rev. Mod. Phys. 40, 1 (1968).
25. G. F. Koster, in Vol. 5 Solid State Physics, F. Seitz and D. Turnbull (eds.) Academic Press (1957).
26. J. F. Cornwell, Group Theory and Electronic Energy Bands in Solids, North Holland (1969).
27. T. Timusk and M. Buchanan, Phys. Rev. 164, 345 (1967).
28. M. Buchanan and E. J. Woll, Can. J. Phys. 47, 1757 (1969).
29. W. E. Bron, Phys. Rev. 185, 1163 (1969).
30. D. Kühner and M. Wagner, Phys. Stat. Sol. 40, 617 (1970).
31. M. J. Sangster and C. W. McCombie, J. Phys. C: Sol. State Phys. 3, 1498 (1970).
32. N. B. Manson, Phys. Rev. B4, 2645 (1971).
33. M. J. Sangster, Phys. Rev. B6, 254 (1972).
34. H. G. von Häfele, Ann. Phys., Lpz., 7, 321 (1963).
35. B. Piriou and F. Cabannes, Compt. Rend. 264, 630 (1967).
36. N. B. Manson, W. von der Ohe, and S. L. Chodos, Phys. Rev. B3, 1968 (1971).
37. G. Peckham, Proc. Phys. Soc. 90, 657 (1967).
38. M. J. Sangster, G. Peckham, and D. H. Saunderson, J. Phys. C: Sol. St. Phys. 3, 1026 (1970).
39. R. K. Singh and K. S. Upadhyaya, Phys. Rev. B6, 1589 (1972).
40. R. Loudon, Proc. Roy. Soc. 84, 379 (1964).
41. M. D. Sturge, Phys. Rev. 130, 639 (1963).
42. G. F. Imbush, W. M. Yen, A. L. Schawlow, D. E. McCumber, and M. D. Sturge, Phys. Rev. 133, A1029 (1964).
43. B. DiBartolo and R. Peccei, Phys. Rev. 137, A1770 (1965).
44 S. E. Stokowski, S. A. Johnson, and P. L. Scott, Phys. Rev. 147, 544 (1966).
45. G. E. Devlin, J. A. Ditzenberger, and M. D. Sturge, Bull. Am. Phys. Soc. 7, 258 (1962).
46. M. J. Weber and R. F. Schaufele, unpublished data, private communication.

LUMINESCENCE AND SPECTROSCOPY OF SMALL MOLECULAR IONS IN CRYSTALS

K. K. Rebane, L. A. Rebane

Institute of Physics, Academy of Sciences

of the Estonian SSR, 202400 Tartu, USSR

ABSTRACT

The light molecular ions O_2^-, S_2^-, NO_2^-, PO_2^- and their analogues, inserted into alkali halide crystals as impurities, form luminescence centres of a special molecular type having a number of remarkable properties (a clearly expressed structure of vibronic spectra, rotational-librational motion of the impurity, etc.). In the present paper the following problems are discussed: (1) features of the spectra of the centres formed by small molecular ions; (2) rotation and libration of the impurity ion NO_2^-; (3) luminescence study of the kinetics of the reorientation of the impurity ion S_2^- in the KI crystal.

I. INTRODUCTION

The study of the O_2^-, S_2^-, NO_2^-, and PO_2^- ions as centres is interesting from at least two aspects. First, they serve as convenient probes for investigating the local dynamics of the crystal lattice, i.e., for studying by means of the spectra of molecular centres the crystal lattice vibrations in the centre and in its vicinity, and the details of electron-phonon and phonon-phonon interactions. Secondly, from the point of view of matrix spectroscopy it is important that in the ionic crystal the "frozen gas" of negative molecular ions be stabilized.

II. GENERAL DESCRIPTION OF THE SPECTRA OF MOLECULAR CENTRES IN CRYSTALS

II.A. Structure of Luminescence and Absorption Spectra. Parameters of Potential Curves

The first and the main feature of the spectra of the centres formed by small molecular ions in alkali halide crystals is their clear-cut vibronic structure caused by the presence of one or several high-frequency local modes which are the intramolecular vibrations (somewhat changed by the crystal environment) of the impurity molecule. The intramolecular frequencies of small molecules (about 1000 cm^{-1}) exceed several times lattice vibration frequencies (about 200 cm^{-1}). That results in the fact that intramolecular vibrations remain well localized on the molecule, appearing in the spectra as high-frequency local modes of the whole system - crystal + molecule . The electron vibrational (vibronic) absorption, excitation and luminescence spectra of the molecular centre, have a very characteristic structure due to the excitation of high-frequency local vibrations at electronic transition*. In Fig. 1 the O_2^- luminescence spectrum in the KBr crystal at 90°K is presented. The system of vibronic sub-bands belongs to the electronic transition between two electronic states of the O_2^- molecule ($^2\Pi_u \rightarrow {}^2\Pi_g$); every vibronic sub-band in the luminescence spectrum corresponds to the transition between the lowest vibrational level of the excited electronic state and one of the vibrational levels of the quantum number ν of the ground electronic state. In the presence of more than one local vibration the level will be characterized by a set of quantum numbers ν_1 ν_2...

The areas under vibronic sub-bands enable us to find the distribution of the transition probabilities over vibrational levels ν; these areas are proportional to the probability of the transition to the level with the quantum number ν, which is determined by the corresponding Franck-Condon factor.

On further lowering the temperature a narrow (with the width about one cm^{-1}) no-phonon line and wide phonon wing appear in each vibronic sub-band (see Fig. 2). The spectra of the other molecules mentioned above are in general the same. It should be noted that the vibronic structure of the spectra of molecular centres may

---

* The results of studying the vibronic structure of the spectra of luminescence and absorption of molecular centres of O_2^-,etc., are given in reviews (1) and (2).

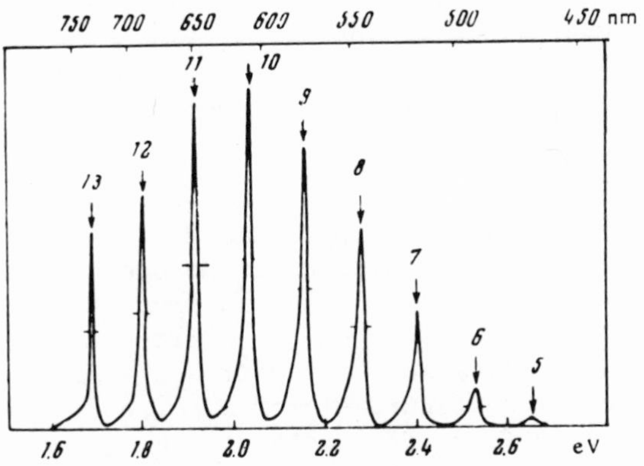

Fig. 1 Luminescence spectrum of molecular ion O_2^- in KBr at liquid nitrogen temperature. The vibrational sub-bands of $0' \rightarrow \nu$ transitions are well separated.

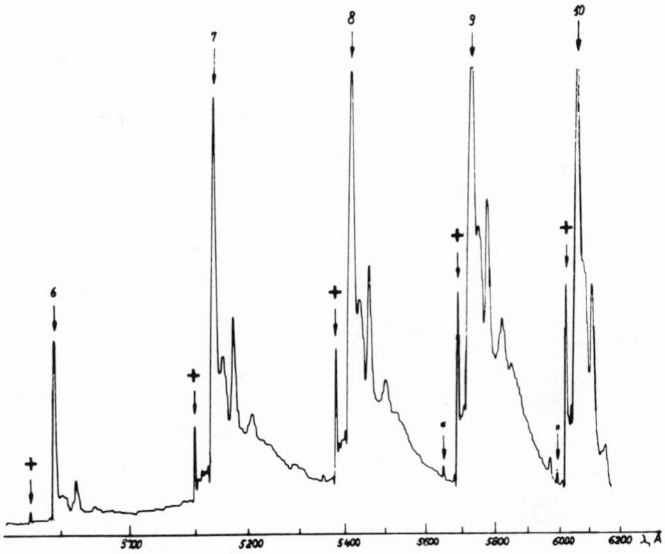

Fig. 2 Luminescence spectrum of molecular ion O_2^- in KBr at 4.2 K. No-phonon lines and phonon sidebands are distinctly separated. Superscript (+) indicates the no-phonon lines of additional series. The similarity law is fairly well fulfilled.

serve as an excellent example of the theoretical picture of the spectrum of the impurity centre in the "ideal" case, i.e., in the case of a strong interaction of the electronic transition with a high-frequency local mode, and a weak or medium interaction with the phonons of crystal modes (3).

The series of no-phonon lines are the replicas of the pure electronic line over frequencies approximately equal to 1,2...-fold local mode frequency. Their position in the spectrum enables us to determine the frequencies of single electron-vibrational transitions with accuracy up to 1 cm^{-1}, and to solve the problems analogous to the ones solved in the spectroscopy of free molecules: to determine the frequencies of pure electronic transitions ν_{00}, the frequencies and anharmonic coefficients of the intramolecular vibrations which are active in the electronic transition, etc. (4) - (8), up to the study of the isotopic composition of the centres (9). Table 1 contains the values of the mentioned parameters of O_2^- and S_2^- in various alkali halide hosts (4) - (6). The spectroscopic constants of the impurity ions NO_2^- and PO_2^- in the ground and lower excited electronic states are given in (7),(8).

Table 1

Spectroscopic Constants of O_2^- and S_2^- Impurity Ions in the Ground Electronic State

Centre	Crystal	ν_{00} cm^{-1}	ν_0 cm^{-1}	$x_0 \nu_0$ cm^{-1}
O_2^-	NaCl	27283	1149	8.4
	NaBr	26672	1139	7.9
	KCl	27559	1153	8.3
	KBr	26979	1141	9.0
	KI	26514	1124	8.8
	RbCl	27483	1150	8.5
	RbBr	27150	1140	8.2
	RbI*	26761	1127	8.7
	CsCl	26527	1136	9.1
	CsBr	26130	1120	9.0
S_2^-	KCl	20970	627.5	2.5
	KBr	20578	616	2.6
	KI	20026	598.6	2.2
	KI*	19452	598	3.0
	RbBr*	20609	614.5	4.5
	RbI	19618	599	2.5

* Data are taken from (6).

The determination of spectroscopic constants, and also the probabilities of vibronic transitions allow one to build the adiabatic potentials of lower electronic states of the impurity with a sufficiently high accuracy (2),(10), e.g., using the Morse potential. It should be stressed that the molecular ions O_2^-, S_2^-, NO_2^- and others are unstable in the free state. Therefore the stabilization of their ground and first excited electronic states in alkali halides offers an essential possibility for their detailed study.

II.B. Electron-Phonon Interaction. Structure of Phonon Wings

The phonon wing measured with a sufficiently high resolution displays a detailed structure characteristic of the impurity itself as well as the host crystal.

The interpretation of broad sidebands on the long-wave side of the no-phonon lines as the phonon wings, i.e., the transitions arising as a result of the interaction of the electronic transition in the molecular centre with the vibrations of the host crystal environment, is based firstly on a good repetition of the band structure in the series, which corresponds to the theoretically predicted similarity law for the spectrum of the impurity centre in the "ideal" case (see Figs. 2 and 3), and secondly, on the great difference of the fine structure of phonon wings in the spectra of the given impurity molecule in various hosts (11)-(13) (see Figs. 2 and 3).

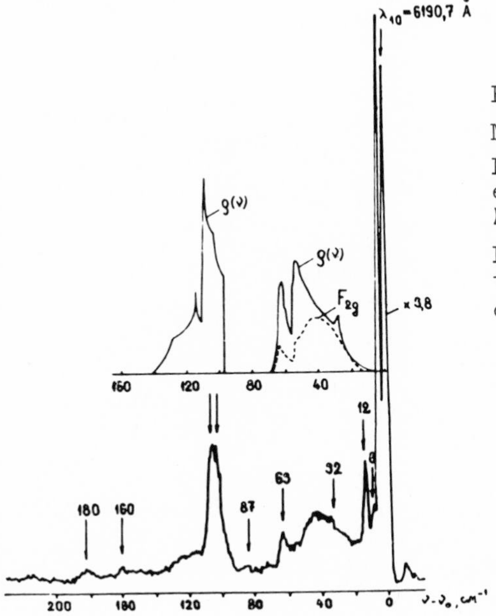

Figure 3

No-phonon line $\nu = 10$ and its phonon sideband in the luminescence spectrum of O_2^- in KI at 4.2°K. The structure of the one-phonon sideband well reproduces the full phonon density function of KI crystals above.

The contribution of multiphonon transitions to the phonon wing is determined by the strength of the coupling with lattice modes characterized by the value of dimensionless Stokes' losses P. P increases along the sequence of the molecular centres O_2^-, S_2^-, NO_2^-, PO_2^-, and the coupling between electronic transition of O_2^- and phonons increases in the sequence of hosts CsCl, KI, NaCl, KBr, KCl, RbBr, RbCl (12), (16).

The analysis of the structure of phonon wings is carried out in (11)-(14). It shows, in particular, that the lattice modes are only slightly distorted by O_2^- ions. For instance, the phonon wing of O_2^- in KI (Fig. 3) reproduces the phonon spectrum of the host crystal very well. The NO_2^- ion, on the contrary, introduces considerable perturbations resulting in a number of maxima which are absent in the phonon spectrum of the host crystal and characteristic of the local dynamics of the impurity centre.

II.C. Electron-Phonon Interaction. Anharmonicity Effects. No-Phonon Lines

In the vibronic series of O_2^- and S_2^- luminescence spectra the deviations from the "similarity law", evidently due to the anharmonic coupling of the local and lattice modes, are considered. The decrease of Stokes' losses in the lattice vibrations with increasing quantum number ν of the intramolecular vibration takes place. It is accompanied by the decrease of the contribution of multiphonon transitions to the phonon wing, and also by the decrease of the width of no-phonon lines in series (15), (16). It is convenient to examine these effects within the frame of the theory of double adiabatic approximation. They may be interpreted as the result of the weakening of the vibronic interaction between the electronic transition and crystal modes with the increase of the degree of excitation of the local mode, and they allow one to estimate some parameters of the anharmonic coupling (16).

The study of no-phonon lines and their temperature behavior offers information about the mechanism of the interaction of vibrations in the centre, and also enables one to investigate the splitting of electronic levels and the librational-rotational motion.

The study of no-phonon lines in the luminescence spectra of O_2^- in KCl, KBr, and KI, and S_2^- in KI show the important role of librational vibrations in the mechanism of their temperature broadening (17). As the decay of the high-frequency local mode quanta into lattice phonons requires the simultaneous creation of a great number of them (e.g., at least 5-6 phonons in case of O_2^- in KCl), the probabilities of such processes are, as a rule, low. This is the main reason why the vibrational broadening of no-phonon lines of molecular centres is connected with modulation processes (18). The case in which the decay of one local mode quantum may be accompanied by the creation of another local mode quantum of a somewhat lower frequency must be an exception.

Such a concrete situation takes place for NO_2^- in KCl. Here the frequencies of local intramolecular vibrations are (7) $\nu_1 = 1326\ cm^{-1}$, $\nu_2 = 803\ cm^{-1}$, $\nu_3 = 1290\ cm^{-1}$. In this way the anharmonic decay of the local mode ν_1 with the creation of the local mode ν_3 and a phonon with the frequency $36\ cm^{-1}$ is possible. The luminescence spectrum of NO_2^- contains vibronic series according to two symmetric vibrations ν_1 and ν_2 . The no-phonon lines of NO_2^- at helium temperatures consist of a small number of rather well-resolved rotational components caused by the quasi-free rotation of the molecule around the axis parallel to the line connecting two oxygen atoms (7). The study of the dependence of the widths of rotational components of the no-phonon lines of quantum numbers 0.4.0 and 1.2.0 (see Fig. 4) on temperature has shown that for the lines related to the excitation of the vibration ν_1 (line 1.2.0) the above mentioned mechanism of anharmonic decay really takes place (19).

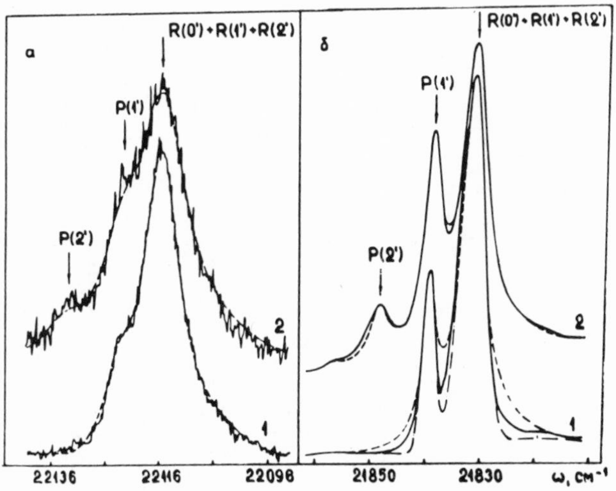

Fig. 4 Contour of no-phonon lines 1.2.0 (a) and 0.4.0 (b) in the luminescence spectrum of NO_2^- in KCl at temperatures 4.2°K (1) and 18°K (2). Approximated contours were obtained by summation over rotational components taken as Lorentz-shape (dashed curve) and Gaussian-shape (dash-dotted curve).

II. D. Radiationless Transitions and Vibrational Relaxation in Molecular Centres

High-frequency local modes play a distinguished role also in the processes of vibrational relaxation and radiationless exchange of the vibronic energy in molecular luminescence centres. The study of the temperature quenching of the luminescence of O_2^- and S_2^- has shown (20) that the radiationless transitions in them take place from more than only one of the lowest intramolecular vibration levels and therefore the temperature dependence of the luminescence yield does not obey the well-known Mott formula.

In case of NO_2^- centres a rapid stepwise decrease of the luminescence yield with the increasing frequency of excitation was observed within a single electronic-vibrational absorption band at 70°K (21), (22). The sharp dependence of the yield on the frequency of the exciting light takes place also at helium temperatures and has a clear stepwise character as well (see Fig. 5 where the luminescence yield spectrum of NO_2^- in KCl is shown at 4.2°K). The decrease in the yield at the excitation of higher vibronic states of the centre is the result of the competition between the probabilities of vibrational relaxation

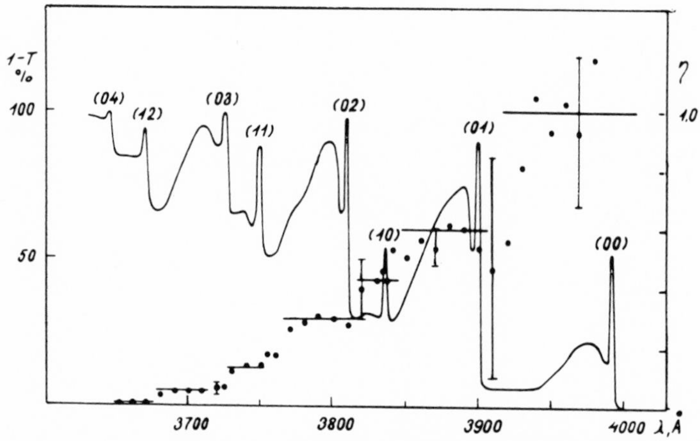

Fig. 5 Absorption spectrum (solid curve) and luminescence yield spectrum (full circles) for NO_2^- in KCl at 4.2°K.

and the radiationless transition, the probability of the latter process increasing much more rapidly with the quantum number of the intramolecular vibration than the probability of the vibrational relaxation. Constancy of the yield within the phonon wing of a fixed no-phonon line shows that the relaxation of the excitation energy of the lattice modes takes place very fast in comparison with the radiationless transition rate into another electronic state.

A rather detailed information about the characteristic times of the relaxation processes in the NO_2^- centre has been obtained on the basis of hot luminescence spectra (23)-(25).

III. ROTATION AND LIBRATION OF THE NO_2^- ION IN POTASSIUM HALIDE CRYSTALS

III.A. Hindered Rotation of the Impurity Molecule in Crystals

Molecular ions are, as a rule, placed in regular lattice sites of the host in such a way that the symmetry planes and axes of the molecule are oriented along the symmetry planes and axes of the crystal. As a result, the molecules in a crystal have several equivalent orientations and transitions take place between them. These transitions are equal to the more or less hindered rotational motion around fixed axes.

The rotation of an impurity molecule in a crystal is in many respects analogous to the internal rotation in multiatomic molecules. The basic features of this phenomenon were cleared up by Pauling (26) who carried out the first quantum mechanical study of the problem in the case of the simplest model: a one-dimensional rotation in a cosine potential field with two minima. The generalization of the problem to the case of the three-dimensional rotation in a potential field, presented as a sum of spherical functions of octahedral symmetry (Devonshire potential) and the case of the three-dimensional rotation of the symmetric gyroscope in this field was carried out in (27),(28).

For molecular centres in crystals, cases exist in which the number of equivalent wells of the potential is $N = 2, 3, 4$, and 6. In Fig. 6 (according to the data of (30)) the energy spectrum of one-dimensional rotation in the field consisting of potential boxes with the depth $V = 4\varepsilon$ is shown (ε is the rotation constant of the free rotator). The levels above the barrier are close to the levels of the free rotator. In case of increasing the height of the potential barriers between the wells, the energy spectrum in the lower part will approach the librator spectrum in the Pauling model, while each librator level, including the zero level, will have a multifold degeneracy equal to the number of equivalent wells. This degeneracy is partly removed by the tunneling reorientation processes of the molecule between equivalent wells. A review of tunneling transitions of impurity molecules is given in (29). In (30) it is shown that the energy spectrum depends weakly on the shape of the potential and is determined mainly by the height of potential barriers between wells and by the relation of barrier widths of the well width.

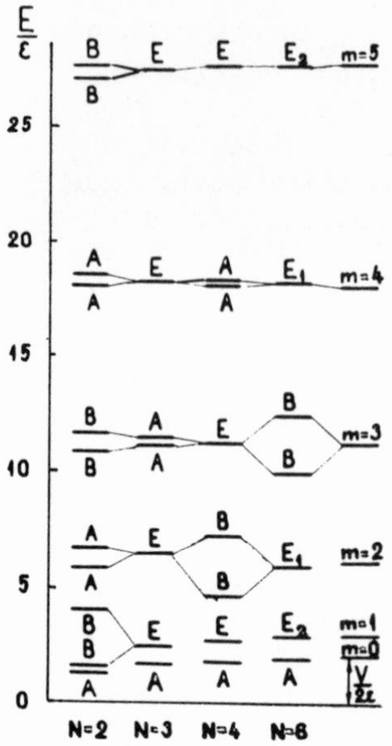

Fig. 6

Levels of a one-dimensional rotator in the field of N = 2,3,4 and 6 potential well with depth $V = 4\varepsilon$. On the right the levels of the free rotator are shown.

III.B. Rotational Structure of NO_2^- Vibronic Spectra

The hindered rotation of impurity molecules in a crystal may be studied by spectroscopic methods on the basis of the data on the polarization and fine structure of no-phonon lines in the vibronic spectra, and also in the infrared absorption spectra and Raman scattering spectra. Let us examine the hindered rotation of NO_2^- in potassium halide crystals by means of the data obtained by different spectroscopic methods: the fine structure of infrared vibrational absorption lines (31)-(33), the fine structure of no-phonon lines in vibronic spectra (34),(35), the polarization and fine structure of the lines in Raman scattering spectra (36)-(38).

The no-phonon lines in the vibronic absorption and luminescence spectra of NO_2^-, corresponding to the excitation of the totally symmetric vibration ν_2, have a similar fine rotational structure in KCl, KBr and RbCl crystals; at 4°K it consists of three main components, the separation between them depending slightly on the host. The no-phonon lines of NO_2^- in the KI crystal are structureless. In Fig. 7 the components of the pure electronic absorption line and 0.3 luminescence line of NO_2^- in KCl are given. On the basis of the position of the components and the dependence of their intensities on temperature (see also Fig. 4) the scheme of rotational levels in the ground electronic state $^1A_1$ and excited electronic state $^1B_2$ was drawn. The rotational terms are in good agreement with the dependence $F(K) = \varepsilon K^2$, where ε is the rotation constant of the one-dimensional rigid rotator. For NO_2^- in KCl ε = 2 and

4.5 cm^{-1} is obtained in the ground and excited electronic states, respectively. The values ε appeared to be close to the rotation constant around axis a for the freely rotating NO_2^- . Therefore the rotation of NO_2^- in KCl, KBr and RbCl crystals may be treated as a nearly free (especially in the excited electronic state) one-dimensional rotation. Electron-rotational transitions obey the selection rule $\Delta K = \pm 1$ in accordance with the selection rule for the rigid rotator in the case where the dipole transition moment is perpendicular to the rotation axis (39). Polarization measurements of NO_2^- spectra in KCl in case of uniaxial stress show (40) that rotation takes place in the crystal field with four equivalent minima. That corresponds to the orientation of the rotation axis a of the molecule parallel to <100> directions. The model of the one-dimensional rotator in the field C_{4v} explains the fine structure of lines ν_1 and ν_2 in the infrared absorption spectrum of NO_2^- in KCl and KBr as well, observed in (31). The absence of the structure of no-phonon lines in the NO_2^- spectrum in KI enables one to conclude that the rotation of NO_2^- around axis a is absent in this crystal. However, we must not forget that the simple model of one-dimensional rotation around axis a can explain the structure of the lines of totally symmetric vibrations ν_1 and ν_2 only, and does not tell anything about the structure of line ν_3 of antisymmetric vibration.

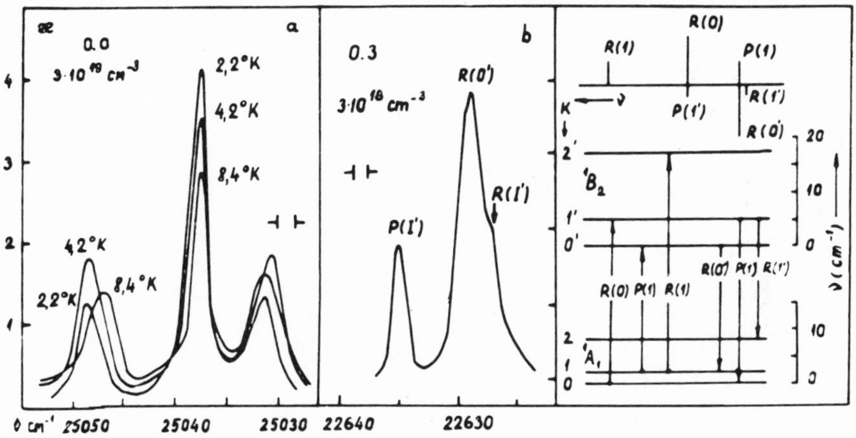

Fig. 7 Rotational structure of no-phonon lines in vibronic spectra of NO_2^- in KCl. (a) Pure electronic transition in the absorption spectrum; (b) no-phonon line 0.3 in the luminescence spectrum at 4.2°K; (c) scheme of rotational terms and interpretation of roto-vibronic transitions.

III.C. Polarization and Structure of Raman Scattering Lines of NO_2^-. The Role of Librations

Let us examine briefly the results obtained from the Raman scattering (RS) spectra of NO_2^-.

The RS spectrum of the impurity crystal could be divided into low- and high-frequency regions. The low-frequency region, adjacent to the Rayleigh scattering line, gives evidence about lattice modes and also about the low-frequency local and pseudo-local vibrations induced by impurities. We shall call this region the phonon wing of the Rayleigh line. In Fig. 8 the corresponding spectra for different orientations of KI-NO_2^- crystals at 5.5°K are given. We can see a pronounced structure having characteristic polarizations.

The high-frequency region includes, first of all, the lines of high-frequency local modes of the impurity and carries information about the symmetry of the centre of a given type and the character of its motion in the crystal. In Fig. 9 the summary of the results obtained in the study of RS spectra of KCl, KBr, KI and RbCl crystals doped with NO_2^- ions at 10°K is given. We are here interested in the lines of NO_2^- only. It can be seen that some of them, ν_1 in KI and partially in KBr, and ν_3 in KCl and KBr, reveal an additional structure. The polarization data are contained in Table 2.

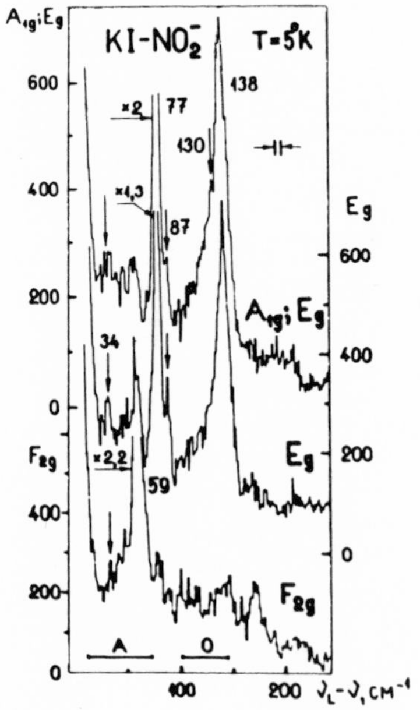

Fig. 8

Phonon sideband of Rayleigh line of KI-NO_2^- at excitation by the argon laser (λ = 4880 Å) and at 5°K.

Table 2

Polarization Degree of Intramolecular Lines of NO_2^- Impurity Ion

Hosts	ν_1	ν_2	ν_3	Orientation of crystals*
KCl	0.3	0.4	0.3	(100)
	-	-	-	(110)
KBr	0.4	0.45	0.25	(100)
	0.5	0.2	0.75	(110)
KI	0.4	0.4	5.0	(100)
	0.4	0.4	1.1	(110)
RbCl	0.7	0.25	0.3	(100)
	0.6	-	0.5	(110)

*(100) orientation denote $e \| [010]$ and $n_\perp \| [001]$
(110) orientation denote $e \| [110]$ and $n_\perp \| [001]$.

For the impurity centres fixed in the crystal the intensity and polarization of lines in the RS spectrum are determined by the scattering tensor of the intramolecular vibrations and the orientation of the centre (41). If NO_2^- is fixed in the orientations with molecular axes a along <100> and b along <110>, the local symmetry of the centre will be C_{2v} (which coincides with the symmetry of the free molecule); the polarization of the lines ν_1 and ν_2 will be mainly parallel, and that of the ν_3 line wholly perpendicular.

It may be recalled that the rotation around axis a leads to the selection rules (according to the rotational quantum number) $\Delta K = \pm 1$, i.e.,to the disappearance of the pure vibrational line ν_3 from the scattering spectrum, and to the appearance of its combinations with rotation. The polarization of the ν_3 line remains perpendicular. Both the observed spectral structure and the polarization of the ν_3 line in KCl, KBr and RbCl crystals do not fit these conclusions. Therefore in case of these crystals it is necessary to take into account the librational motion of NO_2^-.

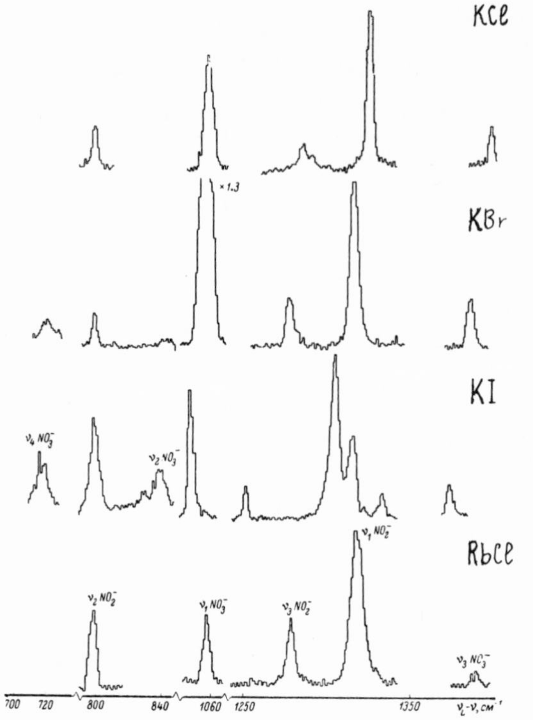

Fig. 9

Raman scattering lines of intramolecular vibrations of ions NO_2^- and NO_3^- in alkali halide crystals at 10^oK; excitation with the 4880 Å argon laser line.

A detailed discussion of librational coordinates and the equations describing the motion of the impurity molecule are given in (42). Librations around axes b and c together with rotation around a lead to the fact that combined frequencies $\nu_i \pm \nu_l$ of vibrations ν_i and librations ν_l must appear, which have different polarizations and new selection rules for rotational quantum numbers (43). So, the line $\nu_3 + \nu_l$ will have selection rules $\Delta K = 0, \pm 2$ and mainly parallel polarization, which is in a better accordance with experiment.

NO_2^- probably performs in the KI lattice a complicated librational motion around axis a as well as around axes b and c.

IV. REORIENTATION OF S_2^- IN KI

The orientation of O_2^- and S_2^- ions has been studied by means of the spin resonance spectra in a well-known series of papers by Kanzig (44)-(47). It was established that O_2^- in the ground electronic state performs tunneling transitions between six equivalent equilibrium orientations of the molecular axis along <110> .

The time of reorientation at helium temperature was found to be a few seconds. The reorientation of the S_2^- molecule needs an activation energy, and at 4°K it is actually frozen in. In (2), (48) it is shown, however, that S_2^- may change its orientation in KI at helium temperature, if the crystal is illuminated in the absorption band of S_2^-. The kinetics of reorientation can then be studied through luminescence measurements (48).

Excitation by polarization light causes depopulation of these orientations in the crystal in which the molecules absorb the light, and the accumulation of molecules in other orientations. Thus, illumination of the crystal with polarization E_y (see Fig. 10) leads to the accumulation of molecules in orientations 3 and 4. Such a non-equilibrium distribution of centres is preserved quite long due to the small probability of tunneling transitions. Exciting then the crystal with polarization E_x and recording luminescence J_{xx} in parallel polarization, it is possible to observe the depopulation of orientations 3 and 4 by the time dependence of luminescence intensity. The curve J_{yx} enables one to observe at first the population and then the depopulation of orientations 1 and 2. It is evident that in the experimental geometry, shown in Fig. 10, orientations 1 and 2, 3 and 4, and 5 and 6 cannot be distinguished, and reorientations are called forth by 60° turns only.

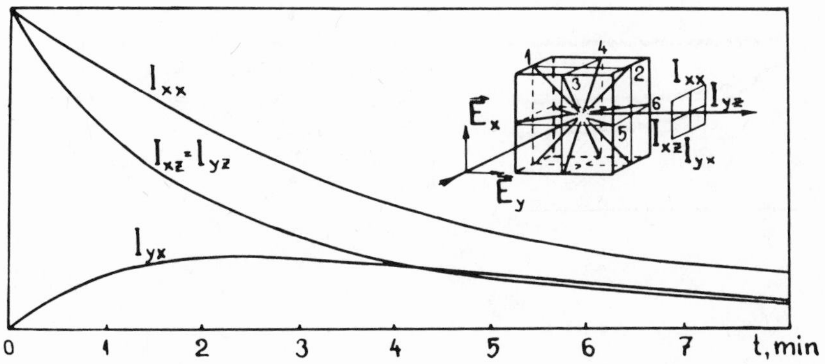

Fig. 10 Time dependence of the intensity of polarized luminescence of S_2^- in KI due to the reorientation of S_2^- under polarized excitation with λ = 410 nm at 4.2°K. The scheme of the experiment is given above.

The agreement between theory and experiment confirms that S_2^- centres in KI are really oriented along <110> and are reoriented as a result of electronic excitation, but the decay curves do not allow one to determine probabilities of reorientation in ground electronic state Q and excited state q separately. The estimates of total probability give $(q+Q) = 10^{-2}$; i.e., the reorientation takes place in one of the 100 excited centres.

Further it was interesting to compare light reorientation and thermal reorientation. For that purpose the crystal was put into the thermostatic chamber of the cryostat, the temperature of which was regulated and kept with a great accuracy, and the decay curves of luminescence were again recorded. The results are shown in Fig. 11. It is clearly seen that in the narrow temperature region from 9 to 12°K a sharp decrease of the reorientation time takes place, and the intensity of the remaining luminescence, which appears to be the level of dynamical equilibrium between orientation under the influence of light and chaotic reorientation, is simultaneously increased.

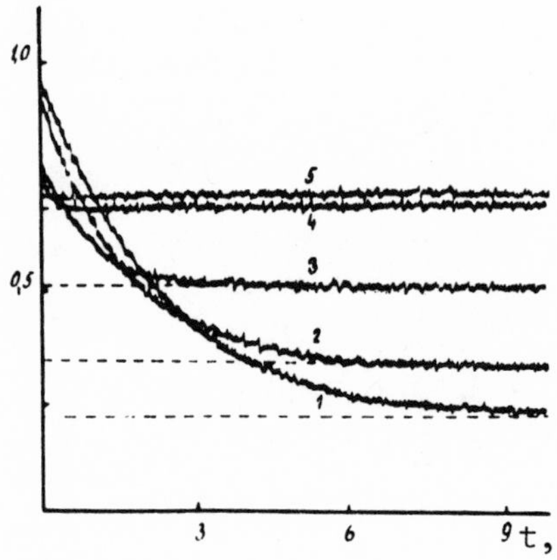

Fig. 11

Time dependence of the intensity of luminescence of S_2^- in KI at different temperatures and reorientational times τ:
1 - T = 4.2-9°K, τ = 133 sec;
2 - T = 10°K, τ = 105 sec;
3 - T = 10.5°K, τ = 60 sec;
4 - T = 11°K, τ = 15 sec;
5 - T = 12°K.

The experimental curves enable us to determine the probabilities of thermal reorientation in the course of electronic excitation of S_2^-, which practically coincide with the ESR results (45). The dependence of the reorientation time on temperature refers to a thermally activated process with the activation energy $\Delta E = 225\ cm^{-1}$.

It was concluded that the thermal reorientation of S_2^- accompanying electronic excitation occurs in the ground electronic state after relaxation over all possible vibrational levels.

Reorientation processes should be taken into consideration not only in recording luminescence spectra but also when investigating other spectra of molecular centres. For example, the reorientation of S_2^- in KI is the cause of unusual effects in the resonant scattering spectra of this crystal: the intensity of the Raman scattering line falls rapidly in lowering temperature below 15°K (49). The reason for such an unusual temperature dependence is the reorientation of S_2^- under laser illumination (50).

REFERENCES

1. L. A. Rebane, Physics of Impurity Centres in Crystals, p. 353 (ed., G. S. Zavt), Tallinn (1972).

2. K. K. Rebane, L. A. Rebane, J. Pure Appl. Chem. 37, Vol. 1-2 (1974).

3. K. K. Rebane, Impurity Spectra of Solids, Plenum Press, New York - London (1970).

4. L. Rebane and P. Saari, Eesti NSV Teaduste Akad.Toimetised, Fuus. Matem. 19, 123 (1970).

5. I. Sildos and L. Rebane, Eesti NSV Teaduste Akad. Toimetised, Fuus. Matem. 20, 354 (1971).

6. M. Ikezawa and J. Rolfe, J. Chem. Phys. 58, 2024 (1973).

7. R. Avarmaa and L. Rebane, Phys. Stat. Sol. 35, 107 (1969).

8. R. Avarmaa, Eesti NSV Teaduste Akad. Toimetised, Fuus. Matem. 17, 78 (1968).

9. L. A. Rebane, Trudy IFA AN ESSR, No. 37, 14 (1968); Eesti NSV Teaduste Akad. Toimetised, Fuus. Matem. 17, 72 (1968).

10. K. K. Rebane, A. I. Laisaar, L. A. Rebane, O. I. Sild, Izv. AN SSSR ser. fiz. 31, 2010 (1967).

11. K. Rebane, L. Rebane, O. Sild, Localized Excitation in Solids, 117, Wallis, R.F., ed. (Plenum Press, N.Y., 1968).

12. L. A. Rebane, and P. M. Saari, Fiz. Tverd. Tela 12, 1945 (1970).

13. L. A. Rebane, Opt. i Spectr. 31, 6230 (1971).

14. J. Rolfe, M. Ikezawa and T. Timusk, Phys. Rev. B7, 3913 (1973).

15. L. A. Rebane, Yu. N. Charchenko, Opt. i Spectr. 28, 943 (1970).

16. L. A. Rebane, O. I. Sild, T. J. Haldre, Izv. AN SSSR, ser. fiz. 35, 1395 (1971).

17. L. A. Rebane, A. M. Freiberg, J. J. Koni, Fiz. Tverd. Tela 15, 3318 (1973).

18. M. A. Krivoglaz, Fiz. Tverd. Tela 6, 1707 (1964).

19. L. A. Rebane, A. M. Freiberg, Fiz. Tverd. Tela (to be published).

20. L. A. Rebane, and R. A. Avarmaa, Trudy IFA AN ESSR, No. 37, 63 (1968).

21. K. K. Rebane, R. A. Avarmaa, and L. A. Rebane, Izv. AN SSSR, ser fiz. 32, 1381 (1968).

22. L. Rebane, P. Saari, and R. Avarmaa, Eesti NSV Teaduste Akad. Toimetised, Fuus. Matem. 19, 44 (1970).

23. K. Rebane and P. Saari, Eesti NSV Teaduste Akad. Toimetised, Fuus. Matem. 17, 241 (1968).

24. P. Saari and K. Rebane, Solid State Commun. 7, 887 (1969); Eesti NSV Teaduste Akad. Toimetised, Fuus. Matem. 18, 225 (1969).

25. P. Saari, Phys. Stat. Sol. (b) 47, K79 (1971); XI European Congress on Molecular Spectroscopy, Abstract No. 104, Tallinn (1973).

26. L. Pauling, Phys. Rev. 36, 430 (1930).

27. A. I. Devonshire, Proc. Roy. Soc. A153, 601 (1936).

28. P. Sauer, Z. Phys. 199, 280 (1967).

29. V. Narayanamurti and R.O. Pohl, Rev. Mod. Phys. 42, 210 (1970).

30. K. Rebane, O. Sild, Eesti NSV Teaduste Akad. Toimetised, Fuus. Matem. 19, 311 (1970).

31. V. Narayanamurti, W. D. Seward and R. O. Pohl, Phys. Rev. 148, 481 (1966).

32. T. Mauring, Eesti NSV Teaduste Akad. Toimetised, Fuus. Matem. 18, 105 (1969).

33. R. Bonn, R. Metselaar and J. Van der Elsken, J. Chem. Phys. 46, 1988 (1967).

34. L. Rebane, Czechosl. J. Phys. B20, 608 (1970).

35. R. A. Avarmaa, Thesis, Tartu (1970).

36. A. R. Evans and D. B. Fitchen, Phys. Rev. B2, 1074 (1970).

37. K. K. Rebane, L. A. Rebane, T. J. Haldre and A. A. Gorokhovski, Advances in Raman Spectroscopy, Vol. 1, p. 379 (Heyden and Son, Ltd., London, 1973).

38. L. A. Rebane, T. J. Haldre, A. E. Novik, and A. A. Gorokhovski, Fiz. Tverd. Tela 15, 3188 (1973).

39. G. Herzberg, Infrared and Raman Spectra of Polyatomic Molecules, New York (1945).

40. R. A. Avarmaa, Opt. i Spectr. 29, 715 (1970).

41. A. A. Kaplyanski, V. K. Negoduiko, Opt. i Spectr. 35, 3 (1973).

42. A. A. Kiselev, and A. V. Ljapzev, Phys. Stat. Sol. (b) 62, 271 (1974).

43. A. A. Kiselev and A. V. Ljapzev, Fiz. Tverd. Tela (to be published).

44. W. Känzig, J. Phys. Chem. Solids 23, 479 (1962).

45. K. Bachmann, W. Känzig, H. R. Zeller and A. Zimmermann, Phys. Kondens. Materie 7, 360 (1968).

46. H. R. Zeller and W. Känzig, Helv. Phys. Acta 38, 638 (1965).

47. H. R. Zeller, R. T. Shuly and W. Känzig, J. Phys. (Paris), Colloq. 28, 81 (1967).

48. A. Treschalov, I. Sildos and L. Rebane, Eesti NSV Teaduste Akad. Toimetised, Fuus. Matem. 22, 451 (1973).

49. W. Holzer, S. Racine and J. Cipriani, Advances in Raman Spectroscopy, Vol. 1, p.393 (Heyden and Son, Ltd., London, 1973).

50. L.A. Rebane, A.B. Treschalov, and T.J. Haldre, Fiz. Tverd. Tela (to be published).

SOME PROBLEMS OF THE VIBRATIONAL STRUCTURE OF OPTICAL SPECTRA OF IMPURITIES IN SOLIDS

K. K. Rebane

Institute of Physics, Academy of Sciences of the Estonian SSR, 202400 Tartu, USSR

ABSTRACT

During the last decade in the study of the impurity spectra of solids, considerable attention has been paid to both the theoretical and experimental problems of no-phonon lines and phonon sidebands. This interest is reasonable: no-phonon lines represent clear-cut narrow spectral lines which are highly sensitive to the slightest alterations of the physical situation in the impurity centre. The $\omega:\Delta\omega$ relation for the frequency ω broadening $\Delta\omega$ is often about 10^4 and may be even more.

Two problems are discussed in this treatment: (1) the no-phonon lines and phonon wings in various impurity spectra of solids; (2) the inhomogeneous broadening of luminescence spectra.

I. NO-PHONON LINES AND PHONON WINGS (SIDEBANDS)

I.A. Vibronic Spectra of Absorption and Luminescence

The no-phonon line is the optical analogue of the Mössbauer γ-line (1)-(5)*. The spectroscopy of sharp no-phonon** lines gives reliable data about the electron energy of the excited state, about the frequencies of local vibrations and their

*At the same time there exists an analogue to the no-phonon neutron scattering line and no-phonon x-ray scattering lines.

**We consider as "no-phonon" lines the pure electronic line as well as its high-frequency local vibration replicas.

alterations caused by different reasons (isotopic shifts, splittings and shifts by external fields, etc.). The no-phonon lines in extremely weak hot luminescence spectra serve as a reliable basis when interpreting hot luminescence transitions.

The phonon wing has its characteristic shape and often rather clear pronounced details of structure offering valuable information on the local dynamics of the lattice in the vicinity of the impurity, i.e., on the lattice vibrations of the host crystal and on their perturbation by the impurity, on local and pseudolocal vibrations, on librational vibrations and rotations of the impurity.

No-phonon lines were observed and interpreted (but without paying necessary attention to their new quality as narrow spectral lines in solids in the presence of the electron-phonon coupling) in a number of low-temperature spectra of atomic and ionic impurity centres in crystals (4)-(8), especially in the case of rare-earth impurities where the electron-phonon coupling is very small (8). The characteristic temperature behavior of no-phonon lines was not noticed and investigated* before the Mössbauer effect was discovered and the very remarkable analogy between it and the vibronic optical transitions was established (1)-(5). The paper by Gross et al. (2) was the first one in which the typical dependence on temperature --- a rapid decrease of the integrated intensity of the no-phonon (pure electronic) line with increasing temperature without a considerable broadening --- was demonstrated experimentally in the example of the I_1 -centre in CdS crystals (Fig.1).

The best systems for investigating vibronic no-phonon lines and the structure of phonon wings, and for obtaining very detailed information about the electronic states, local dynamics and processes in the centre of luminescence turned out to be small molecular ions O_2^-, S_2^-, NO_2^- and their analogues inserted in alkali halide crystals as impurity centres (see the article by K.K. Rebane and L. A. Rebane in this book).

I.B. Infrared Absorption Spectra

The infrared absorption lines, caused by high-frequency local vibrations in impurity centres, have been well-known for a couple of decades. The U-centre, i.e., the H^- ion in the anion position in alkali halides is one of the best investigated systems (10),(11). Another group of well-investigated systems are the same alkali halides activated by molecular ions OH^-, SH^-, NO_2^-, NO_3^- and the

*In a theoretical paper of Krivoglaz and Pekar (9) the main formula was obtained, but not enough attention was paid to the interpretation.

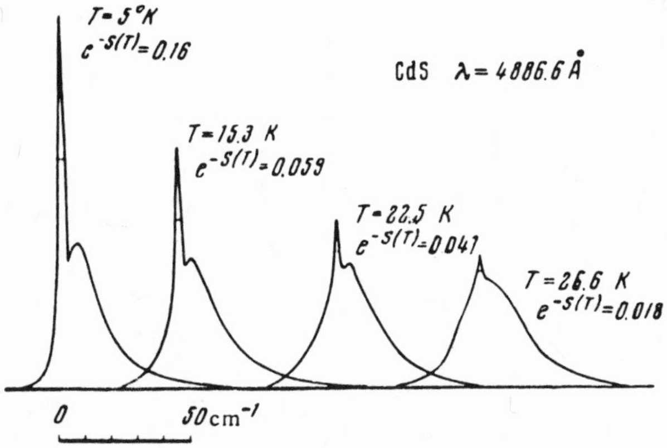

Fig. 1 Experimental data of Gross et al. (2) on the temperature dependence of the spectral region of CdS luminescence near the line with λ = 4888.6 Å. Intensity of the purely electronic line decreases rapidly with temperature, but the width of this line remains unchanged.

others having considerable dipole moments and producting therefore a rather strong impurity infrared absorption (12).

In 1965 Fritz (10) found that in the low-temperature spectra of absorption lines which are rather sharp ($\Delta\omega$ being of the order of 1 cm^{-1} and less) are accompanied by broad sidebands (tens of cm^{-1}) having properties characteristic of the phonon wings in the vibronic absorption spectra. The phonon wings at the infrared absorption lines of local vibrations of impurities are caused by the anharmonic coupling terms between the local vibration and the lattice modes, and as that coupling is weak the wings are very weak too. The same reason leads to the fact that the broadening of the no-phonon line (instead of the decrease of the integrated and peak intensities) is the dominant effect of increasing the temperature of the crystal.

The "double adiabatic approximation" (13)-(14) offers a very simple theoretical pattern for understanding the "no-phonon line - phonon wing" situation in the infrared spectrum. In the double adiabatic approximation the vibrations of the crystals are considered for the second time* on the basis of the adiabatic approximation:

*For the first time the Born-Oppenheimer adiabatic approximation is applied to separate the electron state from the vibrational ones.

the high-frequency local mode is considered to be a fast sub-system which establishes its quasi-stationary states for each momentary configuration of the slow sub-system of lattice modes. So a full analogy is achieved between the vibronic spectra and the infrared absorption spectra of high-frequency local vibration, the states and potential curves of the local vibration taking the place of the electronic states and the corresponding potential curves (surfaces) in the ordinary picture of vibronic transitions. The characteristic feature of the potential curve picture in the double adiabatic approximation is that the shifts and the alteration of the shape of the potential curves for different quantum states of the local vibration are very weak. So the "no-phonon line - phonon wing" picture of infrared high-frequency local mode impurity absorption is that of extremely small Stokes' losses and Debye-Waller factors almost equal to unity, respectively, and that is the reason why the phonon wings are weak.

I.C. Light Scattering Spectra

In light scattering spectra the no-phonon line is the Rayleigh line - the light scattered without any changes in frequency; i.e., it has exactly the frequency of the exciting light (elastic scattering We have to distinguish between two different kinds of phonon wings: the phonon wing at the Rayleigh line and the one at the Raman lines of high-frequency local vibrations.

The phonon wing at the Rayleigh line represents mainly the first order Raman scattering spectrum. In fact, it is a superposition of various first order spectra from the host lattice and a great variety of intrinsic defects and impurities in the crystal. When the first order Raman spectrum of the lattice vibrations of the host is not forbidden it makes an overwhelming contribution to the phonon wing. If this is not the case, and the first order spectrum of the host is forbidden (as in the case of alkali halides) the phonon wing at the Rayleigh line is actually a mix-up of the second order scattering spectrum of the undisturbed host crystal and several first order spectra of various areas of the crystal disturbed by different impurities or defects. In that case one must be very careful in interpreting the experimental structure of the wing.

The phonon wing at the Raman line of the high-frequency local vibration is a result of at least the second order scattering processe (a local mode quantum and at least one lattice mode must be simultaneously created or annihilated). So there are three factors leading to extremely low intensities of those phonon wings: 1) light scattering, especially the non-resonant scattering, is a weak process 2) the concentration of impurities in crystals is low; 3) the second-order process in lattice vibrations is required. On the other hand, a phonon wing of this kind is very individual, carrying information about the impurity centre to which the local mode belongs.

The phonon wings at the Rayleigh line were investigated in numerous papers (see (15)). Sometimes they offer information about the local dynamics in impurities (see, for instance, (16)). The second order individual spectrum offers information about librations and rotations of the impurity (see (17) and the article by K. K. Rebane and L. A. Rebane in this book).

I.D. Hot Luminescence Spectra

Phonon wings were observed in the hot luminescence spectra of the NO_2^- molecular ion in alkali halide hosts (18) (see Fig. 2). That offered additional help in the interpretation of the spectra. The fact that the shapes of the phonon wings do not differ considerably from those of ordinary luminescence may serve as an experimental evidence of the rapid establishment of thermal equilibrium between the lattice modes well before the next vibrational relaxation step of the local mode takes place.

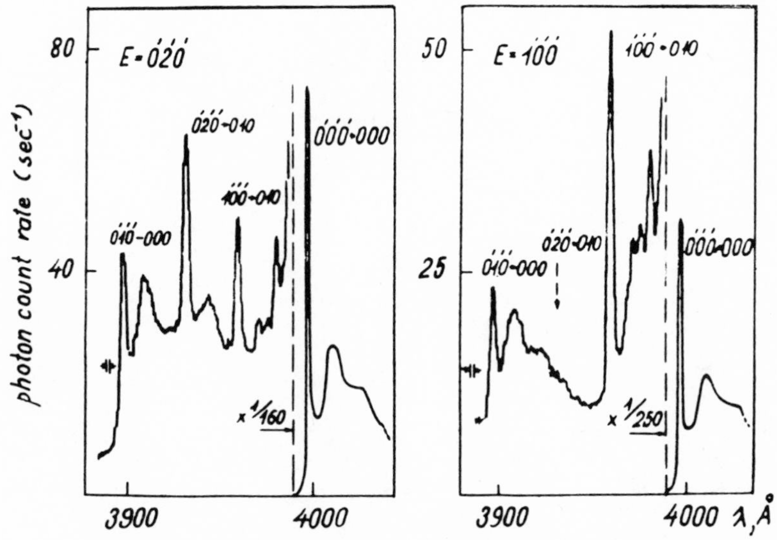

Fig. 2 Hot luminescence spectrum of KCl-NO_2^- crystal at 4.2°K as compared with the purely electronic line of ordinary luminescence is given in two cases of the exciting frequency E: E(1'0'0') < E(0'2'0') causing the removal of the 0'2'0' → 010 line from the HL spectrum and appearance of the phonon sideband of the 0'1'0'→000 line (18).

I.E. Shpolsky Spectra

A few years ago there existed problems connected with the nature of Shpolsky spectra --- the sharp vibronic absorption and luminescence lines of large molecules (mainly of aromatic compounds) inserted in suitable solid matrices (mainly paraffins). The theoretical work (3) was undertaken to interpret the Shpolsky lines as typical vibronic no-phonon lines (see also (4)) in case of a small or medium coupling between the electronic transition and lattice modes (small or moderate Stokes' shifts). But there was no experimental evidence of the existence of phonon wings. The problem arose whether all the numerous Shpolsky systems have extremely small Stokes' shifts or the theory was wrong.

When experiments were designed specially to clear up the situation it was found that very typical "no-phonon line - phonon wing" pictures may be obtained for the perylene molecules frozen in normal hexane at low temperatures (4.2°K) (see (19) and Fig. 3). After that the phonon wings were found for a number of other Shpolsky systems as well (20). But there remained another and very important problem for the practical spectroscopy of big molecules - why can we not have clear-cut no-phonon lines and characteristic phonon wings in the low-temperature luminescence spectra of every (or almost every) molecule, including rather complicated big molecules, inserted into solid matrices, if only the Stokes' losses are not too big? It turns out to be first of all the problem of homogeneity of the impurity molecule system in a solid matrix.

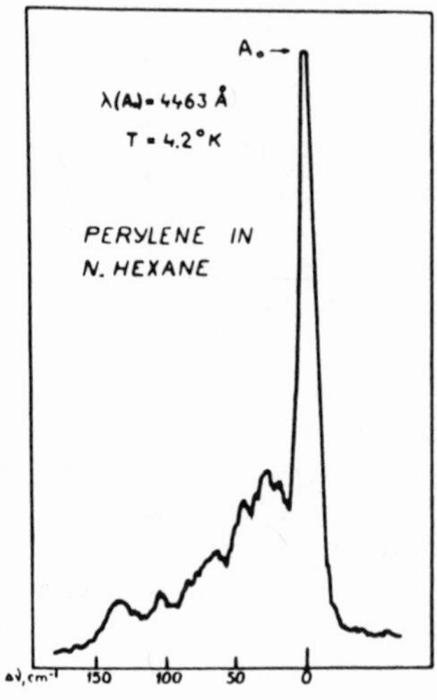

Fig. 3 One of the no-phonon lines of the 0-0 "multiplet" of the fluorescence spectrum of the perylene molecule frozen in n-hexane and the phonon sideband at the no-phonon line at 4.2°K (19).

II. INHOMOGENEOUS BROADENING OF LUMINESCENCE SPECTRA

Let us turn again to Fig. 3 which represents a very typical "no-phonon line - phonon wing" impurity luminescence spectrum. This particular picture belongs to a system of rather big molecules but its main features are the same for a number of various impurity systems (ionic, atomic, small molecules in alkali halides, etc.) provided the coupling of electron transition to crystal modes is of the same order of magnitude.

The halfwidth of the no-phonon line, which is about 4 cm^{-1}, is caused by inhomogeneous broadening (3), (4). The molecules are arranged in the matrix in slightly different conditions (different distances to defects and other impurity molecules, different stress-fields, and so on) and this causes slight differences (up to some wave-numbers) in the energies of electronic transitions.*

It is important that the homogeneous ("real") widths of no-phonon lines may be much smaller; in particular, in the case of pure-electronic no-phonon lines when the real halfwidths are, in principle, determined by the radiation lifetime and may be 0.01 to 0.001 cm^{-1} or even smaller.

Recently Szabo (21) eliminated the inhomogeneous broadening of a pure-electronic no-phonon line by exciting the luminescence of ruby by means of a narrow line of a ruby laser, and got the half-width of the no-phonon line of the order of 0.01 cm^{-1}. It was homogeneous broadening interpreted as being caused by nuclear spin interactions.

Personov et al. (22) used cadmium laser excitation in the pure-electronic line of the perylene molecule inserted in some matrices where the Shpolsky effect does not take place, that is, where perylene molecules have only broad-band spectra without any vibronic structure. The wavelength of the pure-electronic 0-0 line of pery-

*It should be noted that the Shpolsky systems may have,besides the impurity molecules inserted homogeneously in "Shpolsky positions", a considerable part of impurity molecules of the same kind introduced into the matrix rather inhomogeneously In (26) it was reported that in the frozen solutions of perylene in n-hexane (the best Shpolsky system for perylene) only a part (and by far the smaller one) of the perylene molecules is placed in "Shpolsky positions", the other part is distributed rather inhomogeneously having the no-phonon lines broadened by tens of cm^{-1} and causing the broad background in absorption and peculiarities in the secondary radiation (mainly luminescence) spectra.

lene is known from the systems where the Shpolsky effect takes place. It turns out that the wavelength of the cadmium laser line (λ = 4415.6 Å, linewidth 0.05 cm^{-1}) falls into the 0-0 line region of perylene. Under the laser excitation in resonance with the pure-electronic line the results were rather remarkable. Such an excitation seems to open new ways for getting nice "no-phonon line - phonon wing" luminescence spectra for a great variety of molecular systems.

In Fig. 4 the luminescence spectra of perylene, inserted into ethanol are given. The upper spectrum corresponds to ordinary excitation. We can see that no linear structure is presented and the spectrum carries little information about the molecule and the interactions.

The lower spectrum of luminescence occurs under cadmium laser excitation. The same system gives now a prominent liner (or, more exactly, quasilinear (4)) spectrum with a number of no-phonon lines. We can see that in the case of laser excitation we can get much more informative spectra of molecular systems. The linewidths are about 1 cm^{-1} and even smaller.

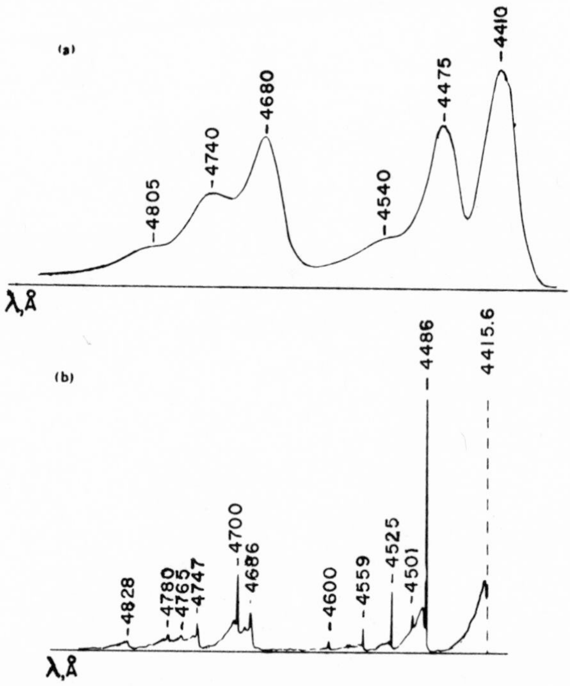

Fig. 4 Perylene fluorescence spectra in n-undecane at 4.2°K: (a) usual excitation (λ = 365 nm, DPW 1000), (b) laser excitation (λ = 4415.6 Å) (Personov et al. (22)).

To prove that they really had no-phonon lines here Persoņov et al (22) carried out temperature dependence measurements (see Fig. 5). The temperature dependence of the intensity which is characteristic of no-phonon lines is very clearly observable. The intensity of the sharp line decreases rapidly with increasing temperature; at $40^{o}K$ the lines are not observable any more.

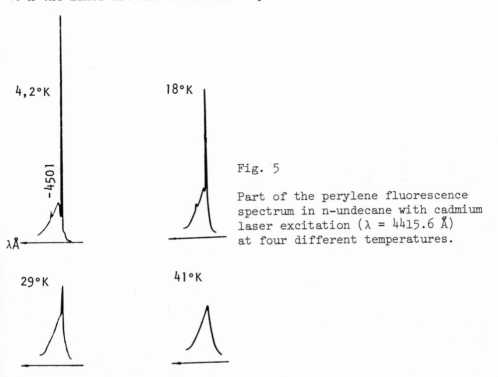

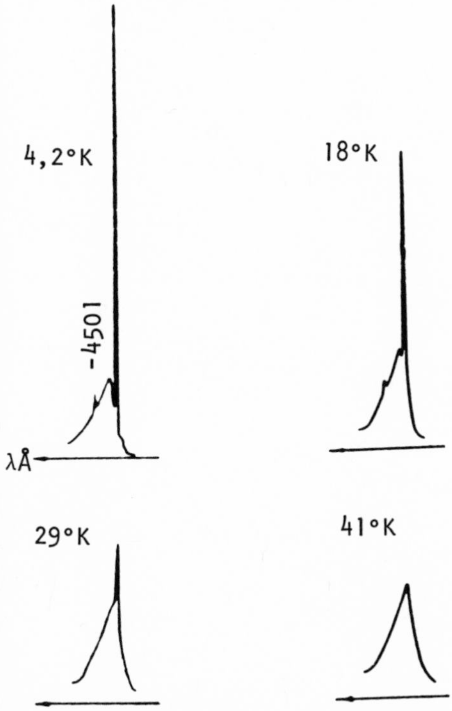

Fig. 5

Part of the perylene fluorescence spectrum in n-undecane with cadmium laser excitation (λ = 4415.6 Å) at four different temperatures.

As the inhomogeneous broadening of no-phonon lines in an arbitrary solid matrix may be tens and even hundred of cm^{-1}, it is possible to get laser excitation in the pure electronic lines of other systems. Some of them (phenanthrene in ethanole, 3, 4, 8, 9-dibenzpyrene, chlorophyll b) were investigated by Personov et al. (22), (23). In the paper (24) the results of obtaining sharp vibronic lines in the luminescence spectra of chlorophyll molecules frozen in ether, toluol, pyridine, ethanole, benzene and acetone matrices are reported.

The interpretation of the described results is the following.

Let us have typical no-phonon lines and phonon sidebands of the impurities which have (due to the inhomogeneities of the matrix) slightly different frequencies of electronic transitions. It is easy to see that in such a case the summing up of the spectra of different impurity molecules leads to the increase of the total intensity of the phonon sideband proportional to the number of molecules N*. The broadening caused by inhomogeneities is small as compared with the width of the phonon wing. In the region of no-phonon lines (or other sharp lines) the main effect is the broadening of the total line, whereas the peak intensity increases with the number of molecules much slower than the intensity of phonon wings (proportional to αN, where α is the homogeneous width of the line divided by the inhomogeneous one). This is the reason why under the ordinary excitation (spectral width largely exceeding the widths of spectral lines) of inhomogeneous impurity systems the absorption and luminescence spectra are without sharp lines even at helium temperatures.

Now, when the sharp excitation line falls into the no-phonon line (the linewidth of excitation being many times smaller than the total inhomogeneous linewidth of the no-phonon line) the picture of the luminescence spectrum is entirely changed: only a small number of the impurities gets excited, but all of the excited centres have the same frequency as the no-phonon line and that results in a cancelling of the broadening caused by inhomogeneities. The peak intensity of the no-phonon line increases now as βN, where β is the homogeneous width of the no-phonon line divided by the spectral width of excitation (only if the latter exceeds the first).

A more detailed theoretical treatment of the problem was given recently by Avarmaa (25). I would like to stress one of his results: it turns out that under certain and not very special circumstances there may be an additional (sometimes rather sharp) maximum ("pseudoline") in the luminescence spectrum caused entirely by inhomogeneities, which does not exist in the real spectrum of a single luminescence centre.

The effect of the appearance of sharp lines and other details of vibronic structure in the luminescence spectra under laser excitation opens new and promising possibilities for fine spectroscopic investigations of molecules (including rather complicated ones) inserted as impurities into different crystalline and glassy matrices. It may turn out that there exist only a few big molecules which do not give structural vibronic spectra in luminescence and under all circumstances have only broadband vibronic spectra.

---

*To be more precise: it is exactly so if one speaks about the absorption coefficient and it remains sufficiently correct for luminescence when the absorption in the sample is small. The intensity of luminescence depends on the quantity of excitation energy absorbed by certain kinds of luminescence centres, and non-linearities occur in the case of strong absorption.

REFERENCES

1. E. D. Trifonov, Dokl. Akad. Nauk SSSR 147, 826 (1962).

2. E. F. Gross, B. S. Razbirin and S. A. Permogorov, Dokl. Akad. Nauk SSSR 147, 338 (1962).

3. K. K. Rebane and V. V. Hizhnyakov, Opt. i Spectr. 14, 362 (1963); K. K. Rebane, Opt. i Spectr. 16, 594 (1964).

4. K. K. Rebane, Impurity Spectra of Solids, Elementary Theory of Vibrational Structure, Plenum Press, New York - London (1970).

5. R. H. Silsbee and D. B. Fitchen, Rev. Mod. Phys. 36, 433 (1964); J. J. Hopfield, Proc. Internat. Conf. Semicond. Phys., Exeter (1962), p. 75.

6. K. K. Rebane, N. N. Kristoffel, E. D. Trifonov and V. V. Hizhnyakov, Isv. Akad. Nauk ESSR, ser. fiz.-mat. i tekhn. nauk 13, 87 (1964).

7. A. A. Maradudin, Repts. Prog. Phys. 28, 331 (1965); Scientific Paper 65-9F5-442-P5, Westinghouse Research Laboratories, Pittsburgh (1965).

8. M. A. El'yashevich, Rare Earth Spectra, Gostekhizdat (1953).

9. M. A. Krivoglaz and S. I. Pekar, Trudy Inst. Fiziki Akad. Nauk Ukr. SSR, No. 4, 37 (1953).

10. B. Fritz, J. Phys. Chem. Sol. Suppl. 1, 485 (1965); B. Fritz, V. Gross, D. Bauerle, Phys. Stat. Sol. 11, 231 (1965).

11. T. Timusk, M. V. Klein, Phys. Rev. 141, 664 (1966); H. Dötsch, S. S. Mitra, Phys. Rev. 178, 1492 (1968); R. W. Mac Pherson, T. Timusk, Can. J. Phys. 48, 2176 (1970).

12. M. A. Cundill, W. F. Sherman, Phys. Rev. 168, 1007 (1968); R. K. Eijnthoven, J. van der Elsken, Phys. Rev. Lett. 23, 1455 (1969); T. Mauring, Isv. Akad. Nauk ESSR, ser. fiz.-mat. nauk 20, 232 (1971); 21, 115 (1972).

13. K. Rebane, O. Sild, Izv. Akad. Nauk ESSR, ser. fiz.-mat. i tekhn. nauk 15, No. 2, 299 (1966); O. Sild, Izv. Akad. Nauk ESSR, ser. fiz.-mat. nauk 17, No. 2, 203 (1968).

14. I. P. Ipatova, A. A. Maradudin, A. V. Subashiev, Fiz. Tverd. Tela 11, No. 8, 2271 (1969); I. P. Ipatova, A. V. Subashiev, A. A. Maradudin, Ann. Phys. (USA) 53, No. 2, 376 (1969).

15. "Light Scattering Spectra of Solids," Proceedings of the International Conference on Light Scattering Spectra of Solids, ed., G. B. Wright, Springer-Verlag, New York, Inc. (1969).

16. K. K. Rebane, L. A. Rebane, T. J. Haldre, A. A. Gorokhovski, Advances in Raman Spectroscopy, Vol. 1, p. 379, Heyden and Son, Ltd., London (1973).

17. L. A. Rebane, T. J. Haldre, A. E. Novik, A. A. Gorokhovski, Fiz. Tverd. Tela 15, 3188 (1973).

18. K. K. Rebane, P. M. Saari, T. H. Mauring, Luminescence in Crystals, Molecules and Solutions, Proc. of the Int. Conf. on Luminescence held in Leningrad, USSR, August 1972. Ed. by Williams, Plenum Press, New York - London (1973), p. 690.

19. K. Rebane, P. Saari, T. Tamm, in "Fifth Molecular Crystals Symposium, Program," University of Pennsylvania, Philadelphia (1970), p. 121; Izv. Akad. Nauk ESSR, ser. fiz.-mat. nauk 19, 251 (1970).

20. R. I. Personov, I. S. Osad'ko, E. I. Alshits, Fiz. Tverd. Tela 13, 2653 (1971).

21. A. Szabo, Phys. Rev. Lett. 27, 323 (1971).

22. R. I. Personov, E. I. Alshits, L. A. Dykovskaya, Pisma v JETP 15, 609 (1972); Optics Commun. 6, 169 (1972).

23. R. I. Personov, E. I. Alshits, L. A. Dykovskaya, B. M. Kharlamov, Zh. Eksper. Theor. Fiz. 65, 1825 (1973).

24. R. A. Avarmaa, Izv. Akad. Nauk ESSR, ser. fiz.-mat. nauk 23, 89 (1974).

25. R. A. Avarmaa, Izv. Akad. Nauk ESSR, ser. fiz.-mat. nauk (to be published).

26. K. Rebane, T. Tamm, XI European Congress on Molecular Spectroscopy, Tallinn USSR, May 28-June 1 (1973), Abstracts, Academy of Sciences of Estonian SSR, Abstract No. 172.

SPECTROSCOPY OF MAGNETIC INSULATORS

Donald S. McClure

Department of Chemistry, Princeton University

Princeton, New Jersey 08540

ABSTRACT

The basic principles of electron exchange are covered first and applied to optical transitions in ion pairs. The Frenkel exciton theory of molecular crystals is developed and applied to magnetic crystals. The electronic and magnetic structure of magnetic insulators is reviewed, including aspects of most importance to the optical properties of these substances. Finally, a review of several systems such as MnF_2 is given.

I. INTRODUCTION

There are new features in the absorption spectra of a collection of interacting absorbers which do not appear when the absorbers are not interacting. The simplest examples might be molecular crystals. A single excited molecule would have a sharp electronic absorption line, but a crystal could have several lines at the approximate position of the molecular transition, depending on the structure of the unit cell. Each of the crystal lines is a part of a quasi-continuous band of energy levels. These bands are called Frenkel-exciton bands and the multiplet arising from a single molecular line is called Davydov splitting.

Additional features occur in the spectra of crystals having magnetic ground states. The ground state in these crystals is degenerate because of the multiplicity of spin orientations possible, and when electron exchange between the atoms is possible, the degeneracy is spread out into a quasi-continuous band of states. The excitations of the ground state are called spin-waves.

The optical transitions in magnetic crystals show, in addition to the usual phonon sidebands, spin-wave sidebands. These bands occur at energies characteristic of the spin-waves at the edge of the first Brillouin zone.

Both excitons and spin waves have a dispersion, or a dependence of their energies (or frequencies) on wavelength (or wavenumber). The combined dispersions of these two types of excitation produce spectral band shapes from which interesting information can be extracted.

The transition probabilities in a magnetic crystal are also different from those of non-magnetic crystals. New types of transitions become possible as a result of the exchange coupling between ions in a magnetic crystal.

Research on these systems has been under way for about ten years and in that time many of the basic ideas have been established, but there are many examples which have not been analyzed. We can expect new knowledge about excited state exchange and anisotropic exchange, energy transfer and the Jahn-Teller effect from a study of these cases.

II. INTERACTING IONS

In order to understand the problems of the spectroscopy of magnetically ordered crystals, we first examine the problem of a pair of interacting atoms. We will do this in three stages:

1. Non-overlapping atoms
2. Overlapping atoms, no spin
3. Effect of adding spin.

II.A. Two Interacting Electrons

As a zeroth order problem, consider two electrons, each in separate boxes separated by a great distance. If the box walls are infinitely high, the wavefunctions of the electrons cannot overlap and electron 1 stays in box A, electron 2 in box B. The two electrons are distinguishable and have separate Schrödinger equations.

$$\mathcal{H}_A A^i(1) = E_A^i A^i(1)$$

$$\mathcal{H}_B B^j(2) = E_B^j B^j(2) \qquad (1)$$

Here i, j denote states of the separate systems. The systems may be thought of as identical or not and both cases will be considered.

1. Non-Overlapping Wavefunctions. Even if the wavefunctions do not overlap, the Coulomb interaction can cause the two systems to interact via the Coulomb force. To handle this situation, it is convenient to define a Hamiltonian

$$\mathcal{H} = \mathcal{H}_A(1) + \mathcal{H}_B(2) + e^2/r_{12} \tag{2}$$

A basis set of wavefunctions consisting of products of the wavefunctions at A and B provides a good approximation to the solution of (2) when the interaction is weak. Thus the basis is

$$A^0B^0,\ A^0B^1,\ A^1B^0,\ A^0B^2 \ldots A^iB^j \ldots \tag{3}$$

where o means lowest state, and successively higher unperturbed states are 1, 2, etc. If A and B have the same potential, then A^iB^j and A^jB^i are degenerate. The energy levels are obtained from a secular equation

$$\begin{vmatrix} K_{00} + E_A^0 + E_B^0 - E & C_{00}^{01} & C_{00}^{10} \\ C_{00}^{01} & K_{01} + E_A^0 + E_B^1 - E & X_{01} \\ C_{00}^{10} & X_{10} & K_{10} + E_A^1 + E_B^0 - E \end{vmatrix} = 0 \tag{4}$$

where

$$K_{ij} = \langle A^i(1)\, B^j(2)\, |\frac{e^2}{r_{12}}|\, A^i(1)\, B^j(2)\rangle$$

$$X_{ij} = \langle A^i(1)\, B^j(2)\, |\frac{e^2}{r_{12}}|\, A^j(1)\, B^i(2)\rangle$$

$$C_{ij}^{k\ell} = \langle A^i(1)\, B^j(2)\, |\frac{e^2}{r_{12}}|\, A^k(1)\, B^\ell(2)\rangle \quad .$$

The most important parts of eq. 4 are the K integrals, called Coulomb integrals, and the X integrals, called excitation exchange integrals. If $E_A^i = E_B^i$, then the first excited state will be split by an amount $2X_{10}$. Thus there can be exchange of excitation energy without overlap or electron exchange. It is caused by the Coulomb interaction. This is about the only interesting thing that happens without overlap.

If we add spin, nothing much of interest happens in this case. Electrons 1 and 2 can have either spin, but the Coulomb interaction is entirely insensitive to spin as can be seen by reducing a Coulomb integral. For example:

$$K_{01} = \int A^0(1)^* \alpha(1)\ B^1(2)^* \beta(2)\ \frac{e^2}{r_{12}}\ A^0(1)\alpha(1)\ B^1(2)\beta(2)$$

$$\times\ dv_1\ dv_2\ d\sigma_1\ d\sigma_2 \quad .$$

The spin integrations factor out since there are no spin-dependent parts of the Hamiltonian

$$\int \alpha(1)\ \alpha(1)\ d\sigma_1 = 1, \text{ etc.}$$

and the integral is left unchanged.

If we have two electrons per system, we can have singlet and triplet states when spin is considered, and in this case excitation energy, involving the singlet ground state, can be transferred only through the singlet states.

If the interaction remains weak, the C-terms don't do much, but even a small X can have spectacular effects since energy transfer can occur if $\hbar/X$ is less than the lifetime of the state.

In the presence of spin-orbit coupling, singlets and triplets on one atom become mixed and weak excitation exchange splitting can occur in the triplet levels.

To summarize, if electron exchange is prevented, there is still the possibility of excitation exchange, and there are effects of spin when the energy between different spin states is large enough (i.e., so that experiments can be performed which distinguish properties of the different states).

2. Finite Overlap. If we now let the barrier between the two systems become finite, the two electrons can interchange by tunneling. The wavefunctions of the system would have to be written to show that over a long period of time, either electron could be in either box with equal probability. Therefore, we have to give equal weight to the two distributions

$$A(1)\ B(2) \quad \text{and} \quad A(2)\ B(1) \quad . \tag{5}$$

Normalized probability amplitudes which do this are

$$\frac{1}{\sqrt{2}}\left[A(1)\ B(2) \pm A(2)\ B(1)\right] \tag{6}$$

giving a probability distribution

$$\frac{1}{2}\left[A^2(1)\ B^2(2) + A^2(2)\ B^2(1)\right] \pm A(1)\ B(2)\ A(2)\ B(1) \tag{7}$$

If the distributions A^2 and B^2 do not overlap appreciably, the $\pm$ term is zero. Then the sum of the probability amplitudes is physically the same as in the previous example, but the electrons are not distinguishable by their location any longer. The physical meaning of the $\pm$ sign will become clearer when the energy is calculated with the wavefunctions (6).

The Hamiltonian must now be written so that the potential at both centers can act on each electron

$$\mathcal{H} = T(1) + T(2) + V_A(1) + V_A(2) + V_B(1) + V_B(2) + \frac{e^2}{r_{12}} \tag{8}$$

where T = kinetic energy, V = potential energy.

The basis set for the interacting two-electron problem is the same as in (3) except that the members must be made symmetric or antisymmetric under electron exchange as in (6). The matrix element for the ground state becomes

$$\mathcal{H}_{00} = K_{00} \pm J_{00} \tag{9}$$

where $$J_{00} = \langle A^0(1)\ B^0(2)\ |\frac{e^2}{r_{12}}|\ A^0(2)\ B^0(1)\rangle \tag{10}$$

and $$K_{00} = \langle A^0(1)\ B^0(2)\ |\frac{e^2}{r_{12}}|\ A^0(1)\ B^0(2)\rangle\ . \tag{11}$$

(Note that there are also contributions to J and K from one-center terms such as $\int A(1)\ B(1)\ V_B(1)\ dv_1 \int A(2)\ B(2)\ dv_2$.)

There are two "ground-state" energies depending upon the symmetry of the wavefunction. The integral J_{00} is called an exchange integral, but to distinguish it from excitation exchange, we should call it an electron exchange integral.

The diagonal matrix element of a singly excited state

$$\left[A^i(1)\ B^o(2) \pm A^i(2)\ B^o(1)\right]/\sqrt{2} \tag{12a}$$

is

$$\mathcal{H}_{io} = K_{io} \pm J_{io} \ . \tag{12b}$$

The definition of the excited state integrals is analogous to the ground state.

The excited state with excitation interchanged is

$$\left[A^o(1)\ B^i(2) \pm A^o(2)\ B^i(1)\right]/\sqrt{2} \tag{12c}$$

and would have a diagonal matrix element

$$\mathcal{H}_{oi} = K_{oi} \pm J_{oi} \ . \tag{12d}$$

Finally, there must be an excitation exchange process which would split these two types of states. The off-diagonal matrix element between states (12a) and (12c) of the same electron exchange symmetry is

$$\mathcal{H}_{oi,io} = X_{oi} \pm Q_{oi} \tag{13}$$

where

$$X_{oi} = \langle A^o(1)\ B^i(2)\ |\mathcal{H}|\ A^i(1)\ B^o(2)\rangle \tag{14}$$

$$Q_{oi} = \langle A^o(1)\ B^i(2)\ |\mathcal{H}|\ A^i(2)\ B^o(1)\rangle \ . \tag{15}$$

The part of the matrix involving two excited states having the same electron interchange symmetry is

$$\begin{vmatrix} K_{io} \pm J_{io} - E & X_{oi} \pm Q_{oi} \\ X_{oi} \pm Q_{oi} & K_{oi} \pm J_{oi} - E \end{vmatrix} = 0 \ . \tag{16}$$

When the potentials V_A and V_B are the same, then $2\left(X_{oi} \pm Q_{oi}\right)$ is the splitting caused by the combined excitation and electron exchange.

The four integrals can be compared in terms of the type of exchange they represent:

	Electron Exchange	Excitation Exchange
K	0	0
J	+	0
X	0	+
Q	+	+

Thus when the potentials are identical, the splitting of an excited atom depends both on the electron exchange and on the excitation exchange. As the two potentials become different, the importance of excitation exchange diminishes.

3. Effect of the Spin. Now introduce the spin and the principle that the wavefunction must be antisymmetric with respect to interchange of identical particles. Eq. 6 for the wavefunctions of two different exchange types is now to be expanded to:

Singlet $\frac{1}{\sqrt{2}}\left[A(1)\,B(2)+A(2)\,B(1)\right]\frac{1}{\sqrt{2}}\left[\alpha(1)\,\beta(2)-\alpha(2)\,\beta(1)\right]$

$\frac{1}{\sqrt{2}}\left[A(1)\,B(2)-A(2)\,B(1)\right]\frac{1}{\sqrt{2}}\left[\alpha(1)\,\beta(2)+\alpha(2)\,\beta(1)\right]$

Triplet " $\left[\alpha(1)\alpha(2)\right]$

" $\left[\beta(1)\,\beta(2)\right]$. (17)

The other four combinations have to be thrown away. The singlets and triplets do not interact in the absence of spin-dependent forces, and so the matrix elements (10), (11), (14), (15) will apply to the singlet when the sign is + and to the triplet when the sign is - .

For the ground state we have

$$E_S = K_{oo} + J_{oo}$$

$$E_T = K_{oo} - J_{oo} \quad . \tag{18}$$

One can now make the following somewhat inaccurate statement:

If J is positive, then the ground state is "ferromagnetic," while if it is negative, the ground state is "antiferromagnetic." Terms involving magnetic interactions should strictly be reserved for cooperative systems, however, and not be applied to two atoms. Extending the reasoning to many atoms will lead to the magnetic state of the crystal.

The requirements of antisymmetric wavefunctions for electrons resulted in the correlation of spin to interchange symmetry as has just been shown. To make this correlation even more graphic, Dirac invented a spin operator which gives the eigenvalues of eq. 18 and enables us to avoid using the antisymmetrized wavefunctions of eq. 17. This is

$$\mathcal{H} = K_{oo} - J_{oo}\left(\frac{1}{2} + 2\,s_i \cdot s_j\right). \qquad (19)$$

This works because $S^2 = s_i^2 + 2\,s_i s_j + s_j^2$ and $S^2 = \underline{S}(\underline{S}+1)$, $s^2 = \frac{1}{2}\left(\frac{1}{2}+1\right) = \frac{3}{4}$, and $S = 0,1$ for singlet, triplet.

In a sense, the antisymmetry principle produces a powerful spin-spin interaction, and this leads to the possibility of magnetic ground states. The direct magnetic dipole-dipole interaction between spins is far weaker than the exchange energy in almost all cases.

The Dirac operator can be used for any number of electrons. For example, if one electron interacts with a closed shell of two electrons we have

$$E = 2K - J_{a1}\left(\frac{1}{2} + 2\,s_a \cdot s_1\right) - J_{a2}\left(\frac{1}{2} + 2\,s_a \cdot s_2\right).$$

Since $s_1 + s_2 = 0$ in a closed shell and $J_{a1} = J_{a2}$, the energy becomes

$$E = 2K - J.$$

The energy matrix including everything can now be visualized. The ground state part is eq. 18, which is really the same as eq. 9 except that the spin of the states has now been identified. An excited state block is like eq. 16. If we have identical potentials, the energies in this 2 x 2 block would become

$$E\,(\text{singlet}) = K_{oi} + J_{oi} \pm \left(X_{oi} + Q_{oi}\right)$$

$$E\,(\text{triplet}) = K_{oi} - J_{oi} \pm \left(X_{oi} - Q_{oi}\right). \qquad (20)$$

Thus if $J > 0$, the antisymmetric or triplet state is lower; also its excitation exchange splitting differs from that of the symmetric or singlet state. These energies and that of the ground state are shown in Fig. 1.

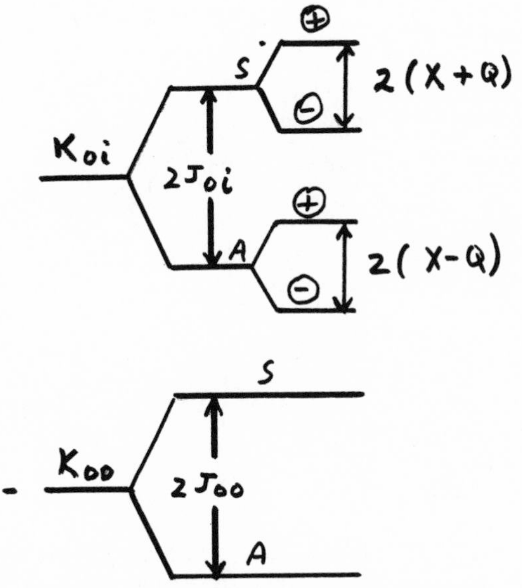

Figure 1

Electron exchange splitting of ground state (below) and exchange or Coulomb splitting of singly excited state of a pair of atoms (above). S, A refer to electron interchange symmetry, +, - to excitation exchange symmetry.

4. Biquadratic Terms. We have ignored the C matrix elements in eq. 4 up to this point, noting only that they can have only a small effect on the ground state. Their contribution to the ground state energy can be displayed by a perturbation expression

$$E_S = K + J - C^2/\Delta E; \qquad E_T = K - J - C^2/\Delta E \,. \qquad (21)$$

In order for an excited state to be able to perturb the ground state, the matrix element C must have certain spin orientations. Explicitly,

$$C^i = \langle A^o(1)\,B^o(2)\,|\mathcal{H}|\,A^o(1)\,B^i(2)\rangle \pm \langle A^o(1)\,B^o(2)\,|\mathcal{H}|\,A^o(2)\,B^i(1)\rangle \,. (22)$$

The second, or exchange component of C, will have the usual spin dependence, and when expressed in terms of the Dirac operator, we have

$$C = K^{oi}_{oo} - \frac{1}{2} J^{oi}_{oo} \left(1 - 4\, s_A \cdot s_B\right) \,. \qquad (23)$$

Now C^2 has a constant part which only shifts the levels, a spin-dependent $s_A \cdot s_B$ term which can be added to the $J\, s_A \cdot s_B$ term from the ground state interaction giving $J' s_A \cdot s_B$, and a term in $\left(s_A \cdot s_B\right)^2$ which is something new. This is a biquadratic exchange term and will modify the structure of the exchange multiplet in the spectra of interacting many-electron ion pairs. It is still an isotropic term however.

Actually this biquadratic term has been shown to be smaller than a term with the same form arising from magnetostrictive exchange.

II.B. Many-Electron Problems (1)

Within the atom or ion a many-electron wavefunction is written as a Slater determinant, D, or as a linear combination of these. Each determinant is equivalent to a sum of N products of the form

$$\pi = u_a(1)\, \chi_{\sigma_a}(1)\, u_b(2)\, \chi_{\sigma_b}(2) \ldots u_h(N)\, \chi_{\sigma_h}(N) \,. \qquad (24)$$

Here we have shown one product in standard order. It turns out that matrix elements can be correctly written using a single product instead of a determinant if certain rules are followed. The $u_a(1)$, etc., are orthogonal to each other as are the $\chi_\sigma(1)$, etc. The matrix of a sum of one- and two-electron operators is needed

$$Q = \sum_i f_i + \sum_{i<j} g_{ij} \,. \qquad (25)$$

Because of the orthogonality, the matrix of f is trivial to calculate and the matrix of g is not much harder

$$\langle D'|Q|D\rangle = \left\langle \pi' \left| \sum_i f(i) \right| \pi \right\rangle + \left\langle \pi' \left| \sum_{i<j} g(i,j)(1-P_{ij}) \right| \pi \right\rangle . \quad (26)$$

If π and π' differ by more than one u or by any χ, the first term in (26) is zero, and if they differ by more than two, the second term is zero. The factor $(1-P_{ij})$ means that a matrix element with and without a permutation of electrons i and j is to be taken. A convenient form for this matrix element is

$$\langle D'|Q|D\rangle = \left\langle \bar{\pi}' \left| \sum_i f_{ii}(\gamma',\gamma) \right| \bar{\pi} \right\rangle + \left\langle \bar{\pi}' \left| \sum_{i<j} K_{ij}(\gamma',\gamma) \right| \bar{\pi} \right\rangle$$

$$- \left\langle \bar{\pi}' \left| \sum_{i<j} J_{ij}(\gamma',\gamma) \frac{1}{2}\left(1+4\,S_i\cdot S_j\right) \right| \bar{\pi} \right\rangle \quad (27)$$

where $\bar{\pi}$ means the spin product only. The f, K and J are the one-electron Coulomb and exchange integrals, respectively. The indices γ,γ' are just to remind us which determinants were involved in the matrix elements.

An effective Hamiltonian can be written for the case where the orbitals are of one given set, and the π's differ only by the spin assignments

$$\mathcal{H}_{eff} = \sum_i f_{ii} + \sum_{i<j} K_{ij} - \sum_{i<j} J_{ij} \frac{1}{2}\left(1+4\,S_i\cdot S_j\right) . \quad (28)$$

This is to be evaluated between spin function products $\bar{\pi}'$ and $\bar{\pi}$. For a given spin arrangement the result reproduces the matrix elements of Q between D and D'.

II.C. The Heisenberg Hamiltonian

A common generalization of eq. 28 is the Heisenberg Hamiltonian

$$\mathcal{H}_H = -J_{12}\, S_1\cdot S_2 \quad (29)$$

where S_1, S_2 are the total spins of two interacting ions rather than being the individual electron spins. This Hamiltonian is used to describe magnetic crystals and then is summed over the entire lattice.

The Heisenberg Hamiltonian provided a simplified interaction operator for describing magnetic ground states and made possible a great deal of progress in the understanding of magnetic materials. The tendency to regard this operator as being generally valid, however, actually interfered with the understanding of excited states of magnetic systems. It is important therefore for us to understand the conditions under which the Heisenberg Hamiltonian is valid.

For illustrative purposes, consider a pair of interacting two-electron atoms A and B for which we can write $S_A = s_1 + s_2$ and $S_B = s_3 + s_4$. The eigenfunctions of the pair belong to a total spin operator S where $S^2_{op}\,\psi = S(S+1)\,\psi$ and

$$S^2_{op} = S_A^2 + S_B^2 + 2\,S_A \cdot S_B = (s_1 + s_2 + s_3 + s_4)^2 \ .$$

Therefore we have

$$S_A \cdot S_B = s_1 \cdot s_3 + s_1 \cdot s_4 + s_2 \cdot s_3 + s_2 \cdot s_4 \ . \qquad (30)$$

The Heisenberg Hamiltonian, $J_{AB} S_A \cdot S_B$, therefore corresponds to having the same coefficient in front of every electron-electron interaction between atoms A and B.

For the case of a half-filled shell we can write each spin as $s_i = S/n$ where n is the number in the half-filled shell. Therefore an expression such as $\sum J_{ij}\, s_i \cdot s_j$ can be converted into

$$\sum J_{ij}\left(\frac{S_A}{n}\right)\left(\frac{S_B}{n}\right) = \left(\frac{1}{n^2}\sum J_{ij}\right) S_A \cdot S_B$$

and we have an effective exchange integral $J_{eff} = \frac{1}{n^2} \sum_{ij} J_{ij}$.

Thus the Heisenberg Hamiltonian is valid for this case even when the J_{ij}'s are not all the same.

When the Heisenberg Hamiltonian is valid, the eigenvalues of $S_A \cdot S_B$ give the correct energy spacings of the exchange levels. These can be obtained by working out the matrix elements of

$$S_A \cdot S_B = S_A^z S_B^z + \frac{1}{2}\left(S_A^+ S_B^- + S_A^- S_B^+\right)$$

or by the vector coupling formula

$$2\, S_A \cdot S_B = S(S+1) - S_A\left(S_A+1\right) - S_B\left(S_B+1\right)$$

where $\qquad S = S_A + S_B,\ S_A + S_B - 1 \cdots\cdots S_A - S_B\,.$

II.D. Direct Exchange or Superexchange?

We cannot go into a discussion of exchange mechanisms in such a brief article. So far we have proceeded as if the orbitals of the interacting atoms overlap directly. In ionic crystals this is seldom the case, as the strongest exchanges are usually those in which an intervening atom transmits spin information from one magnetic atom to another. This is the process called superexchange. The formalism of the Dirac operator or the Heisenberg Hamiltonian applies in either case, and only the magnitude of J is different in the two cases.

II.E. Transition Probabilities in Coupled Systems

Soon after the discovery of simultaneous transitions in pairs of coupled rare earth ions by Dieke and Varsanyi, Dexter (2) worked out a simple perturbation scheme to explain their existence. The existence of spin wave sidebands in electronic spectra is another example of simultaneous transitions in a pair of coupled ions, and Dexter's scheme applies to these also. Thus we can explain why a pair of ions absorb light at approximately the sum of their separate transition frequencies, and why forbidden transitions become much stronger in a magnetic crystal.

We can begin by considering pair functions as in eq. 3, and their unperturbed energies as shown in eq. 31.

$$\mathcal{H} = \mathcal{H}_A + \mathcal{H}_B + \mathcal{H}_{AB}, \quad \Psi = \psi_i \varphi_j \tag{31}$$

Types of States		Zero-Order Energy
ground	$\psi_o \varphi_o$	$E_o^o + \mathcal{E}_o^o$
A - exc.	$\psi_a \varphi_o$	$E_a^o + \mathcal{E}_o^o$
B - exc.	$\psi_o \varphi_b$	$E_o^o + \mathcal{E}_b^o$
A + B - exc.	$\psi_a \varphi_b$	$E_a^o + \mathcal{E}_b^o$

In the presence of exchange interactions between the ions A and B, the zero order states mix as given by the solution of secular equations such as (4) and (16). The corrected wavefunctions can be expressed accurately enough by first-order perturbation expressions

$$\chi_{oo} = \psi_o \varphi_o - \sum_{a' \neq o} \sum_{b' \neq o} \frac{\langle oo | H_{AB} | a'b' \rangle}{\delta_{a'} + \epsilon_{b'}} \psi_{a'} \varphi_{b'}$$

$$\delta_{a'} = E^o_{a'} - E^o_o \qquad \epsilon_{b'} = \mathcal{E}^o_{b'} - \mathcal{E}^o_o$$

$$\chi_{ao} = \psi_a \varphi_o - \sum_{a' \neq a} \sum_{b' \neq o} \frac{\langle ab | H_{AB} | a'b' \rangle}{\delta_{a'} - \delta_a + \epsilon_{b'}} \psi_{a'} \varphi_{b'}$$

$$\chi_{ab} = \psi_a \psi_b - \sum_{a' \neq a} \sum_{b' \neq b} \frac{\langle ab | H_{AB} | a'b' \rangle}{\delta_{a'} - \delta_a + \epsilon_{b'} - \epsilon_b} \psi_{a'} \varphi_{b'} \, . \qquad (32)$$

By using the corrected wavefunctions in the expression for the transition moment, one can derive the pair moment, M_{ab}, or the cooperative-transition moment

$$M_{ab} = \int \chi_{oo} \left(M_A + M_B \right) \chi_{ab} \, dv$$

$$= - \sum_{a' \neq o} \frac{\langle oo | H_{AB} | a'b \rangle \langle a' | M_A | a \rangle}{\delta_{a'} + \epsilon_b}$$

$$- \sum_{b' \neq o} \frac{\langle oo | H_{AB} | ab' \rangle \langle b' | M_B | b \rangle}{\delta_a + \epsilon_{b'}}$$

$$- \sum_{a' \neq a} \frac{\langle ab | H_{AB} | a'o \rangle \langle a' | M_A | o \rangle}{\delta_{a'} - \delta_a - \epsilon_b}$$

$$- \sum_{b' \neq b} \frac{\langle ab | H_{AB} | ob' \rangle \langle b' | M_B | o \rangle}{- \delta_a + \epsilon_{b'} - \epsilon_b} \, . \qquad (33)$$

This expression can be written specifically for the simultaneous excitation of a pair of ions or for the spinwave sideband. For the first case, a, b represent electronic states of ions A and B, whereas in the second a represents a spin deviation on A and b represents an electronic excitation on B. The primed states are intermediate states which can make allowed transitions to states a, b or o. The integrals $\langle oo|H_{AB}|a'b\rangle$, etc., are Coulomb and exchange contributions. In antiferromagnets the exchange part is of great importance because it makes spin-forbidden transitions possible. The spin dependence of these integrals could be expressed by the Dirac operator as has been done by Tanabe and Gondaira to display the spin dependence of the transition moment contributions in antiferromagnets. A recent example of a study of an interacting pair of ions is van der Ziel's work on Cr^{+++} pairs in ruby (3).

Figure 2 diagrams the states involved in the exchange coupling to give the transition moment for the case of the last term in eq. 33.

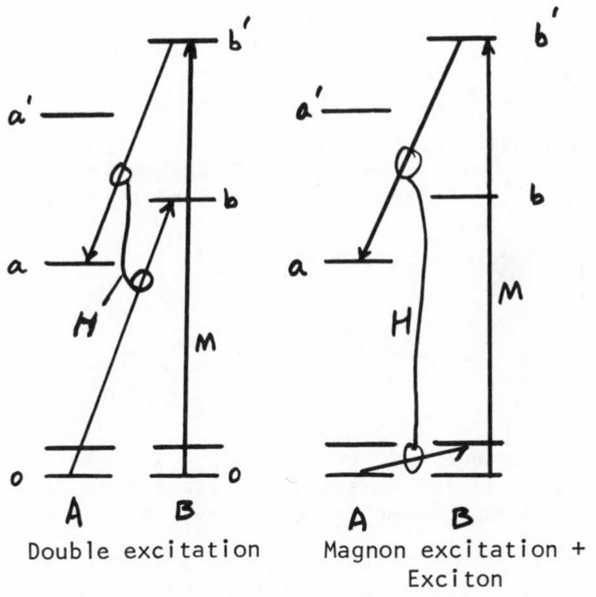

Fig. 2 Illustrating Exchange Induced Dipole Moment

III. CRYSTALLINE SYSTEMS

The major premise in treating magnetic insulator crystals is that the magnetic ions interact only weakly and have properties similar to those of isolated impurity ions in crystals. This means that the Frenkel exciton model should be a good starting point for a description of the excited states of a magnetic crystal. We will begin this section by developing the Frenkel model for a non-magnetic molecular crystal and then show how to make the slight modifications necessary to describe a magnetic crystal.

III.A. Frenkel Excitons and Davydov Splitting (4)

First consider a crystal having one molecule per unit cell. There are three basic translation vectors t_a, t_b, t_c which can be used to carry one molecule into another. The position of any molecule is given by $at_a + bt_b + ct_c$, where a, b, c are integers including 0, 0, 0, the origin.

The wavefunction of the ground state is just the product of the single molecule wavefunctions of the crystal

$$\psi_g = \prod_{i=1}^{N} \varphi_i \quad . \tag{34}$$

This state can be considered to be a Slater determinant, but only a single such determinant is needed, as we assume that the ground state is a filled state, invariant under all electron exchanges.

The singly excited states are written as

$$\psi_e = \frac{1}{\sqrt{N}} \sum_{n=1}^{N} C_n \varphi_1 -- \varphi_n^e -- \varphi_N \tag{35}$$

where the excited molecule with wavefunction φ^e can be at any site n, and overlap in the normalization factor is ignored. The coefficients C_n are determined by symmetry; all C_n must have the same absolute magnitude since all product functions differing only in the position of the excitation must have equal probability. The C_n can be shown to be

$$C_n = C(a,b,c) = \exp(ik \cdot r_n) \tag{36}$$

where
$$k = \alpha k_a + \beta k_b + \gamma k_c$$

$$r_n = a't_a + b't_b + c't_c \tag{37}$$

and a', b', c', α, β, γ are integers, $t_a \cdot k_a = \frac{2\pi}{Na}$, etc.

The wavefunctions are thus of the form

$$\psi_k = \frac{1}{\sqrt{N}} \sum_n^N \exp\left[ik \cdot r_n\right] \varphi_1 -- \varphi_n^e -- \varphi_N \tag{38}$$

The energy is obtained by calculating the diagonal elements of these functions with the Hamiltonian:

$$\mathcal{H} = \sum_i H_i + \sum_{i<j} g_{ij} \quad . \tag{39}$$

The one-center part of $\mathcal{H}$ cannot lead to a k-dependent term in the energy, since this can only come from interactions between different electrons, so the first part of $\mathcal{H}$ gives the same number for every $\psi(k)$. The second term has a k-independent part

$$\int \psi_k^* g_{ij} \psi_k \, dv = \sum_m \langle \varphi_n^e(i) \varphi_m(j) | g_{ij} | \varphi_n^e(i) \varphi_m(j) \rangle$$

and a k-dependent part in which different terms of the summation contribute. If only nearest neighbors are considered, this term could be

$$T = \frac{1}{N} \langle \varphi_n^e(i) \varphi_{n+1}(j) | g_{ij} | \varphi_n(i \text{ or } j) \varphi_{n+1}^e(j \text{ or } i) \rangle$$
$$\times \exp\left[ik\left(r_n - r_{n+1}\right)\right] \quad . \tag{40}$$

This is an excitation exchange term since the interaction between electrons on the two centers has caused the excitation to move from n to n+1.

The wavefunctions φ_n, etc., may be many-electron functions and the excitation exchange integral may or may not involve electron exchange. So as to avoid the problem of having to consider anti-symmetrization of ψ with respect to electrons on different centers (where it is assumed that the individual centers are already anti-symmetrized), we could carry out the summations over the electrons in the orbitals of the different centers using Dirac's operator; so the above term would be

$$T = \left[K_{n,n+1} - J_{n,n+1}\left(\frac{1}{2} + 2\,s_n \cdot s_{n+1}\right)\right] \exp\left[ik\left(r_n - r_{n+1}\right)\right] \quad (41)$$

where K and J have the same meaning as before.

If the ground and excited states have the same spin multiplicity, both the K and J parts of the excitation exchange term are effective, and usually K is larger than J. But if they have different multiplicity, the spin integration in the Coulomb integral gives a zero, while that in the exchange integral does not necessarily give zero. There are many examples in which the exchange integral causes energy migration, as in the triplet exciton bands of organic crystals and transfer of energy in ruby.

Returning to the solution of the Schrödinger equation, we can write all the interactions between two molecules as h_{ij}, so that i,j here mean molecules instead of electrons. We can also to a good approximation neglect all but interactions between nearest neighbors, so that the sum $\sum h_{i,j}$ is $\sum_n \left(\sum_{nn} h_{n,nn}\right)$.

When the complete Hamiltonian operates on the wavefunction (38) and the diagonal matrix element is taken, we get

$$E = \langle\psi_k|\mathcal{H}|\psi_k\rangle = E^e + C + E(k) \quad (42)$$

where E^e is the molecular excitation energy, C is the k-independent part of the interaction energy and $E(k)$ is the k-dependent part of the interaction energy

$$E(k) = 2\,J \sum_{nn} \cos k \cdot r_{nn} \quad (43)$$

where J is the integral in (40) and r_{nn} is one of the vectors to the nearest neighbors. This form of the k-dependent energy gives the dispersion relation for the exciton and a band of energies up to $4\,J$ in width.

The spectroscopic result of the presence of an exciton band of this form is somewhat disappointing. The transitions from the ground state are determined in one-photon spectroscopy by the electric dipole matrix element

$$\int \psi_g \sum_i er_i \psi_e dv_i = \frac{e}{\sqrt{N}} \langle\varphi_n|r|\varphi_n^e\rangle \sum_n \exp\left[-ik\cdot r_n\right]. \quad (44)$$

The sum is always zero unless $k = 0$, and then it equals N. So in the case $k = 0$ only we have a transition moment

$$M_o = e\sqrt{N} \langle\varphi_n|r|\varphi_n^e\rangle \quad . \quad (45)$$

The intensity of the absorption is NM^2, but the intensity per molecule is just the same as for a free molecule.

Thus even in the presence of the exciton band, we would only get a single line having the same strength per molecule as the free molecule.

If we consider the effect of phonons or spin waves, then the band width due to the exciton energy becomes apparent because the selection rule limiting transitions to $k = 0$ is broken (see Sec. III. C.).

The more interesting case occurs when there are two or more molecules per unit cell. Then there are molecules which are not related by a lattice translation and two or more transitions will occur for each molecular transition. This case is handled at the beginning as in the first example; for each site of the unit cell product functions and functions belonging to representations of the translation group of the crystal are formed. These latter are the one-site excitons analogous to eq. 38

$$\psi^i(k) = \frac{1}{\sqrt{N}} \sum_n \exp\left[ik\cdot r_n\right] \varphi_1^i -- \varphi_n^{ie} -- \varphi_N^i \prod_{j\neq i} \varphi^j \quad (46)$$

where i labels the site in the unit cell, and π takes the product over the unexcited molecules on all the other sublattices.

Next the one-site excitons are combined according to the symmetry operations of the group of the wave vector. When $k = 0$, the space group of the crystal is used; otherwise the lower symmetry of the wave vector group is the best that can be used. If the wave vector lies in a symmetry direction, the functions are simplified. When the wave vector lies in a general direction, the space group operations applied to one $\psi^i(k)$ generate an entire set of other $\psi^i(k')$ and the other site functions are also generated. We will see what to do with these later but for the moment will consider the $k = 0$ one-site excitons. The space group in this case can be replaced by the factor group which carries out operations interchanging the sites. A projection operator for each representation of the factor group can be constructed and applied to one of the $\psi^i(0)$ to generate wavefunctions belonging to factor group representations. Alternatively, this can be done more or less intuitively. For two molecules per cell one would get the functions

$$\psi_A = \frac{1}{\sqrt{2}}\left(\psi^1(0) + \psi^2(0)\right)$$

$$\psi_B = \frac{1}{\sqrt{2}}\left(\psi^1(0) - \psi^2(0)\right) . \tag{47}$$

Functions A and B each belong to an exciton band. The selection rule for transitions $\Delta k = 0$ would give two transitions in this case.

The energy difference between the two transitions is called the Davydov splitting, and it is just twice the matrix element of the Hamiltonian connecting the two inequivalent molecules

$$\Delta E_D = 2\langle\psi^1|\mathcal{H}|\psi^2\rangle . \tag{48}$$

Since the molecules are related by symmetry, the results of the one-molecule problem apply to each site. We can set up a 2 x 2 matrix in the ψ^1, ψ^2 basis for a given k

$$\begin{array}{c|cc|} \psi_1(k) & 2J_{11}\sum_{nn}\cos k\cdot r_{nn}^{11} & 2J_{12}\sum_{nn}\cos k\cdot\tau \\ \psi_2(k) & 2J_{12}\sum_{nn}\cos k\cdot\tau & 2J_{22}\sum_{nn}\cos k\cdot r_{nn}^{22} \end{array} \tag{49}$$

where $J_{11} = J_{22}$ is the exchange of excitation integral on one sublattice, J_{12} is that between sublattices, and τ is the non-primitive translation connecting the two lattices. The Davydov splitting at $k = 0$ is just $4zJ_{12}$, where z is the number of nearest neighbors.

For more complicated crystal structures more "unit cell bands" will appear, one for every atom in the cell. The transition probability to these bands at $k = 0$ will be a vector sum over the molecular moments, with the sign of each moment determined by the symmetry of the factor group state.

Specifically, one can take a moment vector on one of the molecules and apply the projection operator for one of the factor group representations to it. This will result in a combination of molecular moment vectors which belong to the particular factor group representation. Certain factor group representations contain the allowed electric or magnetic moments, and thus one can identify the linear combination of molecular moments which give the crystal moment for a particular symmetry direction.

The projection operator is $\sum_R \chi_R^j R$ where χ_R^j is the coefficient of operation R in representation j. Then we have

$$P_{nm}^{j} = \sum_{R} \chi_R^j R P_{nm} \tag{50}$$

where P_{nm} is any one molecular moment for molecular transition $n \rightarrow m$, and P_{nm}^j is the crystal moment for representation j and molecular transition $n \rightarrow m$.

III.B. Effect of Vibration on the Davydov Splitting and Exciton Dispersion

One cannot assume that the atomic coordinates in a crystal remain unaffected by electronic transitions, since any rearrangement of the electrons must cause some change in bonding. The exchange-of-excitation integrals can be written in terms of Born-Oppenheimer wavefunctions instead of purely electronic wavefunctions. That is, we will use a product wavefunction

$$\psi_e\left(r, Q_o\right) \psi_v\left(Q - Q_o\right) \tag{51}$$

where $\psi_e\left(r, Q_o\right)$ is an electronic wavefunction in which the configuration is fixed at Q_o, and ψ_v is a vibrational wavefunction with Q_o as its equilibrium position. It is a fairly good assumption that the electronic eigenfunctions do not change drastically with changes in Q_o, the equilibrium position, so we will write all electronic wavefunctions for the same configuration, perhaps using perturbation theory to correct them if necessary. The ψ_v must be written so as to show the changes in Q_o explicitly.

The excitation transfer integral involving a ground state $\psi_g\left(r,Q_g\right)\ \varphi_v\left(Q-Q_g\right)$ and an excited state $\psi_e\left(r,Q_g\right)\varphi_w\left(Q-Q_e\right)$ is

$$\langle g_1v_1\ e_2w_2|\mathcal{H}|g_2v_2\ e_1w_1\rangle = \langle g_1e_2|\mathcal{H}|g_2e_1\rangle\langle v_1|w_1\rangle\langle w_2|v_2\rangle = TS^2 \quad (52)$$

where T is the excitation transfer integral and S is the vibrational overlap integral at a single site. The value of S can be obtained from the absorption spectrum. If we are considering the zero phonon line, then S^2 is the ratio of the integrated intensity of that line to that of the entire band. In general, the Davydov splitting of a vibronic level should be proportional to its intensity in the spectrum. This conclusion is based on the foregoing assumptions as to the validity of the Born-Oppenheimer separation; it does not take into account off-diagonal vibronic matrix elements.

III.C. Exciton Dispersion and Davydov Splitting in a Magnetic Crystal

The problem we shall consider here is how does the exciton dispersion and Davydov splitting in a magnetic crystal differ from that in a molecular crystal? These two types of crystal are physically similar in that the excitations of each are localized, but differ in that molecular crystals have closed-shell ground states, while magnetic crystals have open-shell ground states and have an effective magnetic field acting on each ion. Magnetic crystals also have spin wave excitations.

For the purely excitonic levels of a magnetic crystal, we can use Frenkel exciton theory just as in a molecular crystal. The ground state of a given sublattice can be written as a product eigenfunction over the ions in the i-th sublattice

$$\psi_i = \prod_j^{(i)} \varphi_j s_j \quad (53)$$

where $\varphi_j s_j$ means a wavefunction on atom j with orbital part φ_j and spin part s_j. The entire ground state wavefunction is a product of those for all sublattices

$$\Phi = \prod_i \psi_i \ . \quad (54)$$

The ground state, however, can have spin excitations or spin deviations at finite temperatures; these excitations may also be observed in fluorescence experiments. We can write a spin wave as a Frenkel exciton as was done in Sec. III. A.

$$\psi_k^i = \frac{1}{\sqrt{N}} \sum_{\ell}^{N} \varphi_\ell^i \, d_\ell^i \prod_{j \neq \ell}^{(i)} \varphi_j \, s_j \, \exp\left[-ik \cdot r_\ell^{(i)}\right] \quad (55)$$

where d_ℓ^i represents a spin deviation at ion ℓ on sublattice i. This is a special sort of spin wave, since normally they are not confined to one sublattice. The more rigorous description of spin waves is to be found in Sec. IV. A. However, in the spectroscopic transitions the spin wave is often confined to one sublattice, as will be shown later in this section.

The electronic excitations of a sublattice have the same form as for a molecular crystal, eq. 38. The transfer integral is written analogously. Only the spin functions are different from the molecular case, as they correspond to a spin projection rather than to a total spin eigenfunction. Further, there is often a large spin-orbit coupling and a Jahn-Teller effect to complicate the wavefunction. In any case, the transfer integral on a sublattice will be basically the same as in eq. 40. The magnetic structure enters in the possible values of the transfer integral. If ions n and $n+1$ have the same spin projection, m_s, then the transfer integral has the full value allowed by the orbital properties. Both Coulomb and exchange contributions can exist. But if n and $n+1$ belong to sublattices having opposite spins, only the Coulomb contribution remains. For spin-forbidden transitions in an antiferromagnetic lattice such as occur in Mn^{++}, both the Coulomb and exchange contributions are extremely small, and there can be no appreciable Davydov splitting. Various possibilities are shown in the diagrams of Fig. 3. In the presence of spin-orbit coupling, the restrictions on Coulomb and exchange coupling must be modified.

The most spectacular effect of a magnetic crystal on the spectra of the ions which compose it is the appearance of strong spin wave sidebands on spin-forbidden transitions in antiferromagnets, the fourth example in Fig. 3. According to that example there can be no single ion transitions and no transfer of energy in the lattice. However, a relatively high transition probability accompanies spin forbidden transitions when a spin wave is excited simultaneously and a spectrum appears anyway. We will go into the reasons for this later but first will investigate the simultaneous magnon-exciton transition energies.

Assuming that such a compound excitation can exist, we can construct a zeroth order wavefunction for it by multiplying together an exciton and a magnon wavefunction. For a two sublattice antiferromagnet we take the exciton on sublattice A and the magnon on B. Let the non-primitive translation between them be τ, with the origin on A. Then the two separate wavefunctions are

$$S(k_m) = \frac{1}{\sqrt{N}} \sum_j \exp\left[-ik_m\left(R_j+\tau\right)\right] S_j , \quad X(k_e) = \frac{1}{\sqrt{N}} \sum_n \exp\left[ik_e \cdot R_n\right] X_n$$

$$S_j = d_j \prod_{i \neq j} s_i \qquad\qquad X_n = e_n \prod_{i \neq j} o_i \tag{56}$$

where S_j is a spin deviation at cell j of sublattice A, and X_n is an excitation in the nth cell on sublattice B, and k_m and k_e are the magnon and exciton wave vectors. The state of the crystal containing one of each excitation is expressed by the product

$$\psi\left(k_e, k_m\right) = \frac{1}{N} \left(\sum_j \exp\left[ik_m(R_j+\tau)\right] S_j\right)\left(\sum_n \exp\left[ik_e \cdot R_n\right] X_n\right). \tag{57}$$

If we apply an arbitrary lattice translation ρ to this wavefunction, we find

$$T(\rho)\, \psi\left(k_e, k_m\right) = \exp\left[i\left(k_m + k_e\right)\rho\right] \psi\left(k_e, k_m\right)$$

and the wavefunction belongs to the $k_m + k_e$ or K representation of the translation subgroup. In an optical transition from the ground state

$$\psi_G = \prod_i g_i \tag{58}$$

the selection rule on the total wave vector $K = k_e + k_m$, must be

$$K = k_e + k_m = 0 \tag{59}$$

the wavefunctions of interest to us are therefore characterized by a single quantum number which can be called k, where $k = k_e = -k_m$.

With the excitation and spin deviation each transferred to the opposite sublattice, there will be a similar wavefunction characterized by k. When the two are combined in the ways consistent with symmetry, the resulting pair of wavefunctions represent the Davydov components of the magnon-exciton complex.

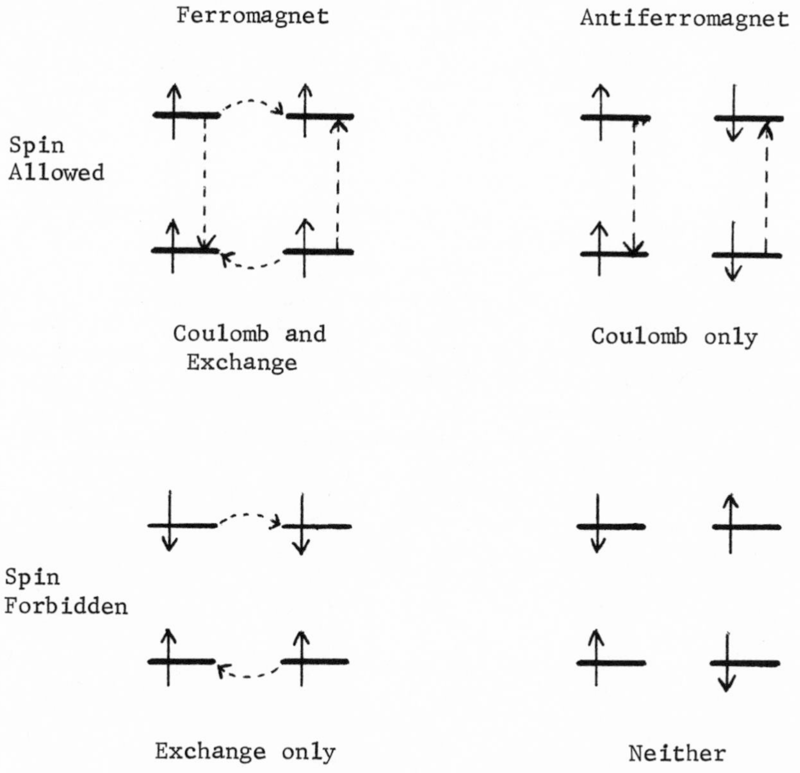

Fig. 3 Exciton Transfer in Magnetic Crystals.

IV. ELECTRONIC STRUCTURE OF MAGNETIC CRYSTALS

Optical spectra are affected by both the static and dynamic aspects of a magnetic crystal, and in this section the most important of these will be discussed. Spin waves appear as side-bands in optical spectra, analogously to phonons. A bound state of a spin wave (or magnon) or an exciton has even been identified. The static structure of a magnetic crystal causes a sort of Zeeman splitting of all levels, and this is ascribed to an exchange and an anisotropy field. The type of spin order, whether ferromagnetic, ferrimagnetic or antiferromagnetic, has an important effect on the spectrum, as already indicated in Fig. 3, and it is worth reviewing the molecular field description of these different systems.

IV.A. Spin Waves (5)

1. <u>The Case of a Chain, $S = 1/2$, Periodic Boundary Conditions</u>. If N spins are present, the ground state eigenfunction for a ferromagnetic chain is

$$\varphi_o = \alpha_1 \alpha_2 -- \alpha_N \ . \tag{60}$$

The next higher state contains one spin deviation

$$\varphi_\ell = \alpha_1 \alpha_2 -- \alpha_{\ell-1} \beta_\ell \alpha_{\ell+1} -- \alpha_N \ . \tag{61}$$

But this function must be degenerate with those for which the spin deviation is on any other site. Therefore some linear combination of φ_ℓ would be a better representation

$$\psi_k = \sum_\ell C_\ell^k \varphi_\ell \ . \tag{62}$$

The Hamiltonian which acts on these functions is the exchange plus the magnetic Hamiltonian

$$\mathcal{H} = - g \mu_B H_o \sum_\ell S_\ell^z - 2 \sum_{\ell > m} J_{\ell m} S_\ell \cdot S_m \ . \tag{63}$$

The second term can be expressed as

$$\mathcal{H}_{ex} = - 2 \sum J_{\ell m} \left(\frac{1}{2} S_\ell^+ S_m^- + \frac{1}{2} S_\ell^- S_m^+ + S_\ell^z S_m^z \right) \tag{64}$$

where S_ℓ^+, S_ℓ^- are raising and lowering operators for the spin. From the theory of angular momentum

$$J_z | jm \rangle = m \hbar | jm \rangle$$

$$J_\pm | jm \rangle = \left[j(j+1) - m(m \pm 1) \right]^{1/2} \hbar | j, m \pm 1 \rangle \ . \tag{65}$$

If we take periodic boundary conditions, this is equivalent to letting the N atoms form a ring, so that atom N is followed by atom 1. If a rotation by $2\pi/N$ is performed on ψ_k, the only change in the ψ_k could be a phase factor of absolute magnitude 1, or $\exp[i(2\pi/N)k]$. If the C_ℓ^k have values like $\exp[-i(2\pi/N)k\ell]$, then a rotation by $2\pi/N$ which shifts atom ℓ to the position occupied by $\ell + 1$, changes C_ℓ^k to $C_{\ell+1}^k$, and so multiplies this and every other C_ℓ^k by just the factor $\exp[i(2\pi/N)k]$. Thus the wavefunctions of this problem are entirely determined by symmetry and need not be constructed after the solution of the secular equation. Thus

$$\psi_k = \frac{1}{\sqrt{N}} \sum_{\ell}^{N} \varphi_\ell \exp[-2\pi i k\ell/N] \tag{66}$$

where $k = 0, 1 \cdots N \cdots 2N$. The diagonal matrix element of this wavefunction with the Hamiltonian is the energy $E(k)$.

The magnetic term of $\mathcal{H}$ simply results in the term $-g\mu_B H_o$ $[(N-1)1/2 - 1/2]$ for the excited state and $-g\mu_B H_o(N/2)$ for the ground state, and the difference is $-g\mu_B H_o$. This is just the magnetic excitation energy.

The exchange term consists of two parts, the parallel exchange $S_k^z S_m^z$, and the transverse $1/2\left(S_\ell^+ S_m^- + S_\ell^- S_m^+\right)$. We limit the exchange to nearest neighbors, so $m = \ell \pm 1$.

Since $S_n^z S_{n+1}^z$ does not change the wavefunction and since φ_ℓ is orthogonal to $\varphi_{\ell+1}$, there are no cross terms in the matrix element $\left\langle \sum_\ell C_\ell \varphi_\ell \sum S_n S_{n+1} \sum_\ell C_\ell \varphi_\ell \right\rangle / N$. Therefore one can cancel both the $\sum_\ell$ with the $1/N$ since there would be N identical terms in the summation and get $\left\langle C_\ell \varphi_\ell \middle| \sum_n S_n S_{n+1} \middle| C_\ell \varphi_\ell \right\rangle$. But $C_\ell^* C_\ell = 1$, so the C_ℓ give unity. Finally, if one subtracts $\left\langle \varphi_o \middle| \sum_n S_n S_{n+1} \middle| \varphi_o \right\rangle$ to get the excitation energy, only the terms in $\sum_n$ for which $n = \ell$ and $n = \ell - 1$ are different in the two states contribute. This contribution is -1, or when the coefficient from $\mathcal{H}_{ex}$ is put in, we get $E = 2J$. Thus the $S_n^z S_{n+1}^z$ term does not give a k-dependent energy. This is the so-called

Ising term. It does not give rise to spin waves.

The other terms in eq. 64 give more interesting results. $S_{\ell}^{-}S_{\ell+1}^{+}$ acts on $\cdots\alpha_{\ell-1}\alpha_{\ell}\beta_{\ell+1}\alpha_{\ell+2}\cdots$ to give $\alpha_{\ell-1}\beta_{\ell}\alpha_{\ell+1}\alpha_{\ell+2}$, so it shifts the spin deviation to the left. $S_{\ell}^{+}S_{\ell+1}^{-}$ acts on $\alpha_{\ell-1}\beta_{\ell}\alpha_{\ell+1}\alpha_{\ell+2}$ to give $\alpha_{\ell-1}\alpha_{\ell}\beta_{\ell+1}\alpha_{\ell+2}$, so it shifts to the right. The only terms in $\sum C_{\ell}\varphi_{\ell}$ which can match the terms shifted by the operator are ones which are similarly shifted. These have a coefficient $\exp[-i\,2\pi k\ell/N]$ for the left shift term whose coefficient is $\exp[+i\,2\pi k(\ell+1)/N]$, so there is a net contribution $\exp[i\,2\pi k/N]$ to the matrix element. The terms in the operator for which $n, n+1 \neq \ell$ give zero because the raising operator annihilates α. For the same reason, the ground state gives zero. When the contributions of the left and right shift terms are added, we have $1/2\ (\exp[i\,2\pi k/N] + \exp[-i\,2\pi k/N]) = \cos 2\pi k/N$, and the term in the energy formula is $-2J\cos 2\pi k/N$. The total energy of excitation is

$$E = g\mu_B H_o + 2J(1 - \cos 2\pi k/N)\ . \tag{67}$$

It is important to realize that it is the transverse exchange components which give rise to a k-dependent energy. The energy formula gives rise to a dispersion curve shown in Fig. 4.

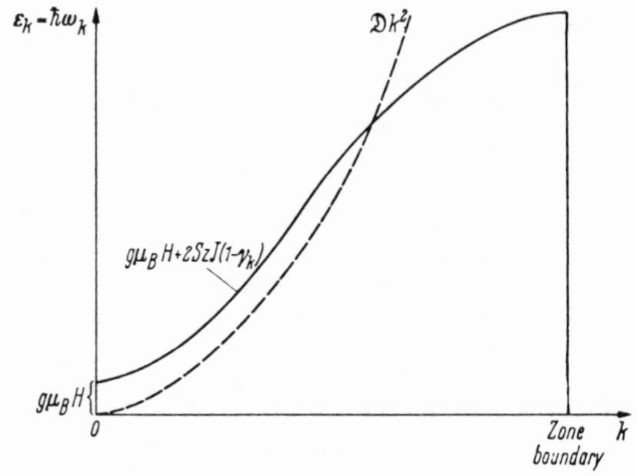

Fig. 4 Magnon Dispersion Curve.

The number set k is usually expressed as a vector having the dimensions of reciprocal length for convenience in three-dimensional problems. In this case we define $k' = 2\pi k/Na$ as the wave vector and the argument of the cosine above becomes $k'a$, where a = the length of the unit cell.

2. The Three-Dimensional Case. The ferromagnetic spin wave energy is

$$\mathcal{E}_k = g\mu_B H_o + 2SzJ(1 - \gamma_k) \tag{68}$$

where z is the number of nearest neighbors, S is the total spin and

$$\gamma_k = z^{-1} \sum_n \exp\left[- ik \cdot r_n\right] \tag{69}$$

and r_n are the vectors to the nearest neighbors, k a vector in reciprocal space.

3. Antiferromagnetic Spin Waves. A simplified Frenkel model was used in Sec. III. C. for a spin wave on one sublattice of an antiferromagnet. This turns out to be justified only when the spin wave is near the edge of the Brillouin zone. The true normal modes of the spin system are shared between the two sublattices at interior zone points. There are always two degenerate modes, one predominantly on one sublattice, one predominantly on the other, except that at $k = 0$ the two modes are equally shared. Their energy in the nearest neighbor approximation is given by

$$\omega_k^2 = \left(\omega_e + \omega_A\right)^2 - \omega_e^2 \gamma_k^2 \tag{70}$$

with $\omega_e = 2zJS$, $\omega_A = 2\mu_B H_A$, and γ_k as in eq. 69, and only the positive root is taken.

In spectroscopic applications one needs to know the density of spin wave states in actual crystals, and in some cases experimental dispersion curves have been found from neutron scattering experiments. The wave functions of the spin waves also have to be known for some problems, and for this one usually goes to the simplified theory which leads to eq. 70.

IV.B. Anisotropy Field and Exchange Field

The first term in eq. 63 was presented as a Zeeman interaction term due to an external H-field. This field makes one spin direction of lower energy than any other; hence it splits the spin degeneracy making $S^z(\max)$ the lowest energy level. The combined effect of the crystal field and the spin orbit coupling can also produce a splitting of a spin state, such that a particular spin level is stabilized, and the spin direction is thereby aligned along the directions of the crystal field. In a tetragonal or trigonal field this effect can be reproduced by an operator term in the Hamiltonian

$$\mathcal{H}_a = D \sum_{\ell} \left[\left(S_\ell^z\right)^2 - \frac{1}{3} S(S+1)\right] \quad . \tag{71}$$

This term is an electric one and therefore does not distinguish $\pm S^z$. It is often the major source of magnetic anisotropy in magnetic insulators such as MnF_2, where the spins are lined up along the C-axis.

The anisotropy term is sometimes replaced by an anisotropy field, a magnetic field of the right magnitude to produce the same splitting as eq. 71. Then eq. 71 would be replaced by

$$\mathcal{H}_a = g\mu_B H_A \qquad (72)$$

as in eq. 63.

When we consider one ion in the crystal, as we do when an optical transition occurs, a reasonable approximation is to think of the action of the exchange plus anisotropy of all the other ions as an effective "exchange field" below the Curie or Néel temperature. The exchange term in eq. 63 is isotropic, but because of the anisotropy energy it is not a bad approximation to replace S by S^z or by the average value $\overline{S^z}$ along the direction of the anisotropy field. The exchange field Hamiltonian is thus

$$\mathcal{H}_{eX} = 2\left(\sum_i J_i S_i^z\right)\cdot S = g\mu_B H_E \cdot S \qquad (73)$$

where S is the spin at the selected site and the sum goes over the other ions in the crystal, giving, if the sum is restricted to the z nearest neighbors

$$H_E = \frac{2\,zJ\,\overline{S^z}}{g\mu_B} \,. \qquad (74)$$

The effective field acting on the ion is the sum of the anisotropy and exchange field, $H_{eff} = H_A + H_E$. In this approximation, the optical properties of an ion in a crystal are just those of an ion in a large magnetic field. The pseudo-Zeeman splittings corresponding to H_{eff} can be $10^5 - 10^7$ gauss so that optical transitions can be thought of as occurring from the lowest Zeeman level of the ground term, all other components lying at energies much greater than kT.

The type of averaging which leads to an effective field could not lead to an explanation of some of the spectral features observed in magnetic materials, but it is often useful in spectroscopy, and also leads to the Weiss theory of magnetism.

IV. C. Weiss Theory of Ferromagnetism (1)

In the Weiss theory of magnetism, the magnetization, or the magnetic dipole moment per unit volume, is

$$M = \bar{N} g \mu_B \overline{S^z} \tag{75}$$

or just the sum of the atomic moments in the crystal. Here $\bar{N}$ is the number of atoms per unit volume and $\overline{S^z} = \sqrt{S(S+1)}$. This magnetization gives rise to a "molecular field"

$$H_E = \lambda M \tag{76}$$

The constant λ is related to the exchange energy (consistent with eq. 74)

$$\lambda = 2zJ/\bar{N} g^2 \mu_B^2 \tag{77}$$

The molecular field acts on the spin of each ion and is the additional aligning force besides the external field. The total field or Weiss field is

$$H_W = H + H_E = H + \lambda M \tag{78}$$

In general, the paramagnetic susceptibility is

$$\chi = M/H = \frac{kT}{H}\frac{\partial Z}{\partial H} = \frac{\bar{N} g \mu_B \sqrt{S(S+1)}}{H} B_S(\alpha) \tag{79}$$

where Z is the partition function, $B_S(\alpha)$ is the Brillouin function and $\alpha = g\mu_B H/kT$. An approximation at low total field strengths (M linear in H) is

$$\chi = \frac{\bar{N} g^2 \mu_B^2 S(S+1)}{3kT} \tag{80}$$

In the presence of the internal field, the magnetization is much larger than in the paramagnetic case, since M results from H_W, not just H. But the susceptibility we measure with an external field is $\chi' = \frac{\partial M}{\partial H}$. We can get χ' by using (78), (79) and (80):

$$\chi = \frac{M}{H_W} = \frac{M}{H + \lambda M}; \quad \chi' = \frac{\partial M}{\partial H} = \frac{\chi}{1 - \lambda\chi} = \frac{1}{\frac{1}{\chi} - \lambda}. \tag{81}$$

This equation can be written using (77) and (80)

$$\chi' = \frac{C}{T - T_c} \quad \text{(Curie-Weiss law)} \tag{82}$$

where

$$C = \bar{N} g^2 \mu_B^2 \, S(S+1)/3k; \quad T_c = 2zJ \, S(S+1)/3k\,. \tag{83}$$

The second formula gives J in terms of T_c. Eq. 82 is valid in the paramagnetic region where M is small and for the beginning of the cooperative process which results in a permanent magnetization. It is not valid in the region $T < T_c$ because of our assumption of small field, small enough to permit the linear approximation (80). It does have the virtue that it explains the magnetic transition; as $T \to T_c$, the susceptibility must become very large since only a very small change in external field can cause a large change in magnetization. The Curie constant, C, can be determined from experiment, as can T_c, and each can be compared to their theoretical expression (83). Below T_c, however, $H \ll H_W$, so the approximation (80) cannot be made and eq. 79 must be used. The equation for χ' becomes a transcendental equation and its solution shows M/M_o as a function of T/T_c in Fig. 5. (M_o is the saturation magnetization for $T = 0$.)

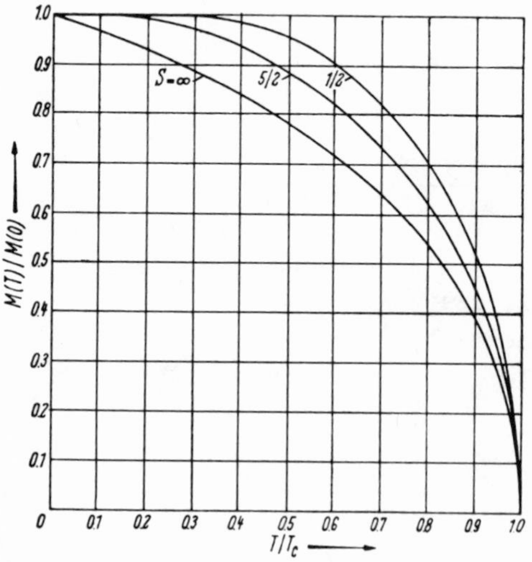

Fig. 5 Molecular field result, as obtained from numerical solution of (79), for magnetization of a ferromagnetic domain as a function of temperature. Curves are shown for $S = 1/2$, $S = 5/2$, and the Langevin limit $S \to \infty$.

IV.D. Antiferromagnets

The Weiss field idea gives an easy way to describe the thermodynamic behavior of antiferromagnets. If there are two equivalent sublattices, their effective fields are

$$H_+ = H - \alpha M_- - \beta M_+$$
$$H_- = H - \alpha M_+ - \beta M_- \qquad (84)$$

For the case of weak fields, the magnetizations are given by eq. 80

$$M_\pm = \frac{1}{2}\,\frac{C}{T}\,|H_\pm| \qquad (85)$$

where

$$C = \frac{\bar{N}\mu^2}{3k}, \qquad \mu^2 = g^2\mu_B^2\,S(S+1) \; . \qquad (86)$$

Substituting eq. 84 into eq. 85

$$M_+(T+\beta C/2) + M_-\,\alpha C/2 = HC/2$$
$$M_-(T+\beta C/2) + M_+\,\alpha C/2 = HC/2 \; . \qquad (87)$$

Solving for $M = M_+ - M_-$, the resultant magnetization gives

$$M = \frac{CH}{T + 1/2\;C(\alpha+\beta)} \qquad \text{or}$$

$$\chi = \frac{C}{T+T_c}, \qquad \text{where} \quad T_c = 1/2\;C(\alpha+\beta) \; . \qquad (88)$$

This equation gives the behavior of the antiferromagnet in its paramagnetic region. The condition for having a spontaneous magnetization on the sublattices can be found from eq. 87 when the external field is set equal to zero. Then the coefficients of M_+, M_- would have to be non-zero, and the condition for the consistency of the two equations becomes

$$\begin{vmatrix} T+\beta C/2 & \alpha C/2 \\ \alpha C/2 & T+\beta C/2 \end{vmatrix} = 0 \qquad (89)$$

which gives $T = 1/2\;C(\alpha - \beta) = T_N$, the Néel temperature. This is the transition temperature.

Its relation to T_c should be noted

$$\frac{T_N}{|T_c|} = \frac{\alpha - \beta}{\alpha + \beta} \tag{90}$$

If the interaction with the other sublattice is greater than with its own $\alpha > \beta$, and since both are positive, $T_N < |T_c|$. This is found experimentally.

For the region $T < T_N$ the full Brillouin function must be used in order to describe the magnetization. If there were no anisotropy energy, the spins would always tend to align themselves perpendicular to an applied magnetic field rather than parallel because $\chi_\perp > \chi_\parallel$. This can be seen in an elementary way by using the diagram, Fig. 6.

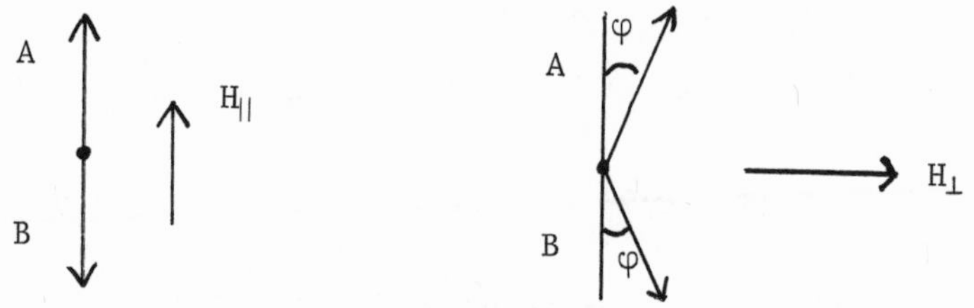

Fig. 6 Illustrating Why $\chi_\parallel < \chi_\perp$ ·

With $H \| M$, the alignment cannot be changed very much by the field, since it is already high and becomes more perfect as $T \to 0$. But when $H \perp M$, the sublattice magnetization vectors can be changed in angle, and the configuration would become more stable as the projections of M on H increase. This energy increment is $M \cdot H = -2M_+ H \sin\varphi$. At the same time the molecular field energy is decreased by the change in the projection of H_+ on M_-

$$\Delta\left(H_+ \cdot M_-\right) = +\beta M_+ \cdot M_- = \beta M_+^2 \cos 2\varphi \cdot$$

Minimizing the total energy:

$$E = -2M_+ H \sin\varphi + \beta M_+^2 \cos 2\varphi \quad \text{gives} \quad \sin\varphi = H/2\beta M_+ \cdot$$

Since $\chi_\perp = \frac{\partial M}{\partial H_\perp}$ and $\delta M = 2M_+ \sin\varphi$, then $\chi_\perp = \frac{2M_+ \sin\varphi}{H} = \frac{1}{\beta}$

Since β is one of the exchange constants, $\chi_\perp$ is nearly independent of temperature. The behavior of the susceptibility as a function of temperature is shown in Fig. 7.

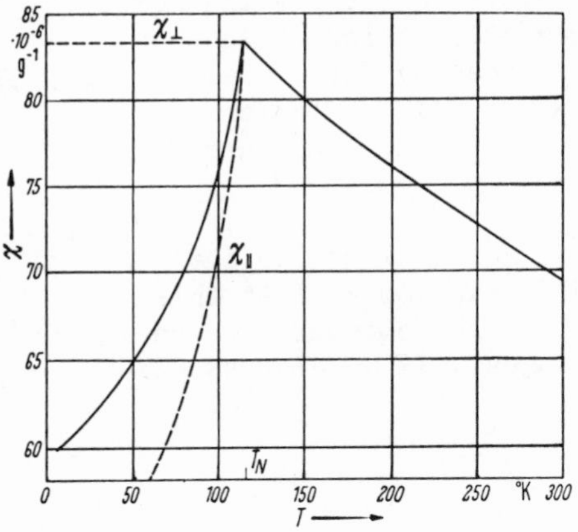

Fig. 7 Solid line, magnetic susceptibility of powdered MnO, as measured by Bizette, Squire, and Tsai. Dotted lines, molecular field theory of single crystal perpendicular and parallel susceptibility. Compare with

$$\chi_{powder}(0^oK) = \frac{2}{3}\chi_\perp = \frac{2}{3}\chi_{powder}(T_N).$$

IV.E. Effects of External Magnetic Fields

From eq. 70, by adding the Zeeman term, the spin wave frequencies near $k = 0$ in a magnetic field H_o are

$$\omega_o = 2\mu_B\left[\left(2H_AH_E + H_A^2\right)^{1/2} \pm H_o\right] . \qquad (91)$$

H_o is supposed to be oriented along the direction of H_A. The anisotropy field H_A can be large compared to the exchange field H_E in compounds of Fe^{++} or Co^{++} where the ground states have large orbital contributions. The spin-orbit splitting essentially provides the large anisotropy in these cases. For other systems $H_A \ll H_E$, and the product H_AH_E gives the main contribution to the radical. These two fields provide the macroscopic precession field in the absence of an external field. When the radical $= H_o$, one frequency goes to zero. At this point the spins "flop" perpendicular to the field. This change lowers the magnetic energy because $\chi_\perp \gg \chi_\parallel$, and the external field can cause greater polarization of the lattice. Thus the critical field for spin flop is

$$H_c = \left[2H_AH_E + H_A^2\right]^{1/2} . \qquad (92)$$

In the spin-flop phase of MnF_2 (H_c = 93 kgauss), the excitons no longer show pseudosplitting. There is a sudden collapse of this splitting from about 18 cm^{-1} to zero as this field is passed.

In ferrous compounds such as $FeBr_2$ the anisotropy dominates over exchange. The "spin" orientation idea is not a valid concept in these cases. $FeBr_2$ is antiferromagnetic at low fields but undergoes a fairly rapid change with increasing field near 40 kgauss to an aligned state. It cannot be called a ferromagnetic state since only external fields can support it. The term aligned paramagnet has been used.

In the case of $FeCO_3$ the optical spectrum can be changed radically by the application of a field near 200 kg. At this field the crystal has changed from an antiferromagnet into an aligned ferromagnet. The spectral changes will be discussed in Sec. V.

IV.F. Group Theory (6)

In this section some special aspects of group theory will be considered.

In a magnetic field the eigenfunctions for spin and angular momentum cannot be combined so as to become real. For an atom the complex φ-dependent part, $\psi = \exp[im\varphi]$, can be factored out where φ measures the angle around the direction of the field. In determining the characters of representations, it doesn't matter which rotation axis is chosen, but for the case of atoms in a magnetic field, it is certainly more convenient to choose the direction of the field. The character is thus obtained from

$$R(2\pi/n)\psi = \chi_n \psi$$

by the operations

$$R(2\pi/n)\psi = \exp[im(\varphi - 2\pi/n)] = \exp[-i\ 2\pi m/n]\ \psi$$

so that

$$\chi_n = \exp[-i\ 2\pi m/n] \quad .$$

Here $2\pi/n$ represents a positive (right-hand rule) rotation about the field. Thus the characters in the rotation group in a magnetic field will usually be complex. For an integral value of angular momentum, n is integral and the characters form a single valued representation of the operations. But when J is half integral, the rotation by 2π gives $P(2\pi)\psi = -\psi$ and the usual isomorphism between products of rotation and products of representations no longer holds. Therefore we invent a symmetry operation of rotation by 4π as the identity. The number of operations is thereby doubled and the groups of these operations are called the double groups.

An example of a double group is Table 1, the cubic group.

TABLE 1

GROUPS $\bar{O}$ AND $\bar{T}_d$ $\qquad O_h = O \times i$

		$\bar{O}$	E	R	$4C_3$ $4C_3^2R$	$4C_3^2$ $4C_3R$	$3C_4^2$ $3C_4^2R$	$3C_4$ $3C_4^3R$	$3C_4^3$ $3C_4R$	$3C_2'$ $3C_2'R$	$\bar{O}$	
		$\bar{T}_d$	E	R	$4C_3$ $4C_3^2R$	$4C_3^2$ $4C_3R$	$3C_4^2$ $3C_4^2R$	$3S_4$ $3S_4^3R$	$3S_4^3$ $3S_4R$	$6\sigma_d$ $6\sigma_dR$		$\bar{T}_d$
	A_1	Γ_1	1	1	1	1	1	1	1	1		
	A_2	Γ_2	1	1	1	1	1	−1	−1	−1		
	E	Γ_3	2	2	−1	−1	2	0	0	0		
	T_1	Γ_4	3	3	0	0	−1	1	1	−1	x, y, z, L_x, L_y, L_z	L_x, L_y, L_z
	T_2	Γ_5	3	3	0	0	−1	−1	−1	1		x, y, z
(E_1)	$E_{1/2}$	Γ_6	2	−2	1	−1	0	$\sqrt{2}$	$-\sqrt{2}$	0		
(E_2)	$E_{5/2}$	Γ_7	2	−2	1	−1	0	$-\sqrt{2}$	$\sqrt{2}$	0		
(G)	$G_{3/2}$	Γ_8	4	−4	−1	1	0	0	0	0		

The first part of the table is simply a repeat of the ordinary group, but the second part shows the representations resulting from transformations of spinors, or half-integral wavefunctions. When a magnetic field is applied along the C_4-axis, some cubic states split; namely, those whose symmetric square contains T_1. The magnetic field destroys the C_3 axes and the C_2 and C_4 axes at right angles to it, but does not destroy the reflection plane at right angles to the field direction or the inversion operation, since it is an axial vector. The resulting group is C_{4h}, shown in Table 2, for both ordinary (vector) and spinor transformations.

TABLE 2

GROUPS $\bar{C}_4$ AND $\bar{S}_4$ $\qquad C_{4h} = \bar{C}_4 \times i$

	$\bar{C}_4$	E	R	C_4	C_4R	C_4^2	C_4^2R	C_4^3	C_4^3R	$\bar{C}_4$	
	$\bar{S}_4$	E	R	S_4	S_4R	C_2R	C_2	S_4^3	S_4^3R		$\bar{S}_4$
A	Γ_1	1	1	1	1	1	1	1	1	z	L_z
B	Γ_2	1	1	−1	−1	1	1	−1	−1	L_z	z
E	Γ_3	1	1	i	i	−1	−1	$-i$	$-i$	$x \pm iy$	$x \pm iy$
	Γ_4	1	1	$-i$	$-i$	−1	−1	i	i	$L_x \pm iL_y$	$L_x \pm iL_y$
$E_{1/2}$	Γ_5	1	−1	ω	$-\omega$	i	$-i$	ω^3	$-\omega^3$		
	Γ_6	1	−1	$-\omega^3$	ω^3	$-i$	i	$-\omega$	ω		
$E_{3/2}$	Γ_7	1	−1	$-\omega$	ω	i	$-i$	$-\omega^3$	ω^3		
	Γ_8	1	−1	ω^3	$-\omega^3$	$-i$	i	ω	$-\omega$		

$\omega = \exp(i\pi/4)$

The cubic double group can be used to classify states in which an odd number of electrons is present but no magnetic field is present. An example is the spin-orbit splitting of a $^4T_1$ or $^4T_2$ state in a cubic field, such as one would find in excited Mn^{++} in $RbMnF_3$. Figure 8 shows on the left the first order spin-orbit splitting of such a state.

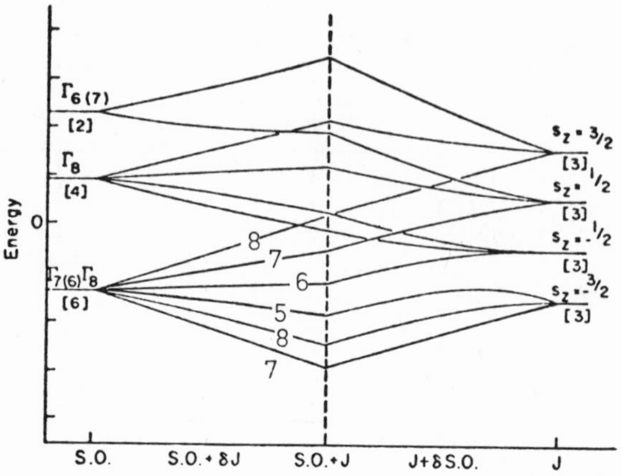

Fig. 8 Correlation between spin-orbit and exchange splitting for a $^4T_{1(2)}$ state. The left corresponds to the spin-orbit effect only (assuming λ to be positive) ; the right-hand side corresponds to a pure exchange splitting; the numbers in brackets give the degeneracy. Inclusion of a Jahn-Teller effect corresponds to a shift from the center line to the right on the diagram. Numbers on the lines are the C_{4h} representations.

Depending on the orbital components, this diagram could be as shown or inverted. $^4T_1(I)$, mostly t^4e, is as shown, but $^4T_2(I)$ is inverted. The accidental degeneracy between Γ_7 and Γ_8 is actually split by higher order interactions.

In a magnetic field along the C_4 axis we would have to use the C_{4h} double group, and then everything would split as the diagram (Fig. 8) shows (7). The left side shows the Zeeman splitting of the spin-orbit components. The right side shows what would happen if the spin-orbit coupling were turned off. This in fact is nearly what happens in the presence of a strong Jahn-Teller effect. The energy levels on the right are just the m_s levels, -3/2, - 1/2, 1/2 and 3/2 due to the exchange field.

The representations to which these levels belong in C_{4h} can be found by reducing the cubic representations of the orbit and the spin part into C_{4h} representations and getting their product, or reducing the double group representations on the left directly. Doing the latter first gives

$$\Gamma_7(E_{5/2}) \rightarrow \Gamma_7^+ + \Gamma_8^+$$

$$\Gamma_6(E_{1/2}) \rightarrow \Gamma_5^+ + \Gamma_6^+$$

$$\Gamma_8(G) \rightarrow \Gamma_5^+ + \Gamma_6^+ + \Gamma_7^+ + \Gamma_8^+ \quad .$$

These give the labels of the Zeeman components on the left. To find out the approximate order of levels, we can assume that the lowest accidentally degenerate pair acts like an angular momentum $J = 5/2$ and see where the m_J components fall in the symmetry group of C_{4h}. If the field direction makes -5/2 lowest, the order of the levels is just the usual one, with 5/2 highest. We get $-5/2 \subset \Gamma_7^+$, $-3/2 \subset \Gamma_8^+$, $-1/2 \subset \Gamma_5^+$, $1/2 \subset \Gamma_6^+$, $3/2 \subset \Gamma_7^+$, $5/2 \subset \Gamma_8^+$. This list contains the correct number of the right kinds of representations for the cubic $\Gamma_7 + \Gamma_8$ level and establishes a probable energy ordering.

At the right side of the diagram we have the -3/2 spin component lowest, and it is split by the orbital partners of T_1. The spin component belongs to Γ_8^+ of C_{4h}, and T_1 breaks into representations of the ordinary part of C_{4h}; namely $\Gamma_1^+ + \Gamma_3^+ + \Gamma_4^+$. These when multiplied by Γ_8^+ give $\Gamma_8^+, \Gamma_7^+, \Gamma_5^+$ respectively. These three correlate with the lowest three Zeeman components on the left side of the diagram.

In $RbMnF_3$ the ground state $M_S = -5/2$ is Γ_7^+ in C_{4h}. The product of this with the excited state symmetries gives the exciton symmetries. In this case it is

$$\Gamma_7^*(\Gamma_8, \Gamma_7, \Gamma_5) = \Gamma_4, \Gamma_1, \Gamma_2$$

where it is important to remember the complex conjugation.

We still have not gone far enough, however, because $RbMnF_3$ is an antiferromagnetic crystal. With spins along [001] (z-axis) there are two Mn^{++} ions per unit cell having opposed spins. The space group can be simplified when $k = 0$ to the factor group; in this case it is isomorphous to D_{4h}. The character table for D_{4h} is shown as Table 3. The operations $E, C_4, C_2, \sigma_v, \sigma_d$, etc., eight in all, do not interchange atoms. The other operations should be written $\{C_2'|\tau\}$, for example, showing that C_2' can only be a symmetry operation when the sublattices are interchanged via the non-primitive translation τ. This factor group makes it easy to take care of the presence of the antiferromagnetic structure of the crystal.

TABLE 3

GROUPS $\bar{D}_4$, $\bar{C}_{4v}$, AND $\bar{D}_{2d}$ $D_{4h} = D_4 \times i$

$\bar{D}_4$	E	R	C_4 C_4^3R	C_4^3 C_4R	C_4^2 C_4^2R	$2C_2'$ $2C_2'R$	$2C_2''$ $2C_2''R$	$\bar{D}_4$		
$\bar{C}_{4v}$	E	R	C_4 C_4^3R	C_4^3 C_4R	C_4^2 C_4^2R	$2\sigma_v$ $2\sigma_vR$	$2\sigma_d$ $2\sigma_dR$		$\bar{C}_{4v}$	
$\bar{D}_{2d}$	E	R	S_4 S_4^3R	S_4^3 S_4R	C_2 C_2R	$2C_2'$ $2C_2'R$	$2\sigma_d$ $2\sigma_dR$			$\bar{D}_{2d}$
$M_1 A_1$ Γ_1	1	1	1	1	1	1	1		z	
$M_2 A_2$ Γ_2	1	1	1	1	1	−1	−1	z, L_z	L_z	L_z
$M_3 B_1$ Γ_3	1	1	−1	−1	1	1	−1			
$M_4 B_2$ Γ_4	1	1	−1	−1	1	−1	1			z
$M_5 E$ Γ_5	2	2	0	0	−2	0	0	x, y, L_x, L_y	x, y, L_x, L_y	x, y, L_x, L_y
$E_{1/2}\Gamma_6$	2	−2	$\sqrt{2}$	$-\sqrt{2}$	0	0	0			
$E_{3/2}\Gamma_7$	2	−2	$-\sqrt{2}$	$\sqrt{2}$	0	0	0			

We now combine the exciton symmetries on the separate sublattices into excitons in the crystal. The up-sublattice excitons are Γ_3^+, Γ_1^+, Γ_2^+, listed in the same order as for the down-sublattice. The wavefunction on the two sublattices must be combined with a + or - sign with respect to one of the symmetry operations, which interchanges sublattices. If we choose C_2' and define it as the [101] axis, we can generate a consistent set of symmetry functions. Perhaps a little neater method, however, is to transform the transition density of one sublattice into that of the other. These are the products of ground and excited state wavefunctions, and we deal with the product symmetry as just found for the two sublattices. The operation $C_2' = C_2^y$ reverses the spin direction and changes the orbital components. We could proceed by classifying the site functions under D_{4h} and then identifying these functions with the two sublattice functions.

There is a fairly obvious correlation between the two groups, however. The M_5 representation of D_{4h} can be reduced in C_{4h} into $M_5 = \Gamma_3 + \Gamma_4$, so the two site functions Γ_3, Γ_4 become the degenerate representation M_5 in D_{4h}. The degeneracy arises because C_2' converts C_4 into C_4^3. The other representation coefficients of D_{4h} are the same as those of Γ_1 and Γ_2 of C_{4h} for the same operations and differ in the signs of the entries for C_2', C_2''. Thus a function of Γ_1 symmetry of C_{4h} is expanded into two functions of symmetry M_1 and M_2 of D_{4h}. Γ_2 is expanded into $M_3 + M_4$.

Magnetic dipole transitions in the crystal can occur along z for M_2 excitons and as right or left circularly polarized transitions for M_5 excitons. Thus the final answer as to the polarization rules is: one transition is z-polarized, one is forbidden, and one is x,y polarized. Since these are magnetic transitions, they would be observed as σ, forbidden, and π respectively, going with the Γ_7, Γ_5 and Γ_8 excited states.

These selection rules are apparently obeyed as far as can be seen in uniaxial stress studies of $RbMnF_3$(8). A strong σ-polarized line is seen when the stress reaches the value needed for spin reorientation, and a weaker π line is observed until it gets lost in the tail of the magnon sideband. The third transition is not observed in agreement with its forbidden nature. The order of the two observed levels is in agreement with the correlation diagram, $E_\sigma < E_\pi$.

The groups used in the foregoing analysis are called the unitary components of the magnetic groups. The non-unitary components contain time inversion as a part of the symmetry operation. For example, we could have used the point group D_{4h} to classify the site functions if we had replaced operations like C_2' with $C_2'T$ where T is the time inversion operator. For most purposes, however, one can stay with the ordinary groups.

The somewhat complicated analysis of the group theory of $RbMnF_3$ excitons is summarized in Fig. 9.

O_h cubic $^4T_1 \rightarrow GXT_1 \rightarrow \underbrace{\Gamma_8 + \Gamma_7}, \Gamma_8, \Gamma_6$

$-5/2, -3/2, -1/2, 1/2, 3/2, 5/2$

C_{4h} Mag. Field

$\Gamma_7 \quad \Gamma_8 \quad \Gamma_5 \quad \Gamma_6 \quad \Gamma_7 \quad \Gamma_8$

C_{4h} Exch. Field $(-3/2)XT_1 \rightarrow \underbrace{(\Gamma_7 \Gamma_8 \Gamma_5)}\downarrow \; ; \; \underbrace{(\Gamma_6 \Gamma_7 \Gamma_8)}\uparrow$
$\downarrow \Gamma_7^* \qquad\qquad \downarrow \Gamma_8^*$

Exciton Rep $(\Gamma_1 \Gamma_4 \Gamma_2)\downarrow \; ; \; (\Gamma_2 \Gamma_3 \Gamma_1)\uparrow$

$M_3 + M_4 - - - - o$

D_{4h} Factor Gp. $M_5 \leftarrow - - \sigma$

$M_1 + M_2 \leftarrow -- \pi$

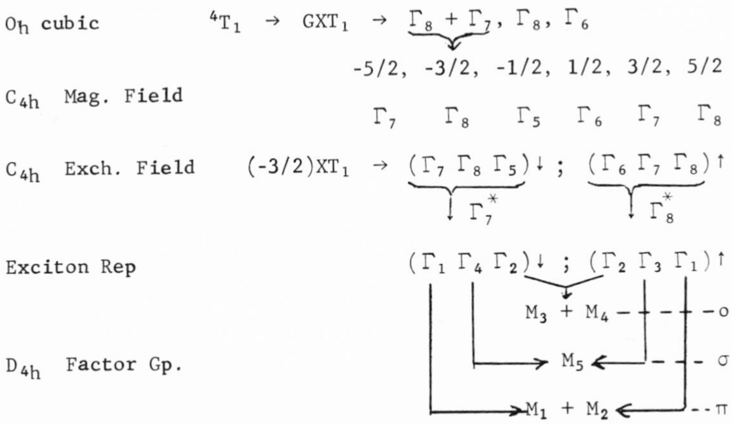

Fig. 9   Decomposition of a $^4T_1$ exciton in D_{4h} magnetic group in 2-sublattice antiferromagnet, $RbMnF_3$.

V. EXAMPLES OF SPECTRA OF MAGNETIC CRYSTALS

The optical spectra of a few magnetic crystals are now fairly well understood, and the interpretation of the data has led to increased understanding of energy transfer, exchange coupling and other interesting quantities. The best studied examples of antiferromagnets are probably MnF_2 and Cr_2O_3, both of which have orbitally non-degenerate ground states. More complex problems are presented by such compounds as CoF_2 and FeF_2 having orbitally degenerate groundstates. Less of a quantitative nature is known about ferromagnetic insulators such as $CrBr_3$ and metamagnetic salts such as $CsMnCl_3 \cdot 2H_2O$. The latter is a nearly one-dimensional magnet in the sense that the J-value is large in only one crystallographic direction. There is considerable interest in these highly anisotropic one- and two-dimensional magnetic structures.

V.A. MnF_2

MnF_2 is one of the nearly ideal antiferromagnets and its magnetic properties were well understood by the time its optical spectrum was studied seriously. Although each ion is in an orthorhombic field, it is nearly an octahedral one. There are two ions per cell and the crystal is tetragonal in overall symmetry, as shown in Fig. 10.

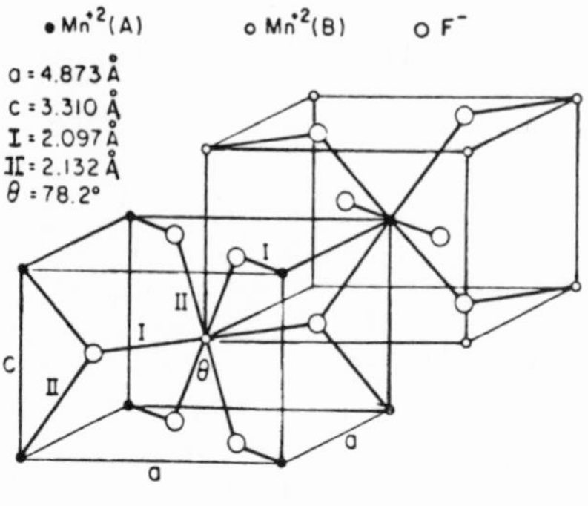

Fig. 10 Crystal structure of antiferromagnetic MnF_2. Symbols I and II refer to the two types of Mn − F bonds. Spins on sublattices A and B are up and down, respectively.

This figure also shows the spin arrangement in the ordered state. A special feature of this structure is that the two ions in the unit cell have opposite spins; therefore, since the two ions were already distinguished in the chemical structure, the spin ordering does not change the number of ions in the unit cell. An example of the other behavior would be $RbMnF_3$ in which there is one Mn per cell in the paramagnetic state, but two in the antiferromagnetic state.

In the spectrum of MnF_2 one sees sharp magnetic dipole lines, which are transitions occurring on single Mn^{++} ions (9). These are the excitons of the system. The selection rules for these are determined by the site group, C_{2h}. The procedure for analyzing them is the same as was discussed in Section IV. F, and they seem to be well understood.

The major features in the optical spectrum of MnF_2 are the spin-wave sidebands. All of the transitions are spin forbidden so the magnetic dipole lines are very weak, even for this type of transition. Electric dipole transitions are forbidden; however, the spin-wave sidebands appear in the spectrum by this mechanism, and therefore both the spin and parity selection rules are broken. This feat is accomplished by the exchange induced electric dipole mechanism, discussed in Section II. E. These cooperative transitions are several orders of magnitude stronger than the magnetic ones.

One of the most interesting spectral regions in MnF_2 is 25,200-25,300 cm^{-1}, the $^6A_{1g} \rightarrow {}^4E$ transition. This spectrum is shown in Fig. 11.

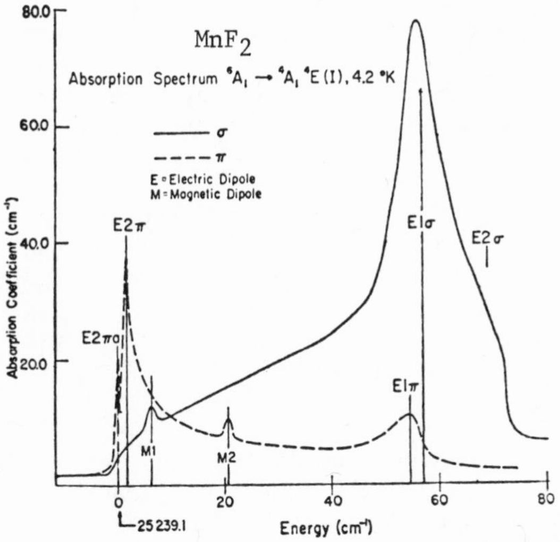

Fig. 11 Absorption near the origin of the transition $^6A_1 \rightarrow {}^4A_1$, $^4E(I)$ at 4.2^0 K.

Two magnetic dipole lines are observed, denoted M1 and M2. There are strong σ-polarized magnon sidebands about 60 cm^{-1} higher. This energy corresponds to the energy at the zone edge of the MnF_2 magnon spectrum, similar to the right side of the diagram of Fig. 4. Because of the selection rule eq. 69, transitions occur at all points in the magnon and exciton zone, but both magnon and exciton wave vectors are equal. This results in the appearance of a broad band of magnon states, which maps out the joint density of magnon-exciton levels. The actual line shape is influenced by the transition probabilities for each value of k in eq. 59. In order to work this out, one must add up the pair moments following Sec. II. E for every near neighbor of an excited Mn^{++} ion; these neighbors have the spin deviations which cause the ion pair transitions as in Fig. 2. The k-dependence has to be found by a Fourier analysis of the pair moments.

The results for an exciton on sublattice A (Fig. 10) are

$$M_1^{\xi} = 4 \sin\left(k_z\ c/2\right)\left[P_1^{\xi} \cos\left(k_x - k_y\right) a/2 + P_2^{\xi} \cos\left(k_x + k_y\right) a/2\right]$$

$$M_1^{z} = 4\ P_2^{z} \cos\left(k_z\ c/2\right) \sin\left(k_x - k_y\right) a/2 \tag{93}$$

where P_1 is the effective moment due to ions in the set of four B sublattice Mn^{2+} in the $-x+y$ plane of Fig. 10, while P_2 are moments caused by the set in the $x+y$ plane. The direction z corresponds to the c axis and ξ to $a_1 + a_2$ (in a plane of set 2). If the exciton had been in the other sublattice, the direction $\eta = a_1 - a_2$ would appear in eq. 93. When the two fixed sublattice exciton-magnon functions are combined so as to produce Davydov components, the moments can lie anywhere in the a_1, a_2 plane, thus showing the behavior appropriate for a D_{4h} crystal structure.

When moments for a given polarization are summed over all transitions, i.e., over all k, an expression for the line profile can be found. The line shape will have the general form

$$\alpha_p(v) = \text{const} \sum_k |M_p(k)|^2\ \delta\left(v - v_k^e - v_k^m\right) \tag{94}$$

where M_p is one of the preceding expressions, eq. 93, depending upon the polarization p, the δ-function fixes the energy at which the k index in the sum is to be taken and k is summed over the first Brillouin zone.

The sideband shape is a convolution of the magnon dispersion and exciton dispersion. Once the magnon dispersion and the k-dependent transition probability have been worked out, it is possible to fit the sideband shape by using exciton dispersion parameters. In this way one can find the excitation exchange rates in various states and for various directions. Normally, the excitons have very little dispersion. This is because the inter-sublattice migration is prevented by the antiferromagnetic structure and because intra-sublattice exchange is small. (In the ground state $J_1 = 0.22\ cm^{-1}$ where J_1 is the nearest neighbor intra-sublattice exchange integral. $J_2 = 1.22\ cm^{-1}$, and $J_3 < 0.035\ cm^{-1}$.) The Franck-Condon factors for many of the states cut down the exciton dispersion by factors of perhaps 10-100 following the analysis of Section III. B. For all these reasons the magnon sidebands reflect the magnon density of states rather faithfully in many cases.

In one sideband of the ^4E state of MnF_2, the exciton dispersion is extraordinarily large. There is a negligible Franck-Condon factor in this state since it has the same electron configuration as the ground state e^2t^3. Also the intra-sublattice exchange in the c-axis direction is extraordinarily large for the $E\theta$ component of ^4E. As a result, the exciton dispersion is large enough to push the magnon sideband to the low energy side of the magnetic dipole line. This is seen as the sharp cutoff at 25,240 cm^{-1} in Fig. 11. This event can only happen when the exciton dispersion is of opposite sign from and greater than that of the magnon. Using a simple exciton dispersion law,

$$E^{ex}_{(K)} = E^{o} + K_1 \cos k_z c$$

the value $K_1 = 37\ cm^{-1}$ fits the data.

Also observed near the low energy edge is a sharp band which has been identified as a magnon-exciton bound state.

From the diagram of the crystal structure, Fig. 10, it is easy to see that the c-axis dispersion could be very large. The nearest neighbors along the c-axis have the same spin orientation, and energy transfer by superexchange or direct exchange is possible. In the $RbMnF_3$ lattice there are no such structural features, and in fact no dispersion of the exciton.

In the presence of a very large magnetic field, excitation exchange between opposite sublattices becomes possible as the spins rotate toward the field direction. In a large enough field, an antiferromagnet would become ferromagnetically aligned. This type of Davydov splitting has apparently been observed by Eremenko in $RbMnF_3$ at 300 kgauss (10).

V.B. Cr_2O_3 (11)

Chromium oxide is an antiferromagnetic insulator with a Néel temperature of 308^0K. It has the corundum structure. Ruby containing Cr^{+3} at Al^{+3} sites in corundum has been compared to Cr_2O_3 and there are many resemblances (12). A striking difference, however, is in the line structure of the $^4A_2 \rightarrow {}^2E$ transition. Whereas in ruby there are two weak lines 29 cm^{-1} apart, in Cr_2O_3 there are four or more in a 200 cm^{-1} range. The large splitting has been interpreted as Davydov splitting resulting from excitation transfer between ions having similar spins.

Ion pairs in corundum have been intensively studied, but only recently have the spectral details of the many types of pairs begun to be correctly interpreted (3). Ultimately it will be possible to use the individual pair interactions in ruby to construct the exchange splitting of Cr_2O_3, and thus to compare these quite similar systems.

V.C. $FeCO_3$ (13)

$FeCO_3$ has the calcite structure: The Fe^{++} are surrounded by six $O^=$ ions from the $CO_3^=$ groups, forming an octahedron with a small trigonal distortion. The magnetic structure below 32^0 K is antiferromagnetic with the spins aligned along C_3. In an external magnetic field along C_3 of 180 kgauss or more, the spins of the sublattices become substantially parallel. The absorption spectrum at 25,200-25,500 cm^{-1} is due to magnetic dipole pure electronic lines and to spin-wave sidebands. The latter appear because of the pair mechanism discussed in Sec. II. D. When the external field approaches 180 kgauss, the spin-wave sidebands nearly disappear because the antiparallel spin arrangement necessary to make spin-forbidden bands appear has been destroyed.

V.D. Other Examples

Many more examples could be mentioned here. Rare earth compounds and rare earth-transition metal compounds show ion pair transitions, spin wave sidebands and interesting effects due to the presence of inequivalent magnetic sublattices (14). It is hoped that this brief review will make it possible for the reader to appreciate these novel phenomena.

REFERENCES

1. H. J. Zeiger and G. W. Pratt, Magnetic Interactions in Solids, Clarendon Press, Oxford, 1973.

2. D. L. Dexter, Phys. Rev. 126, 1962 (1962).

3. J. P. van der Ziel, Phys. Rev. B9, 2846 (1974).

4. D. P. Craig and S. H. Walmsley, Excitons in Molecular Crystals, W. A. Benjamin, Inc., 1968

5. F. Keffer, Spin Waves, in Encyclopedia of Physics, Vol. XVIII/2, S. Flugge, Ed., Springer Verlag, Berlin, 1966.

6. E. P. Wigner, Group Theory, Academic Press, New York, 1959.

7. R. Loudon, Adv. in Physics 17, 243 (1968).

8. E. I. Solomon and D. S. McClure, Phys. Rev. B6, 1697 (1972).

9. R. S. Meltzer, M. Lowe, and D. S. McClure, Phys. Rev. 180, 561 (1969).

10. V. V. Eremenko and V. P. Novikov, Pis'ma Zh. Eksp. Teor. Fiz. 11, 478-82 (1970) CA. 73, 60616 x (1970).

11. J. W. Allen, R. M. MacFarlane, and R. L. White, Phys. Rev. 179, 523 (1969).

12. D. S. McClure, J. Chem. Phys. 38, 2289 (1963).

13. V. V. Eremenko, Yu. G. Litivinenko, and V. I. Myatlik, Pis'ma Zh. Eksp. Teor. Fiz. 12, 66-9 (1970).

14. R. S. Meltzer, Phys. Rev. B2, 2398 (1970).

ENERGY TRANSFER PHENOMENA

R. K. Watts

Texas Instruments Inc.

Dallas, Texas 75222

ABSTRACT

The transfer of optical excitation energy between weakly coupled ions in solids is discussed. The most important coupling mechanisms are treated first from a microscopic point of view, and then statistical averages are taken over the volume of the sample to calculate observable quantities such as the yield of sensitizer luminescence and the time development of the sensitizer luminescence following pulse excitation. Three regimes are considered: negligible sensitizer → sensitizer transfer, transfer between sensitizers comparable in probability with sensitizer → activator transfer, and migration of excitation among sensitizers much more rapid than transfer to activators. The discussion is largely applied to rare earth ions.

I. INTRODUCTION

The study of energy transfer has a long history. For many years non-radiative transfer of excitation energy has been studied in organic liquid and solid systems and only slightly more recently in inorganic systems as well. Often the impetus of investigations of inorganic systems has been the possibility of increasing the efficiency of a phosphor or laser. In general the system of interest has two different types of optically active component. One species is called the sensitizer or energy donor and the other is called the activator or energy acceptor. We shall label these A and B respectively in the following discussion. The two species may in some cases be identical. If absorption of exciting radiation by the sensitizer produces luminescence by the activator, we may infer that energy transfer from sensitizer to activator has occurred. That is, we are assuming that absorption or emission bands may be identified separately with activator or sensitizer. This will be the case only if the coupling between sensitizer and activator is sufficiently weak.

Since non-radiative energy transfer from sensitizer to activator represents another decay path for the excited state of the sensitizer, we may expect the lifetime of the excited state to be less in the presence of activators than in their absence. This is not the case if the transfer is radiative - that is, if a photon emitted by a sensitizer is absorbed by an activator. If the sensitizers and activators are identical, the emitted radiation may be trapped in the sample, leading to an apparently increased lifetime. In such a case the apparent lifetime is dependent on the size and shape of the sample, however. Radiative transfer is generally easily recognized and is of no interest here. Non-radiative transfer to activators depletes the population of the excited state of the sensitizer, and decreases the intensity of light emitted in radiative transitions from this state to a lower state of the sensitizer. Ions will be the only type of sensitizer or activator considered.

When the interaction between two ions in a crystal is so strong that the energy levels of the pair do not resemble the levels of the isolated ions, the two ions must be considered a single system. In the opposite limit of weakly interacting ions the energy levels of the individual ions may be shifted because of the interaction by amounts smaller than the widths of the levels, and this is the regime with which we shall be concerned. The interaction will in general be weaker the greater the separation between ions, but may differ greatly depending on the types of ions concerned. Coupling of near neighbor Cr^{3+} ions in Al_2O_3 leads to energy level splittings (1) of $10^3 cm^{-1}$, while for Nd^{3+} near neighbors in $NdCl_3$ the splittings are three orders of magnitude smaller (2). When the coupling is weak, it is reasonable to speak of non-radiative transfer of excitation energy between ions.

Let us consider a crystal containing two impurity ions labeled A and B separated by the distance R. Each ion has two states a, a' and b, b', as shown in figure 1a. We suppose that R is large enough that the wavefunctions of the two ions do not overlap. The Hamiltonian of the system is

$$H = H_A(\vec{r}_A) + H_B(\vec{r}_B) + H_{AB}(\vec{r}_A, \vec{r}_B) \tag{1}$$

$\vec{r}_A$ represents the coordinates of the electrons of ion A. H_A contains the interactions of ion A with the other ions of the crystal, except ion B, and a similar statement applies to H_B. In the absence of the small interaction term H_{AB} the system has four eigenstates

$$|1\rangle = |a'b\rangle \qquad |2\rangle = |ab'\rangle \tag{2}$$

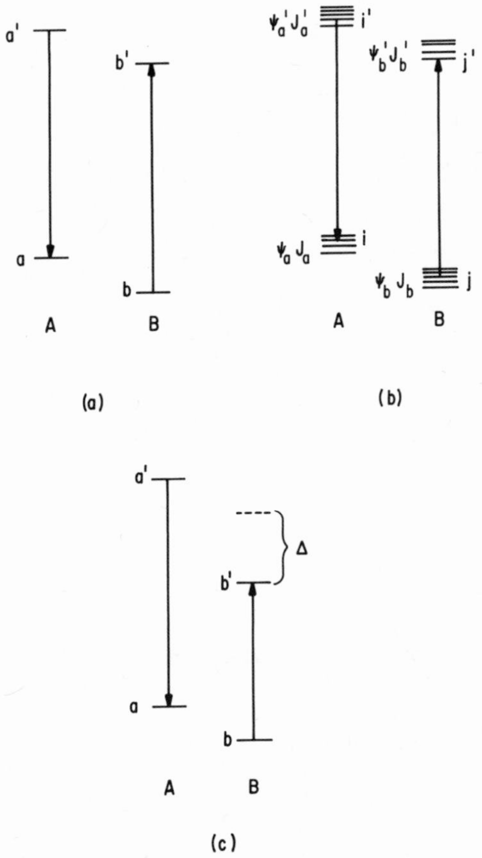

Fig. 1 Non-radiative energy transfer from ion A to ion B. In (a) ion A makes a transition from upper state a' to lower state a and ion B makes a transition from state b to state b'. The transition energies are equal. (b) is a specialization of (a) to rare earth ions. The transitions are from a crystal field component of one multiplet ψJ to a component of another multiplet. In (c) the transition energies differ by the amount Δ, and the transfer is said to be non-resonant. The energy difference Δ is given to lattice vibrational excitation.

$|3\rangle = |ab\rangle \qquad |4\rangle = |a'b'\rangle$

In state $|1\rangle$ ion A is in its upper state a' and ion B is in the lower state b, etc. If H_{AB} is not zero a general time dependent wavefunction of the system can be written as a sum of the form

$$|\psi(t)\rangle = \sum_{j=1}^{4} c_j(t)|j\rangle \exp(-iE_j t/\hbar). \tag{3}$$

The energies E_j are those of the unperturbed states. $E_1 = E_{a'} + E_b$, for example.

If at time t=0 the system is in the state $|1\rangle = |a'b\rangle$, then at a later time t the coefficient $c_1(t)$ will in general be less than one and the other three coefficients will be non-zero. However, since

$$|c_m|^2 = 4|\langle m|H_{AB}|1\rangle|^2(E_m-E_1)^{-2}\sin^2[(E_m-E_1)t/2\hbar], \tag{4}$$

only the coefficient c_2 will be appreciable if $E_{a'}-E_a = E_{b'}-E_b$ as indicated in the figure. We may then limit the sum in equation (3) to the two terms j=1 and 2.

For the sake of simplicity we have made no provision for the decay of excitation. This could be taken into account by including in the Hamiltonian two terms, one to allow for the decay of states of ion A by interaction with the radiation field and one for ion B. But the states of most interest to us are those of iron group or rare earth transition metal ions with only weakly allowed radiative transitions. Thus H_{AB} can be much larger than these two neglected terms if R is not too large and still be very small compared with H_A and H_B.

When $|\psi(t)\rangle$ is inserted in the Schroedinger equation

$$i\hbar\frac{\partial}{\partial t}|\psi(t)\rangle = H|\psi(t)\rangle , \tag{5}$$

and use is made of the equations for the single ions,

$$H_A|a\rangle = E_a|a\rangle , \tag{6}$$

the result is the two coupled equations for the coefficients $c_1(t)$ and $c_2(t)$,

$$\begin{aligned}\frac{d}{dt}c_1(t) &= -iv_1c_1(t) + iuc_2(t)\\ \frac{d}{dt}c_2(t) &= -iv_2c_2(t) + iu^*c_1(t),\end{aligned} \tag{7}$$

where

$$v_1 = \langle 1|H_{AB}|1\rangle/\hbar \quad , \quad v_2 = \langle 2|H_{AB}|2\rangle/\hbar \tag{8}$$

$$u = -\langle 1|H_{AB}|2\rangle/\hbar \quad .$$

The solution of equations (7) is

$$c_1(t) = [\cos(qt) - (i/2q)(v_1-v_2)\sin(qt)]\exp[-i(v_1+v_2)t/2]$$

$$c_2(t) = i(u^*/q)\sin(qt)\exp[-i(v_1+v_2)t/2] \ ,$$

$$q = [uu^* + (1/4)(v_1-v_2)^2]^{\frac{1}{2}} \quad . \tag{9}$$

This is a system of two coupled waves. The excitation oscillates back and forth between states $|1\rangle = |a'b\rangle$ and $|2\rangle = |ab'\rangle$, that is between ions A and B. The exchange of probability is the more complete the smaller the mismatch v_1-v_2. The probability that at time t ion B is excited and ion A is not excited is given by

$$|c_2(t)|^2 = |u/q|^2\sin^2(qt)$$

$$\simeq |u|^2t^2 \ , \text{ for } qt \text{ small.} \tag{10}$$

This model is not realistic, however. It does not correspond to the weakly coupled ion systems which will be discussed in later sections because we have neglected the finite widths of the energy levels due to other interactions and have considered them sharply defined. These widths are not due to interaction with the radiation field. Radiative lifetimes of the ions of interest lie largely in the range 10^{-6}-10^{-3} sec. From the uncertainty principle these correspond to widths of 3×10^{-5} - 3×10^{-8} cm^{-1}, but at room temperature typical observed widths are 1 cm^{-1} - 10 cm^{-1} for rare earth ions, corresponding to the very short relaxation times 3×10^{-11} - 3×10^{-12} sec. These widths are due to ion-phonon interactions of various kinds. The widths are also larger than the shifts $\hbar v_1$ and $\hbar v_2$ produced by the ion-ion interaction. (Inhomogeneous broadening mechanisms which lead, for example, to the very large widths of energy levels of ions in glasses will be assumed to be unimportant.) The energy levels of rare earth ions consist of well separated multiplets, each multiplet being split by static "crystal field" interactions with the other ions of the crystal into a number of sublevels of spacing ten to a few hundred cm^{-1}. Since these spacings are of the order of phonon energies, very rapid transitions in which phonons are created or destroyed occur between the sublevels. These processes lead to thermal equilibrium of the populations and to the observed widths. On the other hand the width of a completely isolated level may be determined by a Raman process in which a phonon is destroyed and another of nearly the same frequency is created. In any event we may assume that at time t=T some such process occurs and interrupts the time development of the phase of the wavefunction, so that the change in $|c_2|^2$ in the small interval (0,T) is given by

$$\delta|c_2|^2 = |u|^2T^2 , \qquad (11)$$

and at the time of interest t, which is much larger than T, $|c_2(t)|^2$ is roughly

$$|c_2(t)|^2 = (t/T)\delta|c_2|^2$$
$$= |u|^2Tt . \qquad (12)$$

The transition probability is then

$$w(1\rightarrow 2) = |u|^2T . \qquad (13)$$

If the level width, or, more properly, the transition linewidth ΔE since the widths of both initial and final levels contribute, is substituted for T through the uncertainty relation $T\Delta E = h$, we obtain

$$w(1\rightarrow 2) = (4\pi^2/h)|\langle 1|H_{AB}|2\rangle|^2(\Delta E)^{-1} . \qquad (14)$$

If the transitions $a'\rightarrow a$, $b\rightarrow b'$ are represented by the line shape functions $g_A(E)$ and $g_B(E)$, each normalized in the sense $\int g(E)dE = 1$, this becomes

$$w_{AB} = (4\pi^2/h)|\langle a'b|H_{AB}|ab'\rangle|^2\int g_A(E)g_B(E)dE . \qquad (15)$$

For a more rigorous derivation of the transfer probability the reader is referred to references (3) and (4), and for a treatment of transfer with stronger ion-ion interactions to reference (5).

II. ION-ION INTERACTIONS

The two coupling mechanisms which must be considered are the electrostatic Coulomb interaction between the electrons of ion A and those of B, and the exchange interaction. Let us first consider the Coulomb interaction and fix the origins of two identically oriented coordinate systems (r_A, θ_A, ϕ_A) and (r_B, θ_B, ϕ_B) at the positions of ions A and B. If the position of ion B, measured in the (r_A, θ_A, ϕ_A) system, is (R, Θ, Φ), the interaction is,

$$H_{AB} = \sum_{s,t} e^2/r_{st} = \sum_{s,t} e^2|\vec{r}_{As} - \vec{r}_{Bt} - \vec{R}|^{-1} . \qquad (16)$$

The sum is over the electrons of the two ions. For R greater than

r_{As} and r_{Bt}, equation (16) can be expanded (6),

$$H_{AB} = e^2 \sum_{\substack{k1,k2\\q1,q2}} R^{-1-k1-k2} G[C^{(k1+k2)}_{q1+q2}(\Theta,\Phi)]^* D^{(k1)}_{q1}(\vec{r}_A) D^{(k2)}_{q2}(\vec{r}_B).$$

$$G = (-1)^{k1}\left[\frac{(2k1+2k2+1)!}{(2k1)!(2k2)!}\right]^{1/2}\begin{pmatrix} k1 & k2 & k1+k2 \\ q1 & q2 & -q1-q2 \end{pmatrix},$$

$$D^{(k)}_q(\vec{r}) = \sum_s r_s^k C^{(k)}_q(\theta_s,\phi_s),$$

$$C^{(k)}_q(\Omega) = (4\pi/2k+1)^{1/2} Y^q_k(\Omega). \tag{17}$$

In the sum k runs from 0 to ∞ and q runs from -k to k. $\begin{pmatrix} \end{pmatrix}$ is the 3j symbol. Dielectric screening factors may be included in equation (17) but seem from magnetic resonance measurements to be unnecessary (7). The matrix element is

$$\langle a'b|H_{AB}|ab'\rangle = \sum_{\substack{k1,k2\\q1,q2}} (e^2/R^{1+k1+k2}) G C^{(k1+k2)*}_{q1+q2} \langle a'|D^{(k1)}_{q1}|a\rangle \times \langle b|D^{(k2)}_{q2}|b'\rangle. \tag{18}$$

When this is squared, the number of indices to be summed over is doubled, but if the wavefunctions are known, $|\langle H_{AB}\rangle|^2$ can be evaluated. A simplified general formula can be obtained by averaging over Θ and Φ. Because of the orthogonality of the C's many cross terms are eliminated. If the other cross terms are neglected, and if $(G)^2$ is averaged over q_1 and q_2, the result is

$$|\langle H_{AB}\rangle|^2 \simeq \sum_{k1,k2} (e^2/R^{1+k1+k2})^2 \frac{(2k1+2k2)!}{(2k+1)!(2k2+1)!} \sum_{q1} |\langle a'|D^{(k1)}_{q1}|a\rangle|^2 \times \sum_{q2} |\langle b|D^{(k2)}_{q2}|b'\rangle|^2. \tag{19}$$

$k_1 = k_2 = 1$ corresponds to electric dipole transitions for both ions. $k_1 = 1$, $k_2 = 2$ corresponds to a dipole transition for ion A and a quadrupole transition for ion B, etc.

Following Kushida (8) and Shinagawa (9), we apply equations (15) and (19) to the case of transfer between two rare earth ions. The transitions to be considered are those from the crystal field components $\psi_a' J_a' i'$ of multiplet $\psi_a' J_a'$ (The prime is meant to apply to the subscripts as well.) to the crystal field components $\psi_a J_a i$ of multiplet $\psi_a J_a$ of ion A with a similar notation for ion B. See figure 1b. ψ represents the other quantum numbers necessary to specify the state. The states $|\psi_a J_a i\rangle$ are linear combinations of the free ion states $|\psi_a J_a M_a\rangle$. In the intermediate coupling regime the state $|\psi_a J_a\rangle$ is in general composed of several states $|SLJ_a\rangle$ of different S and L. The total crystal field splitting of a multiplet is typically a few hundred cm^{-1} or less, and the population of the component levels is given by a Boltzmann distribution $z_i = \exp(-E_i/kT)$, where the energy zero is taken at the position of the lowest level of the multiplet. In this way the distribution of possible initial states is taken into account. Then the transfer rate becomes

$$w_{AB} = (4\pi^2/h) \sum_{\substack{i,i' \\ j,j'}} |\langle i'j|H_{AB}|ij'\rangle|^2 [(z_{i'} z_j / Z_a Z_b) \times \int g_{i'i}(E) g_{jj'}(E) dE]. \quad (20)$$

$|ij'\rangle$ is a short notation for $|\psi_a J_a i, \psi_b' J_b' j'\rangle$, and $Z_a = \sum_{i'} z_{i'}$. If the matrix element is approximated by its average value, (20) becomes,

$$w_{AB} = (4\pi^2/h) N \sum_{k1,k2} (e^2/R^{1+k1+k2})^2 \frac{(2k1+2k2)!}{(2k1+1)!(2k2+1)!} \times \sum_{i,i',q1} |\langle i'|D_{q1}^{(k1)}|i\rangle|^2 \sum_{j,j',q2} |\langle j|D_{q2}^{(k2)}|j'\rangle|^2 \, S, \quad (21)$$

where S is the quadruple sum over the quantity in square brackets in equation (20) and $N^{-1} = (2J_a+1)(2J_a'+1)(2J_b+1)(2J_b'+1)$.

If the ion is located at a lattice site with inversion symmetry, then $\langle D_q^{(k)}\rangle$ is zero for k odd. If there is no inversion symmetry, it will in general not be zero because the states $|\psi J i\rangle$ will contain some small component of opposite parity admixed by the odd-parity components of the crystal field. In fact the transition strength is then usually determined by this forced electric dipole mechanism, and the radiative transition probability for the transition $\psi J \rightarrow \psi' J'$ is given by

$$P_{E1} = (64\pi^4 e^2 \nu^3 / 3hc^3) \chi (2J+1)^{-1} \sum_{i,i'q} |\langle i|D_q^{(1)}|i'\rangle|^2, \quad (22)$$

if the temperature is high enough that all the states $|\psi J i\rangle$ are equally populated. ν is the mean frequency of the inter-multiplet transition. χ is a local-field correction factor. The selection rule for the forced electric dipole transition is $\Delta J \leq 6$. Transitions from a multiplet with J=0 to one with J odd should be very weak.

For the quadrupole transition (k=2) the quantity $\Sigma|\langle D_q^{(2)}\rangle|^2$ must be evaluated. Since the sum is over all the crystal field components of a multiplet it is best to write the states $|\psi Ji\rangle$ in terms of the free ion states $|\psi JM\rangle$

$$|\psi Ji\rangle = \sum_M c(i,M)|\psi JM\rangle \; . \tag{23}$$

Then because of the condition $\Sigma_i c(i,M)c^*(i,N) = \delta_{M,N}$ obeyed by the coefficients c, we have

$$\sum_{i,i'} |\langle\psi Ji|D_q^{(2)}|\psi'J'i'\rangle|^2 = \sum_{M,M'} |\langle\psi JM|D_q^{(2)}|\psi'J'M'\rangle|^2 \; . \tag{24}$$

From the definition of $D_q^{(k)}$ given in equation (17), we see that $D_q^{(k)}$ is a product of a function of angular variables ($C_q^{(k)}$) and r^k, so that

$$\langle\psi JM|D_q^{(k)}|\psi'J'M'\rangle = \langle r^k\rangle\langle\psi JM|\sum_s C_q^{(k)}(\Omega_s)|\psi'J'M'\rangle \; , \tag{25}$$

since all the electrons are in 4f orbitals. $\langle r^k\rangle$ is the expected value of r^k for a 4f orbital. The $C_q^{(k)}(\Omega_s)$ are tensor operators and satisfy the relation

$$\langle\ell m|C_q^{(k)}|\ell' m'\rangle = (-1)^{\ell-m}\langle\ell\|C^{(k)}\|\ell'\rangle\begin{pmatrix}\ell & k & \ell'\\ -m & q & m'\end{pmatrix} \; ,$$

$$\langle\ell\|C^{(k)}\|\ell'\rangle = (-1)^{\ell}[(2\ell+1)(2\ell'+1)]^{1/2}\begin{pmatrix}\ell & k & \ell'\\ 0 & 0 & 0\end{pmatrix} \; .$$

$$\langle f\|C^{(2)}\|f\rangle = -\sqrt{28/15} \; . \tag{26}$$

The tensor operators $u_q^{(k)}$ will satisfy a relation like (26), but with unit amplitude $\langle\|u^{(k)}\|\rangle$, if they are defined by

$$u_q^{(k)} = C_q^{(k)}/\langle\ell\|C^{(k)}\|\ell'\rangle \; . \tag{27}$$

The sum of the one-electron unit tensor operators, $U_q^{(k)}=\Sigma_s u_q^{(k)}(\Omega_s)$, is itself a tensor operator and also satisfies a relation of the same form:

$$\langle\psi JM|U_q^{(k)}|\psi'J'M'\rangle = (-1)^{J-M}\langle\psi J\|U^{(k)}\|\psi'J'\rangle\begin{pmatrix}J & k & J'\\ -M & q & M'\end{pmatrix} \; . \tag{28}$$

From equation (21) the quantity of interest is

$$\sum_{q,M,M'} |\langle\psi JM|D_q^{(k)}|\psi'J'M'\rangle|^2 = \langle r^k\rangle^2\langle f\|C^{(k)}\|f\rangle^2\langle\psi J\|U^{(k)}|\psi'J'\rangle^2 \times \sum_{q,M,M'}\begin{pmatrix}J & k & J'\\ -M & q & M'\end{pmatrix}^2 . \tag{29}$$

The sum of squares of the 3j symbols is one. Combining equations (29) and (24), we obtain,

$$\sum_{i,i',q} |\langle\psi Ji|D_q^{(2)}|\psi'J'i'\rangle|^2 = (28/15)\langle r^2\rangle^2\langle\psi J\|U^{(2)}\|\psi'J'\rangle^2 . \tag{30}$$

The reduced matrix element $\langle\|U^{(2)}\|\rangle$ can be found from the tables of reference (10) if the states $|\psi J\rangle$ are known in terms of pure Russel-Saunders states $|\gamma SLJ\rangle$,

$$|\psi J\rangle = \sum_{\gamma,S,L} d(\gamma,S,L)|\gamma SLJ\rangle . \tag{31}$$

Reference (11) contains a guide to the literature where decompositions of the form of (31) may be found. Since the elements tabulated are $\langle\gamma SL\|U^{(k)}\|\gamma'SL'\rangle$, the additional formula

$$\langle\gamma SLJ\|U^{(k)}\|\gamma'SL'J'\rangle = (-1)^{S+L'+J+k}[(2J+1)(2J'+1)]^{\frac{1}{2}}\begin{Bmatrix} J & J' & k \\ L' & L & S \end{Bmatrix} \times \langle\gamma SL\|U^{(k)}\|\gamma'SL'\rangle \tag{32}$$

is required. Since $U_q^{(k)}$ is an orbital operator, all matrix elements between Russel-Saunders states of different total spin S are zero. $\langle\psi J\|U^{(2)}\|\psi'J'\rangle$ is zero if $|J-J'|$ is greater than 2. When ΔJ is less than 2, $\langle\|U^{(2)}\|\rangle$ may vary by several orders of magnitude from transition to transition. $\langle r^2\rangle$ has been calculated by Freeman and Watson for nine different rare earth ions (12). Since the variation with atomic number is smooth, values for other rare earths may be found by interpolation. $\{\}$ is the 6j symbol (13).

It is usually not necessary to consider terms in equation (19) of higher order than k=2, since they are generally small for separations R of interest. The dependence of w_{AB} on the A-B separation R may be displayed by writing

$$w_{AB} = \alpha^{(6)}/R^6 + \alpha^{(8)}/R^8 + \alpha^{(10)}/R^{10} + \ldots \tag{33}$$

Let us estimate the values of the parameters $\alpha^{(n)}$ for the case where A = Yb^{3+} and B = Er^{3+}. The transfer process will consist of a $^2F_{5/2}\rightarrow{}^2F_{7/2}$ transition for Yb^{3+} and a $^4I_{15/2}\rightarrow{}^4I_{11/2}$ transition for Er^{3+}. If the ions are not at sites with inversion symmetry, quadrupole and dipole transitions are allowed for each. From Weber's data (14) the quantities of interest for Er^{3+} may be calculated. Using also the data of references (8) and (12), we find

$$\sum|\langle{}^4I_{15/2}j|D_q^{(1)}|{}^4I_{11/2}j'\rangle|^2 = 3.7 \times 10^{-21}\text{cm}^2 , \quad \langle r^2_{Er}\rangle = 0.666\text{au}$$

$$\sum|\langle{}^2F_{5/2}i'|\ D_q^{(1)}|{}^2F_{7/2}i\rangle|^2 = 2 \times 10^{-20}\text{cm}^2 , \quad \langle r^2{}_{Yb}\rangle = 0.613\text{au}$$

$$\langle{}^4I_{15/2}\|U^{(2)}\|{}^4I_{11/2}\rangle^2 = 0.0259 , \quad \langle{}^2F_{5/2}\|U^{(2)}\|{}^2F_{7/2}\rangle^2 = 6/49.$$

Taking $S = 0.1/\text{cm}^{-1}$, we obtain $\alpha^{(6)} = 8.3 \times 10^8 \text{ sec}^{-1}\ \text{Å}^6$,

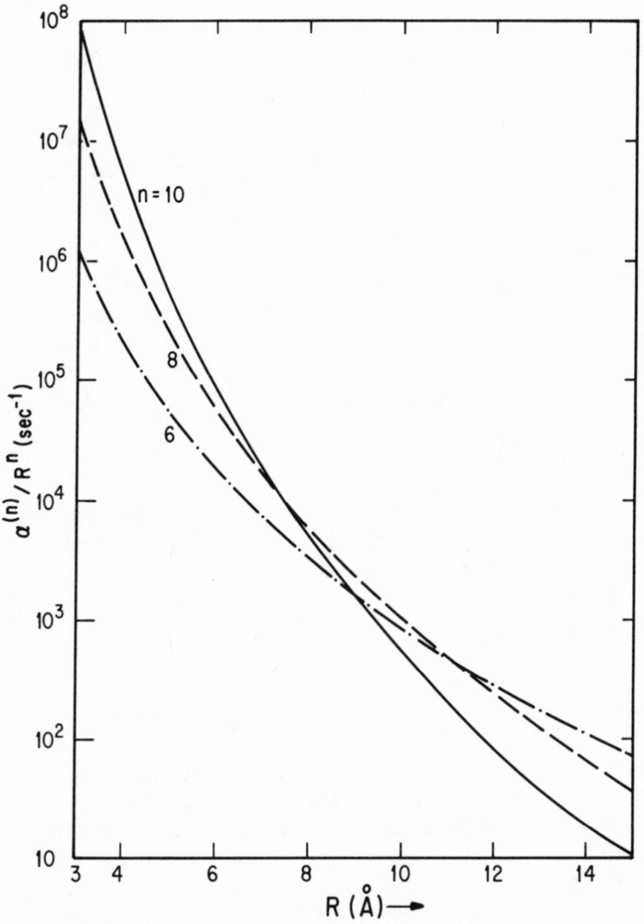

Fig. 2 Contributions to the Yb^{3+}-Er^{3+} energy transfer rate from dipole-dipole (n=6), dipole-quadrupole and quadrupole-dipole (n=8), and quadrupole-quadrupole (n=10) terms vs Yb-Er separation R. The coefficients α are calculated as described in the text. The contribution from the hexadecapole-quadrupole interaction (not shown) is comparable with the dipole-dipole term at R=3Å, but falls off much more rapidly with increasing R.

$\alpha^{(8)} = 9.8 \times 10^{10}$ sec^{-1} Å^8, $\alpha^{(10)} = 5.3 \times 10^{12}$ sec^{-1} Å^{10}. The dipole-quadrupole contribution to $\alpha^{(8)}$ is slightly larger than the quadrupole-dipole contribution. The three terms are plotted in figure 2. Which term dominates the coupling of a particular ion pair depends upon the separation R. Since the transfer is resonant and the lifetime of $^4I_{11/2}$ is generally longer than that of $^2F_{5/2}$, $B \rightarrow A$ back transfer will be important in this case. In most crystals the spontaneous decay rate (measured in a crystal containing only Yb^{3+} impurities) of the Yb^{3+} $^2F_{5/2}$ state is about 10^3 sec^{-1}. For nearest neighbor Yb-Er pairs the transfer rate is much larger than this. For neighbors separated by 10 Å the transfer rate is about equal to the Yb spontaneous decay rate.

Although quadrupole transitions make a negligible contribution to radiative decay probabilities, they can be important for non-radiative energy transfer. To see the reason for this we write the radiative transition probability for the quadrupole transition $\psi J \rightarrow \psi' J'$,

$$P_{E2} = (16\pi^6 e^2 \nu^5/hc^5)(112/225)\chi(2J+1)^{-1}\langle r^2\rangle^2\langle\psi J\|U^{(2)}\|\psi' J'\rangle^2 . \tag{34}$$

Taking the ratio P_{E2}/P_{E1} and the ratio of the dipole-quadrupole and dipole-dipole energy transfer rates w_{AB}^{dq}/w_{AB}^{dd}, we find that, apart from a numerical factor of order one,

$$w_{AB}^{dq}/w_{AB}^{dd} \simeq (\lambda/R)^2(P_{E2}/P_{E1}).$$

λ is the transition wavelength and may be three or four orders of magnitude larger than R. On the other hand radiative magnetic dipole transitions may be observed when the electric dipole transition is forbidden by symmetry, but play a negligible role in optical energy transfer. They are of importance for the analogous intra-multiplet cross-relaxation processes observed in magnetic resonance experiments, however.

If the ions A and B had each only a single active electron, the states a, a', b, and b' would be one-electron states and the matrix element of the electrostatic interaction would be

$$\langle a'(1)b(2)|e^2/r_{12}|a(1)b'(2)\rangle , \tag{35}$$

where the electrons are labeled 1 and 2. When the electron clouds of ions A and B overlap, the initial and final states $|a'b\rangle$ and $|ab'\rangle$ must be made antisymmetric in the coordinates of the two electrons in order to take into account the other contribution to the electrostatic interaction, the exchange term,

$$\langle a'(1)b(2)|-(e^2/r_{12})P_{12}|a(1)b'(2)\rangle = \langle a'(1)b(2)|-e^2/r_{12}|a(2)b'(1)\rangle. \tag{36}$$

P_{12} is an operator which interchanges electrons 1 and 2. The Coulomb term is still given by equation (35). For the general case where a, a', b, and b' are many-electron states, the exchange matrix element is

$$\langle a'b| - \sum_{p,t} (e^2/r_{pt}) P_{pt} |ab'\rangle . \tag{37}$$

Although overlap of the charge clouds makes the condition $R>r_{Ap},r_{Bt}$ invalid, if the overlap is small the Coulomb matrix element will still be approximately given by equation (18). $-\Sigma(e^2/r_{pt})P_{pt}$ can be replaced by the equivalent operator (15),

$$H_{AB}^{(ex)} = -\sum_{p,t}\sum_{\lambda,\lambda'}\sum_{\mu,\mu'} j_{\lambda\lambda'}^{\mu\mu'} C_{\mu}^{(\lambda)}(\Omega_{Ap}) C_{\mu'}^{(\lambda')}(\Omega_{Bt})(\tfrac{1}{2}+2\vec{s}_p\cdot\vec{s}_t)$$

$$= -\sum_{p,t} J_{pt}(\tfrac{1}{2}+2\vec{s}_p\cdot\vec{s}_t) . \tag{38}$$

$H_{AB}^{(ex)}$ operates on the angular and spin dependent parts of the wave functions, the radial integrals being contained in the j parameters. $j_{\lambda\lambda'}^{\mu\mu'}$ is also a function of R, Θ, and Φ. $\vec{s}_p$ is the spin of electron p. When A and B are ions with unfilled f or d shells, the number of independent parameters $\langle J_{pt}\rangle$ will be large. If the symmetry of the pair is high, the number can be reduced somewhat. They will depend exponentially on R when R is large. If all the orbitals had the same asymptotic radial dependence $\exp(-r/r_0)$, then the matrix element of equation (37) would be proportional to $\exp(-2R/r_0)$ in the limit of large R.

For rare earth ions the separation of even nearest cations in a crystal is so large compared with r_0 that direct cation-cation exchange is small. For iron group ions r_0 is not so small. For both types of ion superexchange - the coupling of the cations due to the overlap of their charge clouds with that of an intervening anion and the resulting cation-anion-cation electron exchange - is important. The form of the superexchange interaction operator is the same as in equation (38), but the values of the parameters are more difficult to estimate than for direct exchange. In the case of energy transfer from single Cr^{3+} ions to more strongly coupled Cr^{3+} pairs in Al_2O_3 it was possible to show that electric multipolar coupling was too weak to explain the observed transfer rate and that, therefore, exchange might be the coupling mechanism (16). Lengthy calculations have shown that superexchange provides a coupling of the required order of magnitude to explain the observations (17). Superexchange coupling depends strongly on the crystal structure, on the number and kinds of interaction paths connecting A and B through intermediate anions. For example a Cr^{3+} ion interacts rather strongly with another Cr^{3+} ion at first, second, third, or fourth nearest neighbor sites in Al_2O_3, but in $LaAlO_3$ coupling with a nearest neighbor is much stronger than with second nearest and more distant neighbors (18). For superexchange-coupled Cr^{3+} pairs the parameters $<J_{pt}>$ have been found to depend on the term as well as on the orbitals of the electrons (18). That is, the interaction of two ions must be described by a numerically different set of parameters $<J_{pt}>$ depending whether both ions are in the ground term $^4A_2$ or one is in $^4A_2$ and the other in ^2E. $<J_{pt}>$ is a short notation for the matrix element of J_{pt} taken with respect to the angular dependent (orbital) parts of the wave functions.

The energy transfer we have been discussing is called "resonant" because the transitions of ions A and B occur at very nearly the same frequency. (But the term is not meant to imply the kind of coherent oscillation described by equations (9).) In the non-resonant case, illustrated in figure 1c, transfer can also occur if the energy difference Δ is given to the lattice vibrations. Equation (15) applies to this case also if three modifications are made. The interaction must contain an electron-phonon part. The initial and final states must include the initial and final phonon states which will differ by a number of phonons whose total energy is Δ. The line shape factors must include the phonon sidebands. If one phonon of energy $\hbar\omega = \Delta$ is created in the process of figure 1c, the transfer rate is (19),

$$w_{AB} = (4\pi^2/h)|<a'b|H_{AB}|ab'>|^2 C[n(\omega)+1]\int g_A(E)g_B(E-\hbar\omega)dE. \tag{39}$$

C is a function of electron-phonon coupling parameters, $n(\omega)$ is the number of phonons present of energy $\hbar\omega$, and the matrix element has the same meaning as in equation (15). For the reverse process the square bracket must be replaced by $n(\omega)$ and the sign changed before $\hbar\omega$. If Δ is larger than the maximum phonon energy, more than one phonon will be created in the $A \rightarrow B$ process or destroyed in the reverse $B \rightarrow A$ process. When several phonons are involved, the transfer rate is approximately (20)

$$w_{AB}(\Delta) = w_{AB}(0)\exp(-\beta\Delta). \tag{40}$$

β is a temperature dependent parameter and is related to the analogous parameter in a similar law for the dependence on the energy gap to the next lower level of the non-radiative decay rate of a state by multiphonon emission (20, 21). Since in the reverse process phonons must be destroyed, for sufficiently low temperatures back transfer will be negligible compared with the forward transfer process of figure 1c. The validity of equation (40) has been experimentally verified (22). Phonon-assisted energy transfer is treated in detail in the article by Professor Orbach.

III. STATISTICAL TREATMENT

In order to relate the microscopic transfer rates to

observable quantities we must perform averages over the volume V of the sample. Let us consider a crystal containing a number of sensitizer or donor ions N_A of type A and activator or acceptor ions N_B of type B. Each has two states as shown in figures 1a or 1c. Two quantities which can be measured and which provide some information about A→B energy transfer are the time development of the sensitizer luminescence after pulse excitation and the relative yield of the luminescence. Let a short excitation pulse begin at time $t = -T<0$ and end at $t = 0$, the pulse width T being much less than the time constant of any decay or transfer process of interest. Then at $t = 0$ some randomly situated $N_{a'}(0)$ sensitizers are excited and no activators are excited; $N_{b'}(0) = 0$. Let us find the behaviour of the sensitizer luminescence as a function of time $t>0$. This problem was first treated by Förster (23).

If $\rho_j(t)$ is the probability that sensitizer j at position $\vec{R}_j$ is excited at time t, then $\rho_j(t)$ satisfies the equation

$$\frac{d}{dt}\rho_j(t) = -\frac{1}{\tau}\rho_j(t) - \rho_j(t)\sum_{i=1}^{N_B} w_{AB}(\vec{R}_i-\vec{R}_j). \qquad (41)$$

τ^{-1} is the intrinsic decay rate of state a' in the absence of activators and in general consists of a radiative part P and a non-radiative part W_{nr}: $\tau^{-1} = P+W_{nr}$. It is assumed that no A→A transfer or B→A back transfer occurs. The first condition may imply an upper limit of sensitizer concentration of the order of 1% atomic or less. The second condition will be met for phonon assisted transfer, as shown in figure 1c, if Δ is large enough. It will also be met if the state b' decays rapidly. b' will decay rapidly non-radiatively if there is a state not much farther below it than a few times the energy of an optical phonon. The solution of (41) if $\rho_j(0) = 1$ is

$$\rho_j(t) = e^{-t/\tau}\prod_{i=1}^{N_B}\exp[-tw_{AB}(\vec{R}_i-\vec{R}_j)]. \qquad (42)$$

If the A and B ions are randomly distributed, the probability of finding a given sensitizer excited at time t is $\bar{\rho}(t)N_{a'}(0)/N_a$, where $\bar{\rho}(t)$ is the statistical average of $\rho_j(t)$ over the various possible sensitizer environments. An environment is characterized by a particular distribution of activators around the sensitizer. The number of quanta emitted as luminescence by the sensitizer per unit time is $PN_{a'}(0)\bar{\rho}(t)$. $\bar{\rho}(t)$ is given by the average of

$$\rho(t) = N_A^{-1}\sum_{j=1}^{N_A}\rho_j(t). \qquad (43)$$

Assuming uniform distributions for both sensitizers and activators, we obtain

$$\bar{\rho}(t) = e^{-t/\tau} N_A^{-1} \sum_j \prod_j \int (d^3\vec{R}_j/V)[I(t)]^{N_B}$$
$$= e^{-t/\tau} N_A^{-1} \sum_j (1)^{N_A} [I(t)]^{N_B}$$
$$= e^{-t/\tau} [I(t)]^{N_B} .$$
$$I(t) = V^{-1} \int d^3\vec{R} \exp[-t w_{AB}(\vec{R})]. \qquad (44)$$

The discrete lattice was approximated by a continuum in taking the averages.

We consider the multipolar interactions and assume as a simplification that only one of them is important. $w_{AB}(\vec{R})$ is averaged over angles to give $w_{AB} = \alpha^{(n)} R^{-n}$ with n = 6, 8, or 10, and the constant $\alpha^{(n)}$ is written in terms of a range parameter R_0 given by $\alpha^{(n)} = \tau^{-1} R_0^n$. For R = R_0 then, $w_{AB}(R)$ is equal to the intrinsic decay rate τ^{-1}. Equation (44) becomes

$$\bar{\rho}(t) = e^{-t/\tau} [I_n(t)]^{N_B} ,$$
$$I_n(t) = (4\pi/V) \int_{R_m}^{R_v} R^2 dR \exp[-(R_0/R)^n (t/\tau)]. \qquad (45)$$

R_m is the smallest possible A - B spacing, and R_v is the radius of the large spherical volume V. To evaluate $I_n(t)$ it is convenient to change variables to $x = (R_0/R)^n t/\tau$. Then equation (45) becomes

$$I_n = (3/n) x_v^{3/n} \int_{x_v}^{x_m} x^{-1-3/n} e^{-x} dx .$$
$$x_v = (R_0/R_v)^n t/\tau \quad , \quad x_m = (R_0/R_m)^n t/\tau . \qquad (46)$$

Equation (46) can be integrated by parts. The result for x_v small and x_m large is

$$I_n = 1 - x_v^{3/n} [\Gamma(1-3/n) + (3/n) x_m^{-1} \exp(-x_m)] . \qquad (47)$$

$\Gamma(c)$ is the gamma function, whose value would be (c-1)! if c were a positive integer. For $x_m \gg 1$ the second term in the bracket can be neglected compared with the first. The approximation $x_m \gg 1$ means that $t/\tau \gg (R_m/R_0)^n$. Since R_0 is generally several times as large as R_m, the approximation may still be good even at quite short times t. In this limit, then

$$[I_n(t)]^{N_B} = [1 - (1/N_B) N_B (R_0/R_v)^3 \Gamma(1-3/n)(t/\tau)^{3/n}]^{N_B} . \qquad (48)$$

In the limit that N_B and R_v become infinite, but the activator concentration $c_B = N_B/(4\pi R_v^3/3)$ remains finite, (48) becomes

$$[I_n(t)]^{N_B} \to \exp\{-(c_B/c_0)\Gamma(1-3/n)(t/\tau)^{3/n}\} , \qquad (49)$$

and $\bar{\rho}(t)$ is

$$\bar{\rho}(t) = \exp\{-t/\tau-(c_B/c_0)\Gamma(1-3/n)(t/\tau)^{3/n}\} . \qquad (50)$$

c_0 is a critical concentration defined by $c_0^{-1} = 4\pi R_0^3/3$. In molecular crystals sensitizer decay governed by the dipole-dipole (n=6) interaction mechanism has often been observed (24). For rare earth ions in solids the dipole-quadrupole (n=8) decay law has been seen in some cases also (25).

The decay is non-exponential, beginning as $\exp[-(t/\tau)^{3/n}]$ and becoming more nearly exponential as time progresses and only those sensitizers are left to decay which have no activator nearby. At the very shortest times $t/\tau \ll (R_m/R_0)^n$, which are often not experimentally accessible, the decay is governed by those sensitizers which are separated from an activator by the minimum distance R_m. A similar analysis shows that this initial decay rate, including the intrinsic rate τ^{-1}, is $\tau^{-1}[1+3(n-3)^{-1}(c_B/c_0)(c_m/c_0)^{n/3-1}]$, where $c_m^{-1} = 4\pi R_m^3/3$. Figure 3 is a plot of $\bar{\rho}(t)$ of equation (50) with n=6 for several values of the ratio c_B/c_0. As c_B/c_0 increases, the initial non-exponential part of the decay extends to longer times. For long enough times the decay becomes exponential and all the curves have the same slope.

In the case of the exchange interaction where $w_{AB}(R)$ decreases exponentially with increasing R, the analogue of equation (50) is

$$\bar{\rho}(t) = \exp\{-t/\tau-\gamma^{-3}(c_B/c_0)g(e^{\gamma}t/\tau)\} ,$$
$$g(x) = 6x\sum_{m=0}^{\infty}\frac{(-x)^m}{(m+1)^4 m!} , \qquad (51)$$

and γ is a parameter related to the radial extent of the A and B wavefunctions (26).

If the A ions are excited by a long pulse, rather than a short one, the decay will be somewhat different because during the pulse those A ions with B ions nearby may decay, leading to a different spatial distribution of excitation at the end of the pulse. For a long pulse beginning at $t = -T<0$ and ending at $t = 0$, the function of interest is $\int_{-T}^{0}\bar{\rho}(t-t')dt'$ rather than $\bar{\rho}(t)$.

The relative yield of the sensitizer luminescence η/η_0 is the ratio of the total number of sensitizer luminescence output quanta to the number which would be emitted if the activators were absent.

$$\eta/\eta_0 = \tau^{-1}\int_0^{\infty}\bar{\rho}(t)dt. \qquad (52)$$

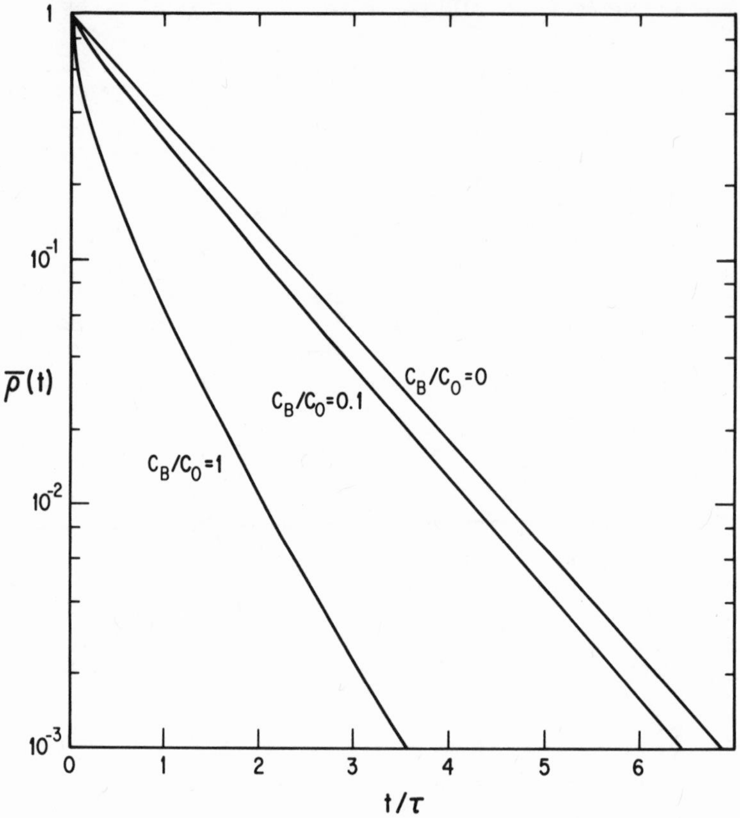

Fig. 3 Decay of sensitizer luminescence vs time after excitation by a short pulse for three values of activator concentration c_B. A dipole-dipole A-B interaction is assumed. For $c_B = 0$ the decay is exponential with time constant τ. The activator concentrations have been divided by the critical concentration c_0 for the sake of generality. Time is measured in units of τ, the lifetime of the excited state of the sensitizer in the absence of activators. As time increases, the slopes of the two lower curves approach that of the curve $c_B/c_0 = 0$.

Equation (52) is valid also for the relative yield under constant excitation as well as under pulse excitation. With increasing activator concentration c_B, the yield η/η_0 decreases, since more energy is transferred to the activators rather than emitted as luminescence.

In many real systems $A \rightarrow A$ transfer cannot be neglected. Since the sensitizers are identical ions, $A \rightarrow A$ transfer is resonant and may be more rapid than $A \rightarrow B$ transfer if the two concentrations are comparable. Excitation energy may then be able to migrate among the sensitizers before passing to an activator. We do not consider here the more complicated case in which the spectra have an inhomogeneous width larger than the homogeneous width. Then the resonance may be destroyed and migration inhibited. This case is discussed in a following section. There are two ways of dealing with the problem of simultaneous $A \rightarrow A$ and $A \rightarrow B$ transfer.

The first method consists in treating the migration as a diffusion process, as in the case of diffusion of spin orientation among magnetic ions. A diffusion equation,

$$\frac{\partial}{\partial t}\rho(\vec{R},t) = [-\tau^{-1}+D\nabla^2-\sum_{i=1}^{N_B} w_{AB}(\vec{R}-\vec{R}_i)]\rho(\vec{R},t) \quad , \qquad (53)$$

is solved for the excitation density $\rho(\vec{R},t)$, and $\phi(t)$, the analogue of $\bar{\rho}(t)$, is obtained.

$$\phi(t) = \int\rho(\vec{R},t)d^3\vec{R} \quad . \qquad (54)$$

Equation (53) has been solved only for the dipole-dipole interaction, $w_{AB}=\tau^{-1}(R_0/R)^6$. For this case the solution is approximately given by (27)

$$\phi(t) = \exp\{-t/\tau-(c_B/c_0)\Gamma(\tfrac{1}{2})(t/\tau)^{\frac{1}{2}}[\frac{1+10.87y+15.5y^2}{1+8.743y}]^{3/4}\}$$

$$y = D\tau R_0^{-2}(t/\tau)^{2/3} \quad . \qquad (55)$$

In the second method w_{AB} is treated as a random variable. As the excitation hops from sensitizer to sensitizer, the value of w_{AB} which determines whether it will be transferred to an activator varies in a stepwise fashion. The problem is similar to that of line broadening in gas spectra where the Doppler shift is a random variable because of collisions. If in the short time interval $(0,t_1)$, the excitation is on sensitizer number 1, the probability of transfer to an activator during this time is $t_1w_1=t_1\Sigma_i w_{AB}(\vec{R}_1-\vec{R}_i)$. At time $t = t_2$ it hops to sensitizer 2 and at $t = t_3$, to sensitizer 3. During the interval (t_1,t_2) the probability of transfer to an activator is $(t_2-t_1)w_2$, etc. If τ_0 is the mean value of this time interval, then from stochastic theory $\phi(t)$ is given by the solution

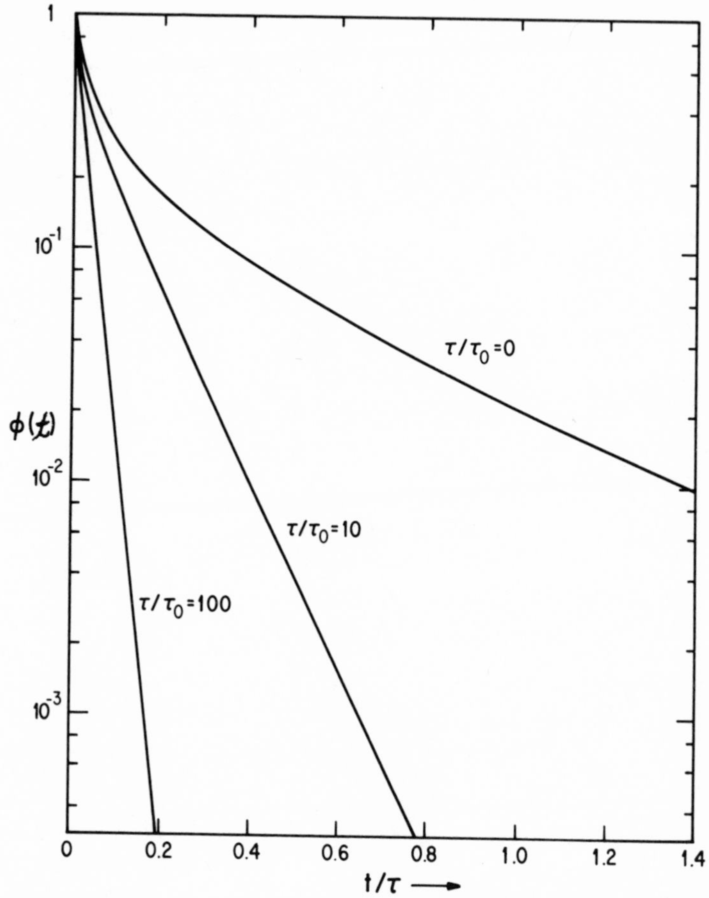

Fig. 4 Decay of sensitizer luminescence after pulse excitation for an A-B dipole-quadrupole interaction and energy migration among sensitizers. $c_B = 2c_0$. The migration is characterized by the hopping rate τ_0^{-1}. As τ_0^{-1} increases, the decay becomes more nearly exponential. The curves were calculated by solving numerically the integral equation (56). In contrast with figure 3 the slopes do not approach the same value as time increases.

of the equation (28),

$$\phi(t) = \bar{\rho}(t)e^{-t/\tau_0} + \tau_0^{-1}\int_0^t \phi(t')\bar{\rho}(t-t')e^{-(t-t')/\tau_0}\,dt'$$

$$= \bar{\rho}(t)e^{-t/\tau_0} + \tau_0^{-1}\int_0^t \phi(t-t')\bar{\rho}(t')e^{-t'/\tau_0}\,dt' \;. \qquad (56)$$

$\bar{\rho}$ is given by equation (50) or (51), for example. Equation (56) is similar in form to a renewal equation, the value of $\phi(t)$ depending upon its value at a previous time t-t'. The solution can be found by numerical methods. Figure 4 shows $\phi(t)$ for the dipole-quadrupole A - B interaction (n=8) for $c_B/c_0 = 2$ and several values of τ_0. In figure 5 the solutions from equations (55) and (56) with $w_{AB} = \tau^{-1}(R_0/R)^6$ are compared and seen to be very nearly identical.

The function $\phi(t)$ is non-exponential for t small enough that migration is unimportant. In this limit it approaches $\bar{\rho}(t)$. For large t, $\phi(t)$ decays exponentially at a rate determined by migration. As migration becomes more rapid, the boundary between these two regions shifts to shorter times until, for sufficiently fast migration, the decay appears to be purely exponential. Figure 4 shows this trend.

Let us look more closely at the behaviour of $\phi(t)$ when A and B interact by the dipole-dipole mechanism. For large t, $\phi(t)$ decays exponentially at a rate $\tau^{-1} + K_M$. The rate K_M is given by

$$K_M(\text{diff.}) = 4\pi D c_B R_s \quad , \quad R_s = (0.676)(\alpha_{AB}/D)^{1/4} \;, \qquad (57)$$

when $\phi(t)$ is the solution of equations (53) and (54). If we imagine a sphere of radius equal to the scattering length R_s centered on the diffusing exciton, K_m can be interpreted as the rate at which B ions enter this sphere. If the diffusion is due to a dipole-dipole A - A interaction, $w_{AA}=\alpha_{AA}/R^6$, then from the theory of exciton diffusion (29) the diffusion coefficient is given by

$$D = \tfrac{1}{2}(4\pi/3)^{4/3} c_A^{4/3} \alpha_{AA} \;. \qquad (58)$$

It has been verified that D is proportional to α_{AA} in experiments on two systems (30, 31). When the sensitizer concentration c_A has been varied, D has been found to increase more rapidly with increasing c_A than $c_A^{4/3}$, however (32, 33). Measured values of D for rare earth ions range from 10^{-14}cm^2/sec. to 10^{-9}cm^2/sec., many orders of magnitude smaller than for excitons in molecular crystals. Substitution of (58) in (57) gives

$$K_M(\text{diff.}) = (21)c_B c_A(\alpha_{AA}^3 \alpha_{AB})^{1/4} \qquad (59)$$

When $\phi(t)$ is found from equation (56), K_M is given by

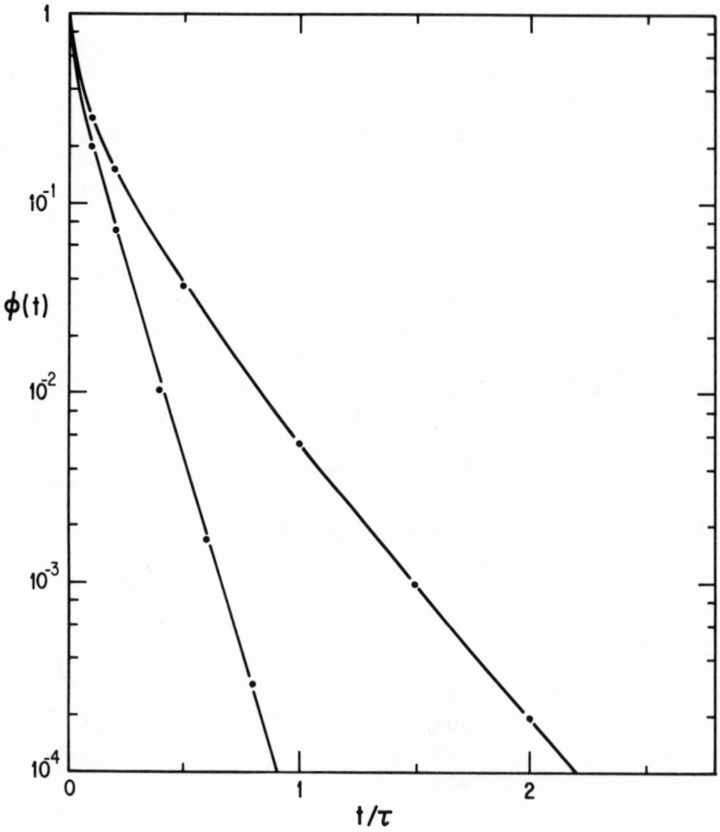

Fig. 5 Decay of sensitizer luminescence after pulse excitation for an A-B dipole-dipole interaction and energy migration among sensitizers. $c_B = 2c_0$. The solutions of equations (56) - solid curves - and (53) - dots - are compared. For the upper curve $\tau/\tau_0 = 1$ and $y = 0.14(t/\tau)^{2/3}$. For the lower curve $\tau/\tau_0 = 10$ and $y = 1.33(t/\tau)^{2/3}$. For $\tau/\tau_0 = y = 0$ both solutions reduce to the same $\bar{\rho}(t)$ (not shown).

$$K_M(\text{hop.}) = (2\pi^2/3)c_B(\alpha_{AB}/\tau_0)^{\frac{1}{2}} . \qquad (60)$$

If another critical separation R_c is defined by $\alpha_{AB} = \tau_0^{-1}R_c^6$, then equation (60) can be written

$$K_M(\text{hop.}) = (\pi/2)(c_B/\tau_0)(4\pi R_c^3/3). \qquad (61)$$

$c_B(4\pi R_c^3/3)$ is the number of B ions in the sphere of radius R_c. K_M is the rate at which B ions enter the interaction sphere, centered on the hopping exciton, and thus R_c is analogous to R_s. If the A ions interact with each other via the dipole-dipole mechanism, the time τ_0 is approximately

$$\tau_0 \simeq \int_0^\infty t\bar{\rho}(t)dt/\int_0^\infty \bar{\rho}(t)dt \simeq (27/8\pi^3)(\alpha_{AA}c_A^2)^{-1}, \qquad (62)$$

where τ_0 is assumed to be much smaller than τ. For very high rare earth sensitizer concentrations τ_0 may be less than 10^{-9} sec. (34). When this is substituted in equation (61), the result is

$$K_m(\text{hop.}) = (20)c_Bc_A(\alpha_{AA}\alpha_{AB})^{\frac{1}{2}} . \qquad (63)$$

Both equations (59) and (63) predict that K_M is proportional to c_B and to c_A, but they differ in the dependence on the interaction constants α_{AA} and α_{AB}.

In the opposite limit $\phi(t)$ (equation (55)) obviously is identical with $\bar{\rho}(t)$ from equation (50) with $n = 6$ for times t short enough that $y \ll 1$, that is, for $t \ll R_s^2/D$. Since $R_s^2/(6D)$ is the average time required for the exciton to traverse a distance R_s, t must be much less than this time in order that the effects of diffusion be negligible in $\phi(t)$.

When migration is not negligible, the relative yield of sensitizer luminescence is given by equation (52) with $\bar{\rho}(t)$ replaced by $\phi(t)$. Figure 6 shows the effect of migration on η/η_0 for the dipole-dipole, dipole-quadrupole, and quadrupole-quadrupole A-B interactions. As the rate of migration increases, the yield decreases because $A \rightarrow B$ transfer becomes more probable.

As the concentration of A ions is increased from a small value, three regimes may be traversed. At large enough A-A separation migration is negligible and the decay of sensitizer luminescence is given by $\bar{\rho}(t)$. As the separation becomes smaller, the decay is limited by the migration rate and is described by a non-exponential $\phi(t)$. With still faster migration, the decay becomes exponential and the rate may be limited by the A-B nearest neighbor transfer rate, leading to an apparent saturation in the rate of increase of the measured diffusion coefficient with increasing A concentration (32).

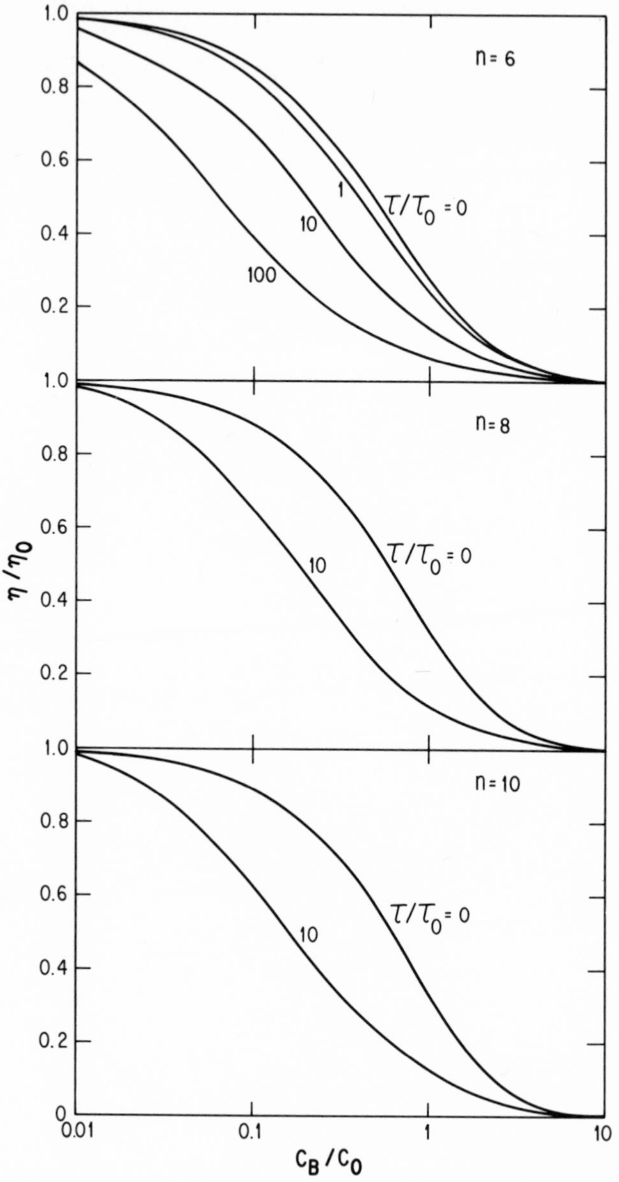

Fig. 6 The effect of migration on the sensitizer luminescence yield for dipole-dipole, dipole-quadrupole, and quadrupole-quadrupole A-B interactions, plotted vs normalized activator concentration c_B/c_0. Each curve is characterized by a value of the migration parameter τ/τ_0. The yield was calculated by integrating the solution of equation (56).

In this case we may write $K_M = Uc_B$, where U is independent of c_A. K_M is still proportional to c_B as discussed in the paragraph following equation (50). Or, if most of the B ions are excited by transfer, the decay of the excited state b' of B could be the limiting step in the process. To take account of this possibility we set K_M proportional to n_b, rather than c_B, where n_b is the number of B ions in state b per unit volume. When migration is sufficiently fast, all A ions decay at the same rate $\tau^{-1}+K_M$. The rapid migration has the effect of averaging the environments of the A ions. Then in the concentration regime where fast migration occurs, the time devolopment of $n_{a'}(t)$, the concentration of excited A ions, satisfies the rate equation (34),

$$\frac{d}{dt} n_{a'}(t) = -\tau^{-1} n_{a'}(t) - U n_b(t) n_{a'}(t) . \qquad (64)$$

Rate equations of this form have been successfully used to explain energy transfer in rare earth phosphors when the concentration is high. Some of the most interesting cases will be discussed in the next chapter. Sometimes a rate equation description, although not strictly justified, may be the only practical theoretical way of treating a system whose behaviour is complicated by a large number of energy levels and possible interaction paths. Then the decay times may be arbitrarily taken as the time for an observed luminescence to decay non-exponentially to e^{-1} of its initial value. In such cases rate equations may be useful for a gross description and may provide some insight into the physical phenomena.

The analysis of the time development of the sensitizer luminescence is not trivial. Good data spanning several orders of magnitude of intensity are required. For determination of the transfer mechanism it is advantageous to vary sensitizer concentrations as well as activator concentrations.

Although energy transfer between rare earth ions has been emphasized, other systems have also been studied. A few examples are A = Cr^{3+}, B = Cr^{3+} - Cr^{3+} pair (36); A = Mn^{2+}, B = Er^{3+} (37); A = Eu^{3+}, B = Cr^{3+} (30); and A = Pb^{2+}, B = Mn^{2+} (38). Radiative recombination of electrons and holes loosely bound in the hydrogen-like states of donor-acceptor pairs in semiconductors (39) can be viewed as due to energy transfer by exchange coupling (40). Transfer from organic molecules to ions has been observed (41), and there is an extensive literature on molecule-molecule energy transfer (24).

IV. INHOMOGENEOUS BROADENING

To this point we have assumed that the broadening of all energy levels of interest is due to interaction with lattice vibrations. This will generally be the case in crystals except perhaps at very low temperatures. Such broadening is called homogeneous because a given level is broadened by the same amount no matter where the ion is located in the crystal. If, on the other hand, the position of the energy level depends upon the location of the ion, then the observed transition linewidth, which consists of contributions from all ions, is broadened inhomogeneously. In this case the transition line shape of each ion i due to homogeneous broadening $g_i(E-E_i)$ is centered about a different transition energy E_i. As long as the spread in the transition energies is much less than the homogeneous width, the inhomogeneous width can be neglected. It is the opposite case - inhomogeneous width larger than homogeneous width - with which we are now concerned.

Inhomogeneous broadening results when the sites occupied by the active ions are inequivalent. In glasses there is considerable disorder and the resulting inhomogeneous broadening for rare earth ions may amount to 100 cm^{-1}. Of course even with a random distribution of sensitizers and activators in an otherwise perfect crystal there are many inequivalent sensitizers: some with an activator as nearest neighbor, some with no activator nearby. The same interactions which are responsible for energy transfer will cause shifts in the positions of energy levels. For rare earth ions these shifts, exemplified by the terms v_1 and v_2 in equation (9), are generally quite small.

Let us consider for simplicity a crystal containing only active ions of a single type. In the absence of inhomogeneous broadening the transitions are resonant. The effect of inhomogeneous broadening is to destroy the resonance, that is to decrease the value of the integral $\int g_i(E-E_i)g_j(E-E_j)dE$. If ions i and j are separated by a large distance, there is no effect, since transfer would be very improbable even with perfect resonance. But when the separation is smaller, transfer may be inhibited by a poor spectral match. The analogue of equation (41) for this case would be

$$\frac{d}{dt}\rho_j(t) = -\frac{1}{\tau}\rho_j(t) - \rho_j(t)\sum_i w_{ji} + \sum_i w_{ij}\rho_i(t) \quad . \qquad (65)$$

The last term couples the equations and prevents straightforward solution and averaging. The transfer rates depend upon the difference in transition energies as well as separation: $w_{ij} = w_{ij}(E_i-E_j, \vec{R}_i-\vec{R}_j)$. Even if the ions are randomly distributed over the available sites, the distribution of excited ions may not be random if the spectral width of the excitation is small compared with the width of the inhomogeneously broadened absorption line.

Figure 7 shows an example of inhomogeneous broadening (42). In $Y_3Al_5O_{12}$:Nd the Nd^{3+} ion replaces Y^{3+} and gives rise to four sharp luminescence lines near 1.07μm. The figure shows the

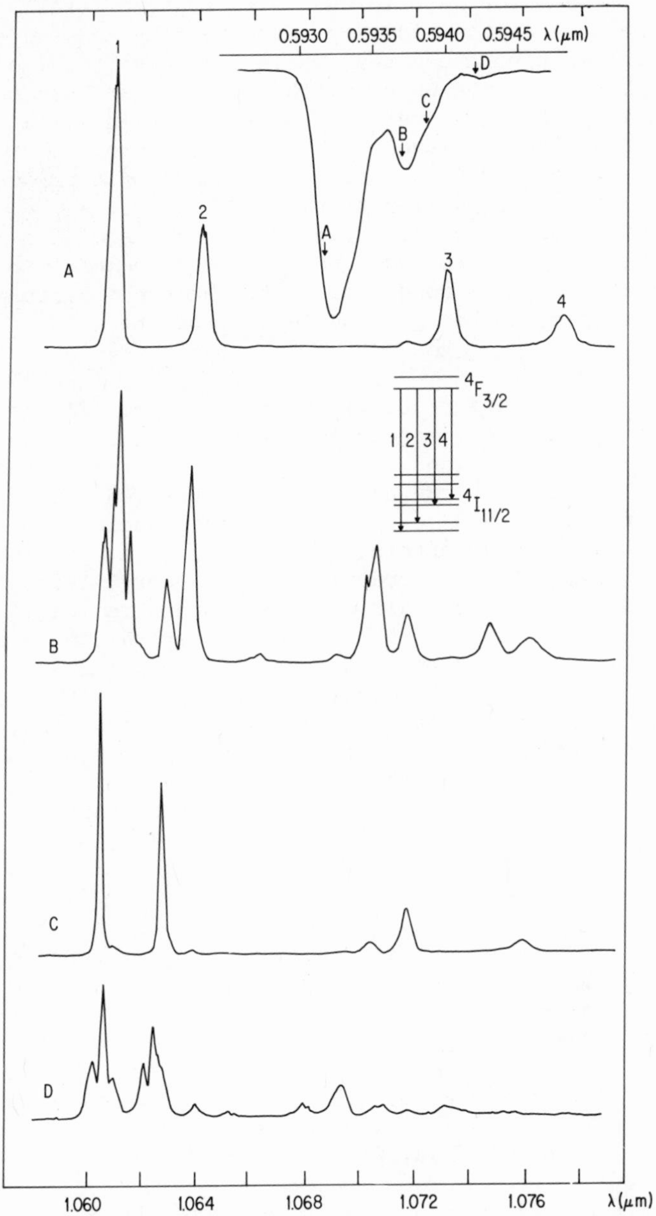

Fig. 7 Luminescence of Nd^{3+} in $Y_3(Al_{0.93}Ga_{0.07})_5O_{12}$ at 2K. As the wavelength of the dye laser excitation source is varied over the width of an absorption line (insert), Nd^{3+} ions with different local environments are selectively excited and the spectra change, since energy transfer is slow compared with the lifetime of $^4F_{3/2}$.

corresponding spectrum for Nd^{3+} in the mixed crystal $Y_3(Al_{0.93}Ga_{0.07})_5O_{12}$. Because of the disorder introduced by the larger gallium ions on some aluminum sites, there are many inequivalent Nd^{3+} ions. At 2 K the homogeneous widths are smaller than the differences in transition frequency of some of the inequivalent ions, and as the excitation wavelength is varied across the absorption line (which has no structure in $Y_3Al_5O_{12}$) from position A to position B to C to D, different types of Nd^{3+} ions are excited. Almost no energy transfer occurs during the lifetime of the $^4F_{3/2}$ state. At higher temperatures homogeneous widths become larger and overlap, energy transfer is more efficient, and the structure in the luminescence spectrum becomes independent of the excitation wavelength.

In glasses the inhomogeneous widths are larger. In certain glass:Nd lasers a "hole" may appear in the luminescence line (43) because the spectral diffusion is relatively slow even at room temperature. Fluorescence line narrowing is observed with a monochromatic excitation source (44). The effect occurs for Nd concentrations less than a certain critical value, as predicted by the theory of Lyo (45). According to this theory, transfer becomes negligible below a certain critical concentration if the ion-ion interaction decreases with increasing separation more rapidly than in the dipole-dipole case and if the strain producing the inhomogeneous broadening is microscopic compared with the mean separation. Energy migration among Eu^{3+} ions in glass, accompanied by the emission of a phonon of energy $\hbar\omega = 60$ cm^{-1}, has recently been observed by Motegi and Shionoya (46). In this case the dipole-dipole mechanism was apparently dominant.

REFERENCES

1. P. Kisliuk, N. C. Chang, P. L. Scott, and M. H. L. Pryce, Phys. Rev. 184, 367 (1969).

2. G. A. Prinz, Phys. Rev. 152, 474 (1966).

3. T. Förster, Ann. Physik 2, 55 (1948).

4. D. L. Dexter, J. Chem. Phys. 21, 836 (1953).

5. T. Förster in Modern Quantum Chemistry, Part III, ed. by O. Sinanoglu (Academic Press, N. Y., 1965), ch. B. 1.

6. B. C. Carlson and G. S. Rushbrooke, Proc. Camb. Phil. Soc. 46, 626 (1950).

7. R. J. Birgeneau, M. T. Hutchings, J. M. Baker, and J. D. Riley, J. Appl. Phys. 40, 1070 (1969).

8. T. Kushida, J. Phys. Soc. Japan 34, 1318 (1973).

9. K. Shinagawa, J. Phys. Soc. Japan 23, 1057 (1967).

10. C. W. Nielson and G. F. Koster, Spectroscopic Coefficients for the p^n, d^n and f^n Configurations (M.I.T. Press, Cambridge, Mass., 1963).

11. G. H. Dieke, Spectra and Energy levels of Rare Earth Ions in Crystals (Wiley Interscience, N. Y., 1968).

12. A. J. Freeman and R. E. Watson, Phys. Rev. 127, 2058 (1962).

13. M. Rotenberg, R. Bivins, N. Metropolis, and J. K. Wooten, Jr., The 3j and 6j Symbols (Technology Press, Cambridge, Mass., 1959).

14. M. J. Weber, Phys. Rev. 157, 262 (1967).

15. P. M. Levy, Phys. Rev. 177, 509 (1969).

16. R. J. Birgeneau, J. Chem. Phys. 50, 4282 (1969).

17. N. L. Huang, Phys. Rev. B1, 945 (1970).

18. J. P. van der Ziel, Phys. Rev. B4, 2888 (1971).

19. R. L. Orbach in Optical Properties of Ions in Crystals, ed. by H. M. Crosswhite and H. W. Moos (Wiley Interscience, N. Y., 1967) p. 445.

20. T. Miyakawa and D. L. Dexter, Phys. Rev. B1, 2961 (1970).

21. L. A. Riseberg and H. W. Moos, Phys. Rev. 174, 429 (1968).

22. N. Yamada, S. Shionoya, and T. Kushida, J. Phys. Soc. Japan 32, 1577 (1972).

23. T. Förster, Z. Naturforsch. 4A, 321 (1949).

24. R. E. Kellogg, J. Luminescence 1, 435 (1970).

25. E. Nakazawa and S. Shionoya, J. Chem. Phys. 47, 3211 (1967).

26. M. Inokuti and H. Hirayama, J. Chem. Phys. 43, 1978 (1965).

27. M. Yokota and O. Tanimoto, J. Phys. Soc. Japan 22, 779 (1967).

28. A. I. Burshtein, Sov. Phys. JETP 35, 882 (1972).

29. M. Trlifaj, Czech. J. Phys. 8, 510 (1958).

30. M. J. Weber, Phys. Rev. B4, 2932 (1971).

31. J. P. van der Ziel, L. Kopf, and L. G. Van Uitert, Phys. Rev. B6, 615 (1972).

32. R. K. Watts and H. J. Richter, Phys. Rev. B6, 1584 (1972).

33. N. Krasutsky and H. W. Moos, Phys. Rev. B8, 1010 (1973).

34. W. B. Gandrud and H. W. Moos, J. Chem. Phys. 49, 2170 (1968).

35. For an alternate derivation of the rate equation see W. J. C. Grant, Phys. Rev. B4, 648 (1971). See also F. K. Fong and D. J. Diestler, J. Chem. Phys. 56, 2875 (1972).

36. G. F. Imbusch, Phys. Rev. 153, 326 (1967).

37. J. M. Flaherty and B. Di Bartolo, Phys. Rev. B8, 5232 (1973).

38. T. P. J. Botden and F. A. Kröger, Physica 14, 553 (1948).

39. F. Williams, Phys. Stat. Sol. 25, 493 (1968).

40. K. Colbow, Phys. Rev. 139A, 274 (1965).

41. M. Kleinerman, J. Luminescence 1, 481 (1970).

42. R. K. Watts and W. C. Holton, J. Appl. Phys. 45, 873 (1974).

43. V. R. Belan, C. M. Briskina, V. V. Grigoryants, and M. E. Zhabotinskii, Sov. Phys. JETP 30, 627 (1970).

44. L. A. Riseberg, Phys. Rev. A7, 671 (1973).

45. S. K. Lyo, Phys. Rev. B3, 3331 (1971).

46. N. Motegi and S. Shionoya, J. Luminescence 8, 1 (1974).

STEPWISE UPCONVERSION AND COOPERATIVE PHENOMENA IN FLUORESCENT SYSTEMS

R. K. Watts

Texas Instruments Inc.

Dallas, Texas 75222, U. S. A.

ABSTRACT

The general discussion of the preceding chapter of energy transfer in the regime of high sensitizer concentration is applied to several cases of stepwise, sequential transfer between rare earth ions, a process which leads to conversion of infrared sensitizer excitation to visible or ultraviolet activator luminescence. The process is compared with the generally less efficient cooperative excitation process in which two sensitizers transfer simultaneously their excitation to an activator. Both types of process are adequately described by rate equations for the populations of the relevant energy levels of the ions. Another class of photon-assisted two-ion processes is briefly discussed on the basis of Dexter's theory.

I. INTRODUCTION

We saw in the preceding chapter how sensitizer → activator energy transfer may be described by rate equations for the populations of the states when the sensitizer concentration is sufficiently high that migration of excitation among the sensitizers is rapid compared with transfer to the activators. Although the lower state of the transition of sensitizer or activator, states a or b of Figure 1 of the previous article is often the ground state, this need not be the case. The theory applies as well when these two are excited states.

A more striking energy transfer phenomenon than quenching of sensitizer luminescence by activator impurities is the stepwise up-conversion, discovered by Auzel (1), in which two or more sensitizers transfer excitation sequentially to an activator, each transfer exciting the activator to a higher energy level. Activator luminescence can then occur at a shorter wavelength, in the visible, than the sensitizer luminescence, which is generally in the infrared (2). There are other ways in which "anti-Stokes," fluorescence can be produced. Those which require a coupling between ions will be discussed in this chapter. They include cooperative excitation, a process in which two sensitizers transfer simultaneously their excitation to an activator. The activator may be excited to an energy at most equal to the sum of the excitation energies of the two sensitizers, as in the sequential transfer process. The study of these processes is of much more recent origin than the general study of sensitized fluorescence.

The systems investigated have consisted of rare earth ions in various host crystals. The positions of the multiplet levels, and, consequently, the gross features of the optical spectra, of rare earth ions are well known and are insensitive to the nature of the crystalline host. The lifetimes of the various states and often even the non-radiative component of the lifetime due to multiphonon emission can be roughly predicted. Thus it is simpler to pick activators and sensitizers from the rare earth group than from other series of transition metal ions with less predictable properties, stronger coupling to the lattice, and lower radiative efficiencies. Much effort has been directed toward increasing the efficiencies of these upconversion processes, generally by variation of concentrations and selection of an optimum host crystal. The efficiencies attainable at moderate excitation are rather large compared with those of other high order optical processes.

It may be worthwhile to mention some other processes which can lead to emission of light of shorter wavelength than the exciting light. Three which are single-ion processes - electronic Raman scattering, two-photon absorption followed by fluorescence, and the single-ion quantum counter process - are not of interest here. In the quantum counter an ion is excited by absorption of radiation at frequency ν_1, and from this excited state it may be further excited to a higher lying level by absorption of photons from a second beam at frequency ν_2, or from the same beam in the degenerate case $\nu_2 = \nu_1$. The main limitation of the efficiency of the quantum counter is the small absorption coefficients of the rare earth ions used. The stepwise upconversion process is more efficient because the exciting radiation is absorbed by sensitizers which may be present in very large concentration and, once excited, may transfer most of their excitation energy to the luminescent activators. It will be more convenient to discuss some specific cases rather than to try to treat the phenomena in a general way. Since all the stepwise and cooperative upconversion processes are described by rate equations, only one of them will be treated in detail.

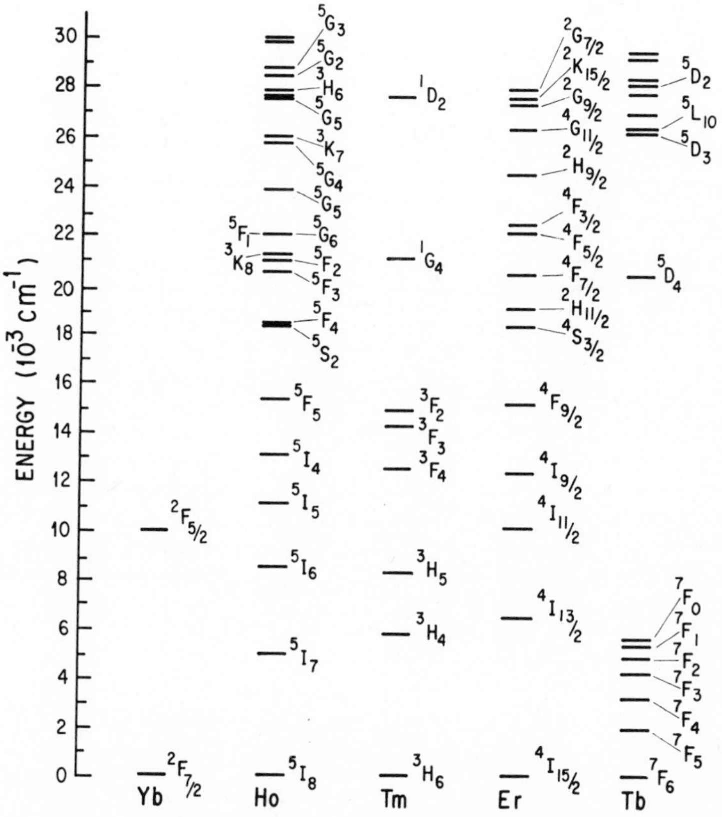

Fig. 1 Energy levels of the trivalent rare earth ions of interest. The multiplet label $^{2S+1}L_J$ is that of the state to which the intermediate coupling state reduces as the spin-orbit interaction goes to zero. These multiplets are labeled ψJ in the text. Each can in general split into a number of sublevels (crystal-field splitting) by interaction with other atoms in a solid. The total crystal field splitting of a multiplet may amount to several hundred cm^{-1} and is not indicated.

II. STEPWISE UPCONVERSION

Stepwise upconversion has largely been studied in crystals containing Yb^{3+} as the sensitizer and Er^{3+}, Ho^{3+}, or Tm^{3+} as the activator. Yb is a convenient sensitizer because, from Figure 1, it has only a single excited multiplet. There is not a large number of excited states which may lead to quenching of the activator luminescence when the Yb concentration has the typical high value of 10% atomic or more. Because of the high sensitizer concentration, rate equations are applicable for the theoretical treatment. Activator concentrations are usually limited to a low value ranging from about 0.1% to a few percent by activator-activator quenching interactions. The Yb^{3+} $^2F_{7/2} \rightarrow {}^2F_{5/2}$ absorption band is also fairly well matched to the emission spectrum of an efficient solid state infrared source, the GaAs:Si diode. Let us look first at the Yb^{3+}, Ho^{3+} system since it is simplest.

From figure 2a an excited Yb^{3+} ion may transfer with phonon assistance to the higher lying $^5I_5$ or lower lying $^5I_6$ levels of Ho^{3+}. We assume that the temperature is low enough that transfer to $^5I_5$ can be neglected. It has been shown that in YF_3 this transfer occurs via the dipole-dipole mechanism. The $^5S_2$ - $^5I_6$ transition is resonant with the Yb^{3+} transition. The $^5S_2$ and $^5F_4$ multiplets are nearly degenerate in energy, but these two will be called $^5S_2$ for short. The $^5I_6$ lifetime is long enough that the probability of a second transfer, exciting the activator to $^5S_2$ is appreciable. This two step process can be written schematically as

$$\begin{aligned} &\text{Yb } {}^2F_{5/2} + \text{Ho } {}^5I_8 \rightarrow \text{Yb } {}^2F_{7/2} + \text{Ho } {}^5I_6 \\ &\text{Yb } {}^5F_{5/2} + \text{Ho } {}^5I_6 \rightarrow \text{Yb } {}^2F_{7/2} + \text{Ho } {}^5S_2 \,. \end{aligned} \tag{1}$$

If the levels $^2F_{7/2}$, $^2F_{5/2}$, $^5I_8$, $^5I_6$, and $^5S_2$ are labeled Y1, Y2, H1, H2, and H3, the rate equations describing the level population densities are

$$\begin{aligned} \frac{d}{dt} n_{Y2} &= X(n_{Y1}-n_{Y2})-Vn_{Y2}n_{H1}-W_{Y2}n_{Y2}-Un_{Y2}n_{H2}+Un_{Y1}n_{H3} \\ \frac{d}{dt} n_{H2} &= Vn_{Y2}n_{H1}-Un_{Y2}n_{H2}+Un_{Y1}n_{H3}-W_{H2}n_{H2} \\ \frac{d}{dt} n_{H3} &= Un_{Y2}n_{H2}-Un_{Y1}n_{H3}-W_{H3}n_{H3} \quad . \end{aligned} \tag{2}$$

X is the rate of sensitizer excitation by the absorption of radiation and is the product of the Yb^{3+} absorption cross section and the incident photon flux. U and V are the transfer coefficients, as in equation (64) of the preceding chapter. For simplicity the forward and back transfer coefficients are set equal for the resonant transfer process. W_i^{-1} is the lifetime τ_i of state i when only one type of ion is present. Parameter values (3) measured in a

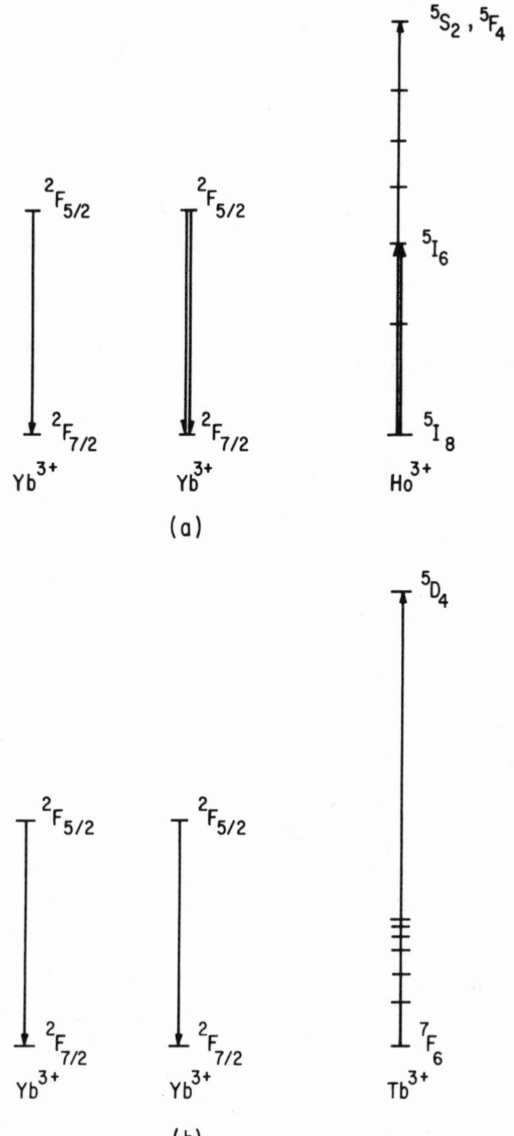

Fig. 2 Transfer by a stepwise process (a) and a cooperative process (b). In (a) the first transfer is non-resonant (double arrows) and the second is resonant. In (b) the $^5D_4$ level has twice the energy of $^2F_{5/2}$. The other multiplets of Ho and Tb are labeled in Figure 1.

crystal of $LiY_{0.799}Yb_{0.2}Ho_{0.001}F_4$ are

$$W_{H2} = 650 \text{ sec.}^{-1} \qquad Vc_H = 90 \text{ sec.}^{-1}$$
$$W_{H3} = 7.46 \times 10^3 \text{sec.}^{-1} \qquad Uc_Y = 1.6 \times 10^6 \text{ sec.}^{-1}$$
$$W_{Y2} = 314 \text{ sec.}^{-1}$$

W_{Y2}^{-1}, W_{H2}^{-1}, and W_{H3}^{-1} are the measured decay rates of the states $^2F_{5/2}$, $^5I_6$, and $^5S_2$ observed in crystals containing only Yb or only Ho respectively. $(W_{Y2} + Vc_H)^{-1}$ is the decay rate of $^2F_{5/2}$ observed in a crystal containing both Yb and Ho. The transfer rate Uc_Y was measured from a comparison of the intensities of the $^5S_2 \rightarrow {}^5I_8$ luminescence under short wavelength excitation in two crystals, one containing only Ho and the other containing both Ho and Yb. In the presence of Yb the luminescence is much less efficient because of Ho → Yb back transfer. The steady state solutions, obtained by setting $(d/dt)n_i = 0$, are for X small (weak excitation)

$$n_{Y2} = Xc_Y(Vc_H + W_{Y2})^{-1}$$
$$n_{H2} = XVc_Yc_HW_{H2}^{-1}(Vc_H + W_{Y2})^{-1}$$
$$n_{H3} = X^2UVc_Y^2c_HW_{H2}^{-1}(Vc_H + W_{Y2})^{-2}(Uc_Y + W_{H3})^{-1} \quad . \qquad (3)$$

For weak excitation n_{Y1} and n_{H1} are nearly equal to the Yb and Ho concentrations c_Y and c_H. From equations (3) when U is large enough that $Uc_Y \gg W_{H3}$, then $n_{H3}/n_{H2} = n_{Y2}/n_{Y1}$. This relation is characteristic of fast resonant transfer. And, as is characteristic of nonresonant transfer, the ratio n_{H2}/n_{H1} may be larger than n_{Y2}/n_{Y1} since $n_{H2}/n_{H1} = (Vc_Y/W_{H2})(n_{Y2}/n_{Y1})$. The power emitted in the green $^5S_2 \rightarrow {}^5I_8$ luminescence transition is given by

$$P = \varepsilon h\nu_{31}W_{H3}n_{H3} \quad . \qquad (4)$$

ε is the radiative quantum efficiency of the transition and ν_{31}, the frequency. For weak excitation P is proportional to the square of the infrared photon flux and to the square of the Yb concentration, from equation (3). It is also proportional to W_{H2}^{-1}, the lifetime of the intermediate state $^5I_6$. The longer this lifetime, the more probable is the second transfer. If V becomes large, n_{H3} and P decrease. In this case the first transfer is too efficient. The Yb^{3+} excited state population is depleted by the first transfer, and there are too few excited activators for the second transfer. Equations (2) can also be solved for the response to an infrared excitation pulse. If the relatively small terms in U in the first two rate equations are neglected, an analytical transient solution may be found. If the excitation rate X is switched from a non-zero value to zero at time t=0, the approximate solutions are

$$n_{Y2}(t) = n_{Y2}(0)\exp(-Wt)$$

$$n_{H2}(t) = Vc_H n_{Y2}(0)(W_{H2}-W)^{-1}[\exp(-Wt) - \exp(-W_{H2}t)]$$

$$n_{H3}(t) = Vc_H U n_{Y2}^2(0)(W_{H2}-W)^{-1}\{(W'-2W)^{-1}[\exp(-2Wt) - \exp(-W't)] - (W'-W-W_{H2})^{-1}[\exp(-Wt-W_{H2}t) - \exp(-W't)]\}. \quad (5)$$

Here $W = Vc_H + W_{Y2}$ and $W' = Uc_Y + W_{H3}$. W' is by far the largest rate, and the terms in $\exp(-W't)$ can be neglected since they go rapidly to zero. Exact solutions for the response to a pulse can be obtained numerically. These are shown in Figure 3 for excitation pulses of various durations (3). $n_{H3}(t)$ reaches a maximum value after the pulse has ended if the pulse is short compared with W_{Y2}^{-1} and W_{H2}^{-1}.

There are two other processes which should be considered in explaining the excitation of Ho^{3+} to the $^5S_2$ level. Since the $^5I_6$ - $^5S_2$ transition is resonant with $^2F_{7/2}$ - $^2F_{5/2}$, the $^5S_2$ state might be directly excited by absorption of the incident photons once $^5I_6$ is populated by transfer. In this case, however, $n_{H3}(t)$ would decrease monotonically as soon as the exciting light were switched off, in contrast with the observed behaviour shown in Figure 3. The other process is the "cooperative" excitation mechanism in which two Yb^{3+} ions simultaneously transfer excitation to Ho^{3+}, exciting it to $^5S_2$ with phonon assistance. This process is written schematically as

$$\text{Yb}\ ^2F_{5/2} + \text{Yb}\ ^2F_{5/2} + \text{Ho}\ ^5I_8 \rightarrow \text{Yb}\ ^2F_{7/2} + \text{Yb}\ ^2F_{7/2} + \text{Ho}\ ^5S_2. \quad (6)$$

The rate equation for n_{H3} would be

$$\frac{d}{dt} n_{H3} = U_c n_{Y2}^2 n_{H1} - W_{H3} n_{H3} \quad . \quad (7)$$

Back transfer has again been neglected since the transfer is not resonant. Solution of the steady state equation again gives n_{H3} proportional to X^2. It seems unlikely that this process plays a dominant role for two reasons. It is nonresonant and should therefore be less probable than a resonant cooperative transfer process, which in other cases is found to be relatively inefficient compared with the stepwise process, as discussed below. In addition the steady state and transient behaviour of the system is well explained on the basis of equations (2) for the stepwise process.

However, the $^5F_3$ state is apparently populated by a resonant cooperative process. The efficiency is much less than for the excitation of $^5S_2$ by sequential transfer. The $^5F_3 \rightarrow {}^5I_8$ luminescence

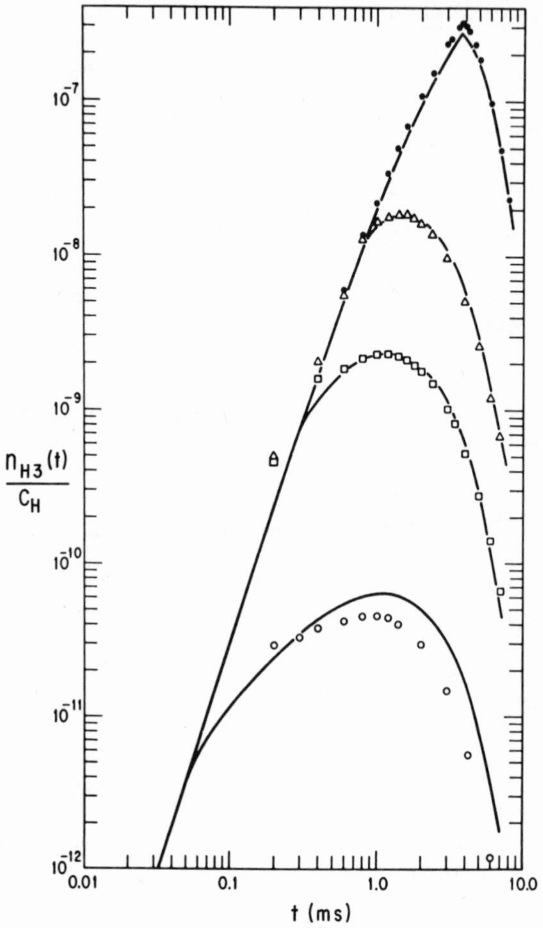

Fig. 3 Time development of the $^5S_2 \rightarrow {}^5I_8$ Ho^{3+} luminescence in $LiYF_4$:Yb,Ho with pulsed infrared excitation of the Yb^{3+}. The solid curves are calculated from equations (2), and the points are experimental data. The pulse begins at t=0 and ends at t=50 μs, 300 μs, 0.86 ms, and 3.8 ms for the four curves in order of increasing peak intensity.

intensity under infrared excitation of the Yb^{3+} is three orders of magnitude less than the $^5S_2 \rightarrow {}^5I_8$ intensity (4). When Ho^{3+} is excited to the $^5F_1$ level by short wavelength light, the excitation rapidly cascades non-radiatively down to $^5F_3$. The $^5F_3$ decay is partly radiative and partly non-radiative, the non-radiative process populating $^5S_2$. From $^5S_2$ the two main de-excitation paths are radiative decay and back transfer to Yb. Under this short wavelength excitation the $^5F_3 \rightarrow {}^5I_8$ luminescence intensity is ten times greater than the $^5S_2 \rightarrow {}^5I_8$ intensity (4). This rules out the possibility that under infrared pumping $^5S_2$ is populated chiefly by the cooperative excitation of $^5F_3$ by two Yb^{3+} ions with subsequent decay to $^5S_2$, since in this case the ratio of the $^5F_3$ and $^5S_2$ luminescence intensities would be the same as under short wavelength pumping.

In crystals containing Yb^{3+} and Tm^{3+} as many as four sequential transfers from Yb to Tm have been observed (5). They are all of the nonresonant type. In the first transfer Tm^{3+} is excited to the $^3H_5$ state. From $^3H_5$ the excitation decays non-radiatively to $^3H_4$. A second transfer further excites the ion from $^3H_4$ to $^3F_2$. There is rapid non-radiative decay to $^3F_4$, from which a third transfer induces a transition to $^1G_4$. From $^1G_4$ a fourth transfer excites the ion to $^1D_2$. Equations similar to equations (2) describe these processes with Tm → Yb back transfer omitted. For sufficiently weak excitation the luminescence intensities of the radiative transitions from $^3H_4$, $^3F_4$, $^1G_4$, and $^1D_2$ are proportional respectively to the first, second, third, and fourth powers of the infrared input flux. However, even at a rather low excitation intensity the rate of the $^3H_4 \rightarrow {}^2F_2$ transition due to the second transfer becomes comparable with the spontaneous decay rate of $^3H_4$, so that as the excitation increases, the population of $^3H_4$ ceases to be proportional to the excitation flux and exhibits a saturation behaviour. This leads to departures from the square, cubic, and quartic dependences on input flux of the populations of the upper levels (6). The ultraviolet emission $^1D_2 \rightarrow {}^3H_6$ is very weak at normally attainable excitation levels. To the eye the infrared-pumped phosphor appears bluish purple because of the red $^1G_4 \rightarrow {}^3H_4$ and blue $^1G_4 \rightarrow {}^3H_6$ emissions.

Green luminescence is efficiently produced in certain crystals activated with Er^{3+} and sensitized with Yb^{3+} when the Yb^{3+} $^2F_{7/2} \rightarrow {}^2F_{5/2}$ transition is pumped in the infrared. The $^4I_{15/2}$ - $^4I_{11/2}$ and $^4I_{11/2}$ - $^4F_{7/2}$ transitions of Er^{3+} are resonant or very nearly resonant, depending on the host crystal, with the Yb^{3+} transition. At very low Yb concentration the intensity of the $^4S_{3/2} \rightarrow {}^4I_{15/2}$ green emission is independent of Yb concentration (7), since the incident radiation is absorbed directly in the $^4I_{15/2} \rightarrow {}^4I_{11/2}$ and $^4I_{11/2} \rightarrow {}^4F_{7/2}$ transitions. $^4F_{7/2}$ decays non-radiatively to $^4S_{3/2}$. For Yb concentrations near 0.6% atomic the emission intensity is linear in the Yb content, as one Yb → Er transfer and one Er absorption dominate. For Yb concentrations greater than about 2% a square law obtains. In this region the behaviour of the system is well described by rate equations with two resonant sequential transfers (8). For materials with concentrations adjusted to maximize the green luminescence the Yb-Er combination yields higher efficiencies (green luminescence intensity per unit of absorbed infrared power) than Yb-Ho. Optimized concentrations of Er are typically ten times those of Ho in such materials. Activator-activator interactions, which quench the green luminescence, determine the upper limit of activator concentration, and these interactions are stronger for Ho than for Er. Examples of such quenching transfer processes are (9, 10),

$$\text{Ho}\ {}^5S_2 + \text{Ho}\ {}^5I_8 \rightarrow \text{Ho}\ {}^5I_4 + \text{Ho}\ {}^5I_7$$

$$\text{Er}\ {}^2H_{11/2} + \text{Er}\ {}^4I_{15/2} \rightarrow \text{Er}\ {}^4I_{9/2} + \text{Er}\ {}^4I_{13/2}\ . \qquad (8)$$

The Er $^2H_{11/2}$ state is populated at room temperature by thermal excitation from $^4S_{3/2}$.

In some host crystals the red Er $^4F_{9/2} \rightarrow {}^4I_{15/2}$ luminescence is more intense than the green under infrared excitation, while in others, generally softer crystals, the green emission dominates. The dominant mechanism in populating the red-emitting $^4F_{9/2}$ levels is apparently different for different host crystals, since different dependences of red luminescence intensity on infrared flux are observed (5, 11). In some cases a two-step process and in others, a three-step process is indicated.

The $^4F_{9/2}$ state may be populated by a three-step process in the following way. After population of the green-emitting $^4S_{3/2}$ state by a two-step process a third transfer may excite the ion from $^4S_{3/2}$ to $^2G_{7/2}$. Decay to $^4G_{11/2}$ may be followed by an Er → Yb back transfer in which the Er^{3+} transition $^4G_{11/2} \rightarrow {}^4F_{9/2}$ occurs. $^4F_{9/2}$ may be populated by a two-step process in several ways: There may be a non-radiative decay from $^4S_{3/2}$ to $^4F_{9/2}$. Or $^4F_{9/2}$ may be populated by the interaction of two Er^{3+} ions, one in the $^4F_{7/2}$ state and the other in $^4I_{11/2}$, the first making the transition $^4F_{7/2} \rightarrow {}^4F_{9/2}$ and the second, the transition $^4I_{11/2} \rightarrow {}^4F_{9/2}$. Or a second transfer from Yb may occur with the Er transition $^4I_{13/2} \rightarrow {}^4F_{9/2}$. In this case $^4I_{13/2}$ might be populated by non-radiative decay from $^4I_{11/2}$ or by direct nonresonant transfer from Yb.

From Figure 1 it is apparent that the profusion of energy levels may provide several excitation paths, and in general a particular energy level may be populated by more than one process. In crystals with large maximum phonon frequencies nonresonant energy transfer rates will be larger because of the lower order of the process. Similarly internal multiphonon decay will also be more rapid in such crystals, and radiative efficiencies will be smaller. The crystal field splitting of the $^{2S+1}L_J$ multiplets also varies from one type of crystal to another, and these differences can make a pair of transitions resonant in one host material and non-resonant in another. Small differences in the match of transition frequencies may be quite important at low temperatures (12).

Since the intensity of the green luminescence is proportional to the square of the infrared excitation intensity, the efficiency of the luminescence is proportional to the excitation intensity, at least in the regime of weak excitation where saturation effects are not important. As the excitation intensity increases, a saturation of the visible output eventually occurs. The detailed saturation behaviour depends upon the crystal host. The ultimate efficiency can be quite high. Experiments in which a YAlG:Nd laser was used as excitation source have shown that for $BaYF_5$ and YF_3 ultimate efficiencies, defined as visible luminescence power emitted divided by incident infrared excitation power, are larger than 15% in the red and 7% in the green (13).

III. COOPERATIVE TRANSFER

A second example of a cooperative excitation process is the simultaneous transfer from two Yb^{3+} ions to a Tb^{3+} ion, written

$$\mathrm{Yb}\ {}^2F_{5/2} + \mathrm{Yb}\ {}^2F_{5/2} + \mathrm{Tb}\ {}^7F_6 \rightarrow \mathrm{Yb}\ {}^2F_{7/2} + \mathrm{Yb}\ {}^2F_{7/2} + \mathrm{Tb}\ {}^5D_4. \quad (9)$$

From Figures 1 and 2b it can be seen that there is no Tb^{3+} energy level to serve as the intermediate state for sequential transfer. $^7F_0$ has a very short lifetime because of rapid non-radiative decay to the nearby $^7F_1$ level. Further, if $^7F_0$ were the intermediate state, in the first transfer an energy of 4×10^3 cm^{-1} would have to be dissipated as phonons, and in the second transfer the phonons would have to supply a similar amount of energy. Thus sequential transfers can be neglected. Since the energy of $^5D_4$ is equal to twice the Yb^{3+} excitation energy, the cooperative transfer is described by the rate equations,

$$\frac{d}{dt} n_{Y2} = X(n_{Y1} - n_{Y2}) - W_{Y2}n_{Y2} - U_c n_{Y2}^2 n_{T1} + U_c' n_{Y1}^2 n_{T2}$$

$$\frac{d}{dt} n_{T2} = -W_{T2}n_{T2} + U_c n_{Y2}^2 n_{T1} - U_c' n_{Y1}^2 n_{T2} . \quad (10)$$

The $^7F_6$ and $^5D_4$ states are written T1 and T2. U_c' is the coefficient for Tb → Yb back transfer. As in the case of the cooperative Yb-Ho process, the efficiency of Tb excitation is very low. If we neglect U_c in the first equation, the steady state solution is

$$n_{T2} = U_c n_{T1}(W_{T2} + U_c' n_{Y1}^2)^{-1} X^2 W_{Y2}^{-2} n_{Y1}^2 \quad . \qquad (11)$$

As the Tb concentration is changed, the lifetimes of the states vary, and in YF_3 the $^5D_4$ luminescence intensity, and hence n_{T2}, is found to be proportional to $(W_{T2} + U_c' n_{Y1}^2)^{-1} W_{Y2}^{-2} n_{T1}$, as expected (14).

The sequential transfer process can be viewed as a series of independent single transfers like those discussed in the preceding chapter. The cooperative transfer process (equations (6) and (9)) is a true three-atom interaction process, however. To describe it equation (15) of the preceding chapter, must be replaced by

$$w_{ABC} = (4\pi^2/h)|\langle a'b'c|H_{ABC}|abc'\rangle|^2 \int dE \int dE' g_A(E') g_B(E-E') g_C(E). \qquad (12)$$

w_{ABC} is the rate of the non-radiative cooperative process in which atoms A, B, and C make simultaneously the transitions $a' \to a$, $b' \to b$, $c \to c'$ respectively. g_A, g_B, and g_C are the normalized homogeneous line shape functions, as before. The interaction H_{ABC} consists of two-atom interactions: H_{AB}, H_{BC}, and H_{AC}. To find the contribution of the multipolar interactions to the matrix element of equation (12) we write the wavefunction of the three-atom system, correct to first order,

$$\langle a'b'c|^{(1)} = \langle a'b'c| + \sum_{\substack{a''\neq a' \\ b''\neq b'}} \frac{\langle a'b'|H_{AB}|a''b''\rangle}{E_{a'}+E_{b'}-E_{a''}-E_{b''}} \langle a''b''c|$$

$$+ \sum_{\substack{b''\neq b' \\ c''\neq c}} \frac{\langle b'c|H_{BC}|b''c''\rangle}{E_{b'}+E_c-E_{b''}-E_{c''}} \langle a'b''c''|$$

$$+ \sum_{\substack{a''\neq a' \\ c''\neq c}} \frac{\langle a'c|H_{AC}|a''c''\rangle}{E_{a'}+E_c-E_{a''}-E_{c''}} \langle a''b'c''| \quad . \qquad (13)$$

Then the matrix element $\langle a'b'c|H_{ABC}|abc'\rangle$ consists of contributions from pairs of two atom interactions:

$$\langle a'b'c|H_{ABC}|abc'\rangle = \sum_{b''\neq b'} \left\{ \frac{\langle b'c|H_{BC}|b''c'\rangle \langle a'b''|H_{AB}|ab\rangle}{E_{b'}+E_c-E_{b''}-E_{c'}} \right.$$

$$\left. + \frac{\langle a'b'|H_{AB}|ab''\rangle \langle b''c|H_{BC}|bc'\rangle}{E_{a'}+E_{b'}-E_a-E_{b''}} \right\} + \sum_{b''\neq b} \left\{ \frac{\langle b''c|H_{BC}|bc'\rangle \langle a'b'|H_{AB}|ab''\rangle}{E_b+E_{c'}-E_{b''}-E_c} \right.$$

$$\left. + \frac{\langle a'b''|H_{AB}|ab\rangle \langle b'c|H_{BC}|b''c'\rangle}{E_a+E_b-E_{a'}-E_{b''}} \right\} , \qquad (14)$$

plus contributions of similar form from $H_{AB} \times H_{AC}$ and $H_{AC} \times H_{AC}$.

If A and B are the same type of ion and if $E_{c'} = 2E_{a'}$, $E_{a'} = E_{b'}$, $E_a = E_b = E_c = 0$, then the first and fourth terms are equal and the second and third are also equal. The intermediate states b" may be other states of the $(4f)^n$ configuration and a quadrupole transition may be allowed, or they may be states of a configuration of opposite parity, allowing an electric dipole transition. Thus there are many virtual transitions to be considered in calculating the matrix element of H_{ABC} as well as several more interaction types (dipole-dipole/dipole-dipole, dipole-quadrupole/dipole-dipole, etc.) than for transfer between two ions. If the wavefunctions of the ions A, B, and C overlap appreciably, then a non-zero matrix element can be obtained with the zero order wavefunctions (the first term in equation (13)). In this case exchange effects are important, and the magnitude of the coupling is more difficult to estimate.

The final expression for w_{ABC} in the case of multipolar interactions contains terms with various dependences on separation of the form $R_{AB}^{-n} R_{BC}^{-m}$, where n and m are 6, 8, or 10, and other similar terms with different combinations of subscripts A, B, and C. w_{ABC} falls off more rapidly with increasing separation of the three ions than the two-ion transfer rate. It may therefore be a good approximation to consider only nearest neighbors. If A and B are Yb ions, C is a Tb ion, and R is the nearest neighbor (cation-cation) separation, then, since the Yb concentration is typically much larger than the Tb concentration, we add the contribution from each pair of Yb neighbors to obtain the rate constant of equation (10),

$$U_c n_{Y2}^2 = (1/2)q(q-1)(n_{Y2}/c_+)^2 w_{ABC}(R). \qquad (15)$$

q is the number of nearest cation sites around a Tb ion. c_+ is the number of cation sites per unit volume. $(1/2)q(q-1)(n_{Y2}/c_+)^2$ is the average number of pairs of excited Yb ions on these q sites. For nearest Yb-Tb neighbors in YF_3 Kushida finds that the dipole-quadrupole/dipole-quadrupole interaction dominates. He calculates an efficiency for the cooperative excitation in good agreement with experiment (15). Namely, to produce the same visible fluorescence power from YF_3:Yb,Tb as from YF_3:Yb,Er at the same activator concentration, the infrared excitation intensity must be greater by a factor 10^2 for Yb,Tb.

IV. ENERGY TRANSFER WITH PHOTON COOPERATION

The energy transfer processes discussed in the preceding sections are studied by observation of an associated absorption or emission of radiation. The absorption may be necessary to prepare the system in an initial state from which transfer can proceed.

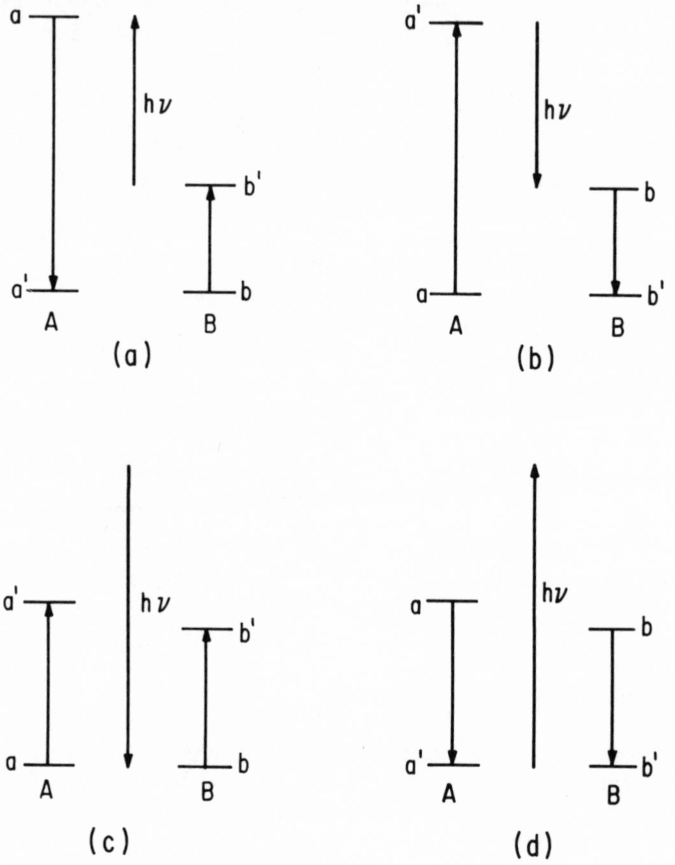

Fig. 4 Energy transfer with photon cooperation, (a) and (b), and cooperative absorption (c) and luminescence (d). An arrow marked "hν" indicates photon absorption in the process if it points down and photon emission if it points up. Process (b) is the inverse of (a) and (d) is the inverse of (c).

The emission generally occurs in competition with a non-radiative transfer process (sensitizer emission) or results from the decay of a state to which an ion has been excited by transfer (activator emission). There is another class of energy transfer phenomena in which the radiation field is necessary for energy conservation. That is, the transfer mechanism is an ion-ion-photon interaction. It is formally similar to the ion-ion-phonon interaction discussed in the first section. The energy transfer rate is equal to the rate at which photons are emitted or absorbed since one photon is either emitted or absorbed during each transfer. Some examples which have been observed experimentally are shown schematically in Figure 4. An arrow labeled "hν" indicates that a photon is absorbed in the process if this arrow points down and emitted if the arrow points up. Only the processes shown in Figures (4a) and (4b) are energy transfer processes in the sense we have been using. In the processes of Figure (4c) and (4d) energy is not transferred from one ion to the other: these processes are called "cooperative absorption" or "cooperative luminescence."

In all these processes atoms A and B are initially in states a and b and finally in states a' and b'. The theory of Dexter applies to all of them (16). The wavefunction for the initial state, corrected to first order to take account of the ion-ion interaction H_{AB}, is

$$\langle ab|^{(1)} = \langle ab| + \sum_{\substack{a''\neq a \\ b''\neq b}} \frac{\langle ab|H_{AB}|a''b''\rangle}{E_a+E_b-E_{a''}-E_{b''}} \langle a''b''| \ . \tag{16}$$

The electric dipole transition matrix elements are $\langle ab|\vec{M}(A) + \vec{M}(B)|a'b'\rangle$ where $\vec{M}(A) = \sum_s e\vec{r}_{As}$ and the sum is over the active electrons. Or in the notation of the preceding chapter the equivalent elements $\langle ab|eD_q^{(1)}(\vec{r}_A) + eD_q^{(1)}(\vec{r}_B)|a'b'\rangle$ may be calculated. With the wavefunctions of equation (16) the element is given by

$$\begin{aligned}\langle ab|D_q^{(1)}(\vec{r}_A) + D_q^{(1)}(\vec{r}_B)|a'b'\rangle = &\sum_{a''\neq a'} \frac{\langle a''b|H_{AB}|a'b'\rangle \langle a|D_q^{(1)}|a''\rangle}{E_{a'}+E_{b'}-E_{a''}-E_b} \\ &+ \sum_{b''\neq b'} \frac{\langle ab''|H_{AB}|a'b'\rangle \langle b|D_q^{(1)}|b''\rangle}{E_{a'}+E_{b'}-E_a-E_{b''}} \\ &+ \sum_{a''\neq a} \frac{\langle ab|H_{AB}|a''b'\rangle \langle a''|D_q^{(1)}|a'\rangle}{E_a+E_b-E_{a''}-E_{b'}} \\ &+ \sum_{b''\neq b} \frac{\langle ab|H_{AB}|a'b''\rangle \langle b''|D_q^{(1)}|b'\rangle}{E_a+E_b-E_{a'}-E_{b''}} \ . \end{aligned} \tag{17}$$

Let us suppose that the states a, a', b, and b' have the same parity - that they are all 4f states, for example. In the first term a" may be a higher lying state of opposite parity. Then if H_{AB} represents a dipole-quadrupole interaction, this term will be of the form $\sim(e^2/R^4)a_0^4/\Delta E$, where a_0 is the Bohr radius and ΔE, the energy denominator. The single-ion forced electric dipole transition matrix element is $\langle a|D_q^{(1)}|a'\rangle \sim a_0(V_{odd}/\Delta E)$. V_{odd} represents the odd-parity crystal field component which admixes a state of opposite parity into a and a' to make the transition weakly allowed. For a quantitative comparison a system must be specified and the omitted numerical factors included.

The process shown in Figure (4a) has been observed on two different systems. Feofilov and Trofimov (17) report a Gd^{3+}-Yb^{3+} interaction in Yb_2O_3 in which states a and a' are the $^6P_{7/2}$ and $^8S_{7/2}$ states of Gd^{3+} and b and b' are the Yb^{3+} states $^2F_{7/2}$ and $^2F_{5/2}$. In $EuAlO_3$:Cr the same sort of process has been observed (18). Here the states a and a' are the ^2E and $^4A_2$ states of Cr^{3+} and b and b' are the $^7F_0$ and $^7F_1$ states of Eu^{3+}. The inverse process, Cr^{3+} $^4A_2 \rightarrow {}^2E$, Eu^{3+} $^7F_1 \rightarrow {}^7F_0$, has also been observed (Figure (4b)). A similar process involving Tb^{3+} and Yb^{3+} has been seen in which the transitions Tb^{3+} $^7F_6 \rightarrow {}^5D_4$, Yb^{3+} $^2F_{5/2} \rightarrow {}^2F_{7/2}$ occur with photon absorption (19). That excitation of the Tb^{3+} $^5D_4$ state might be due instead to cooperative transfer from two excited Yb ions was ruled out by exciting the system with a short pulse. The $^5D_4$ luminescence intensity does not rise to a maximum and then decrease with time after the pulse. That is, the time dependence of the $^5D_4$ luminescence does not consist of two exponential terms, one with time constant half the Yb^{3+} $^2F_{5/2}$ lifetime, as it would if cooperative transfer were dominant.

The cooperative absorption process of Figure (4c) has been observed for Pr^{3+}-Pr^{3+} pairs (20), for Cr^{3+}-Cr^{3+} pairs (21), and for Cr^{3+}-Eu^{3+} pairs (18). The inverse luminescence process of Figure (4d) has also been observed for Cr^{3+}-Eu^{3+} pairs and for Yb^{3+}-Yb^{3+} pairs (22). The magnitudes of the Pr^{3+}-Pr^{3+} pair absorption and Yb^{3+}-Yb^{3+} luminescence have been fairly well explained on the basis of multipolar interactions (15, 23). The processes involving Cr^{3+}-Cr^{3+} and Cr^{3+}-Eu^{3+} pairs, on the other hand require, because of the observed selection rules, an explanation based on super-exchange coupling. The exchange-induced moment can be represented by an effective dipole moment operator,

$$\vec{M}_{eff} = \sum_{p,t} \vec{M}_{pt} \vec{s}_{Ap} \cdot \vec{s}_{Bt}, \tag{18}$$

after the fashion of equation (38) of the preceding chapter. The sum is over the electrons of the two ions. The $\vec{M}_{pt}$ must be treated as empirically determined parameters, since they are as difficult to evaluate from first principles as the J_{pt} of equation (38). For Eu^{3+}-Cr^{3+} the exchange coupling gives rise to pair absorption which is stronger than the single ion absorption, in contrast to rare earth-rare earth pairs coupled by the electric multipolar interaction.

REFERENCES

1. F. Auzel, Comt. Rend. Acad. Sci. Paris 262, 1016 (1966).

2. F. Auzel, Proc. IEEE 61, 758 (1973).

3. R. K. Watts, J. Chem. Phys. 53, 3552 (1970); R. K. Watts and H. J. Richter, Phys. Rev. B6, 1584 (1972).

4. V. V. Ovsyankin and P. P. Feofilov, Opt. and Spectr. 31, 510 (1971).

5. L. F. Johnson, H. J. Guggenheim, T. C. Rich, and F. W. Ostermayer, Jr., J. Appl. Phys. 43, 1125 (1972).

6. F. W. Ostermayer, Jr., J. P. van der Ziel, H. M. Marcos, and J. E. Geusic, Phys. Rev. B3, 2698 (1971).

7. R. A. Hewes and J. F. Sarver, Phys. Rev. 182, 427 (1969).

8. J. D. Kingsley, J. Appl. Phys. 41, 175 (1970).

9. J. F. Porter, Jr. and H. W. Moos, Phys. Rev. 152, 300 (1966).

10. J. P. van der Ziel, F. W. Ostermayer, Jr., and L. G. Van Uitert, Phys. Rev. B2, 4432 (1970).

11. J. P. Wittke, I. Ladany, and P. N. Yocom, J. Appl. Phys. 43, 595 (1972).

12. H. Kuroda, S. Shionoya, and T. Kushida, J. Phys. Soc. Japan 33, 125 (1972).

13. T. C. Rich and D. A. Pinnow, J. Appl. Phys. 43, 2357 (1972).

14. F. W. Ostermayer, Jr. and L. G. Van Uitert, Phys. Rev. B1, 4208 (1970).

15. T. Kushida, J. Phys. Soc. Japan 34, 1327 (1973).

16. D. L. Dexter, Phys. Rev. 126, 1962 (1962); M. Altarelli and D. L. Dexter, Phys. Rev. B7, 5335 (1973).

17. P. P. Feofilov and A. K. Trofimov, Opt. and Spectr. 27, 291 (1969).

18. J. P. van der Ziel and L. G. Van Uitert, Phys. Rev. 180, 343 (1969); B8, 1889 (1973).

19. V. I. Bilak, G. M. Zverev, G. O. Karapetyan, and A. M. Onishchenko, JETP Letters 14,199 (1971).

20. F. Varsanyi and G. H. Dieke, Phys. Rev. Letters 7, 442 (1961).

21. G. G. P. van Gorkom, Phys. Rev. B8, 1827 (1973).

22. E. Nakazawa and S. Shionoya, Phys. Rev. Letters 25, 1710 (1970).

23. K. Shinagawa, J. Phys. Soc. Japan 23, 1057 (1967).

RELAXATION AND ENERGY TRANSFER*

R. Orbach**

Department of Physics, Tel-Aviv University

Ramat Aviv, Tel-Aviv, Israel

ABSTRACT

The theory of phonon induced spin-lattice relaxation of paramagnetic impurities in solids is reviewed and summarized. The frequency and temperature dependences of non-radiative relaxation transitions are catalogued. The electron-phonon interaction is applied to spatial energy transfer, and some interesting aspects of the range dependence examined. Short range transfer forces (e.g., exchange) are discussed, and critical concentrations for energy transport in inhomogeneously broadened systems derived. The relationship between this effect and degradation of fluorescent efficiency is discussed.

* Supported in part by the National Science Foundation and the U.S. Office of Naval Research. Support is also gratefully acknowledged from the John Simon Guggenheim Memorial Foundation in the form of a one year fellowship grant.

**Permanent address:

Department of Physics
University of California
Los Angeles, California 90024
U.S.A.

I. INTRODUCTION

The problem of energy transfer in solids was given its quantitative backbone by Förster (1) and Dexter (2). These approaches can be roughly characterized as dealing with "resonant" transfer. Overlapping emission and absorption band tails of donors and acceptors, at or near the optical absorption frequency in question, allow energy transfer to take place in the absence of phonon assistance. It was noted by a number of authors (e.g., Orbach (3), Birgeneau (4), Miyakawa and Dexter (5), Soules and Duke (6), and Kohli and Huang Liu (7)) that the resonant transfer rate, proportional to the overlap integral

$$\int f_1(E)\; f_2(E)dE, \qquad (1)$$

was negligible for significantly different no-phonon line emission and absorption frequencies. Here $f_1(E)$ and $f_2(E)$ are the appropriate normalized line shape functions corresponding to the no phonon lines of the donor (ion 1) and the acceptor (ion 2). "Significantly different" means separation in energy much larger than the convoluted line width (for Lorentzian lines, the sum of the donor and acceptor widths). In fact, the condition for resonant transfer may even be more limited. If the line shape functions f(E) are inhomogeneously broadened, then the spectral distribution is in reality a spatial distribution. For short range interactions (fall off faster than $1/|\underline{R}_{12}|^3$, where $\underline{R}_{12} = \underline{r}_1 - \underline{r}_2$, the distance between donor and acceptor) the difficulty of finding an acceptor within a homogeneous width energy, at a range such that the transfer rate is significant, governs the transfer probability. Not only will the result be much smaller than if one uses (1) with the inhomogeneous line shape functions, but there may also be no energy transport at all if the concentration of donors and acceptors lies below a critical value. This "Anderson" (8) localization has been shown by Lyo (9) to account for the sharp drop of fluorescent efficiency in ruby for Cr concentrations in excess of ∿ 0.4 %.

Under these conditions, there arises the question of mechanisms which would "make up" the energy difference between donor and acceptor, and thus facilitate energy transport. To our knowledge, this was first formulated in ref. (3), and developed quantitatively in references (5), (6), and (7). The use of lattice vibrations (phonons) to make up for the energy difference has some curious aspects. In the first place, if the temperature T is such that $\Delta E_1 > \Delta E_2$,

$$k_B T < \Delta E_1 - \Delta E_2 \qquad (2)$$

the theoretical donor-acceptor one phonon assisted transfer rate is found to be temperature independent and, at least for $|\Delta E_1 - \Delta E_2| < \hbar\omega_D$, to increase as the energy mismatch. For the reverse,

$$k_B T > |\Delta E_1 - \Delta E_2| \quad . \tag{3}$$

the calculated rate increases as the product of $k_B T$, and is independent of the energy mismatch. Finally, for $\Delta E_1 < \Delta E_2$, and

$$k_B T < |\Delta E_1 - \Delta E_2| \quad , \tag{4}$$

the transfer rate increases as the square of the mismatch, and exponentially with temperature as $\exp\{(\Delta E_1 - \Delta E_2)/k_B T\}$. These results show that energy transfer can be greatest at resonance, $\Delta E_1 = \Delta E_2$, falls off sharply for $\Delta E_1 - \Delta E_2$ in substantial excess of the sum of the homogeneous widths of donor and acceptor; increases as $\Delta E_1 - \Delta E_2$ or as $k_B T$, whichever is larger; and then diminishes as $\Delta E_1 - \Delta E_2$ exceeds the coupling phonon energy, $\hbar\omega_D$, by virtue of the fact that two phonons are now required for energy conservation. As the energy mismatch continues to grow, and the number of phonons required to satisfy energy conservation increases, the transfer rate continues to drop, and can be written as depending exponentially on the energy mismatch $\Delta E_1 - \Delta E_2$ (see ref. (5)):

$$\exp\{-\alpha(\Delta E_1 - \Delta E_2)\} \tag{5}$$

$$= \exp\left\{-\frac{(\Delta E_1 - \Delta E_2)}{\hbar\omega_{Ph}}[\ell n(N/g(n+1)) - 1]\right\}$$

where we have taken the phonon energy $\hbar\omega_{Ph}$ equal to $(\Delta E_1 - \Delta E_2)/N$. Hence, N is the number of phonons needed to make up the energy mismatch. In (5), n is the phonon occupation number, and g an appropriately defined electron-phonon coupling constant. This exponential dependence was first observed in multiphonon non-radiative relaxation by Riseberg et al. (10), and in multi-phonon energy transfer by Reisfeld and her group (11).

In order to formulate a theory which can yield quantitative expressions for the various rates and critical concentrations listed above, we shall first write down expressions for the coupling Hamiltonian between the electronic states and the running vibrational waves, the phonons. One can take advantage of group theory to minimize the number of independent parameters. Their magnitudes are most difficult to calculate, but using the analogy to static crystalline field theory they can be estimated reasonably well. In the strong coupling limit, the transition rates induced by these interactions will be partially quenched by virtue of the displaced vibrational overlaps. Using this formalism, the non-radiative relaxation rates for one, two, and multiphonon processes are calculated. These topics form the body of section II. In section III we divide our treatment of energy transfer into two distinct parts: resonant and non-resonant. The former begins with various inter-ion coupling mechanisms, followed by estimates of their range dependence. The virtual phonon exchange mechanism will be shown to be strongly energy dependent, while those of multipole and exchange character will not. The effect of inhomogeneous broadening on resonant transfer will be sketched, and Lyo's estimates for ruby (9) discussed. The end of section III deals with non-resonant, or more specifically, phonon-assisted energy transfer. Here, we examine the form of the interaction, its temperature, energy (mismatch), and range dependence, reproducing the results sketched above. Finally, the quenching effect of strong coupling, as discussed by Kohli and Huang (7), will be summarized. In section IV, we present our conclusions, and suggest further areas for investigation.

II. NON-RADIATIVE RELAXATION

We divide this section into four parts. The first constructs the electron-phonon coupling Hamiltonian using the donor or acceptor site symmetry to reduce the number of independent variables. Estimates of the magnitude of the coupling coefficients are also given. We indicate briefly in part two how the transition rates are modified if substantial changes in the diagonal vibrational energy occur in the process of relaxation transitions. Part three deals with the one phonon non-radiative relaxation process, important for close lying (within the Debye energy) electronic levels. The last part treats two-phonon (and higher) transitions, leading to the celebrated exponential energy dependence for the non-radiative transition rate of Riseberg and Moos (10) and Miyakawa and Dexter (5):

$$W_{NR} \sim \exp\{\alpha\Delta E\} \quad . \tag{6}$$

Here, ΔE is the net electronic energy change, and α a (negative) constant (compare with (5) for phonon-assisted energy transfer). Eq. (6) is valid for the required number of phonons $N = \Delta E/\hbar\omega_{ph}$ large compared to unity, where $\hbar\omega_{ph}$ represents some average phonon energy.

II.A. Weak Coupling Limit [Most of the results of this section are derived in some detail by Orbach and Stapleton (12)].

1. Crystal Field Theory. We begin this section by very briefly reviewing the static crystalline field which a paramagnetic defect experiences. If one neglects overlap between the paramagnetic center and the surrounding (ligand) charge cloud, the electrons on the center experience electrostatic potential energy

$$V(\underline{r}) = -\int [e\rho(\underline{R})/|\underline{r}-\underline{R}|]\, d^3\underline{R} \tag{7}$$

where the electron is at $\underline{r}$, and the charge density in a volume $d^3\underline{R}$ at $\underline{R}$ is $\rho(\underline{R})$. Using the conventional expansion

$$1/|\underline{r}-\underline{R}| = \sum_{\ell=0}^{\infty} (r_<^{\ell}/r_>^{\ell+1}) P_\ell(\cos\omega)$$

where ω is the angle between $\underline{r}$ and $\underline{R}$; and the spherical harmonic addition theorem,

$$P_\ell(\cos\omega) = \sum_{m=-\ell}^{\ell} C_\ell^m(\underline{r}) C_\ell^{m*}(\underline{R}) \quad ,$$

where $C_\ell^m(\underline{r}) = [4\pi/(2\ell+1)]^{\frac{1}{2}} Y_\ell^m(\underline{r})$ are the Racah normalized spherical harmonics; we can simplify (7) to

$$V(\underline{r}) = -\sum_{\ell,m} r^\ell C_\ell^m(\underline{r}) \{e\int [\rho(\underline{R}) C_\ell^{m*}(R)/R^{\ell+1}]\, d^3\underline{R}\} . \tag{8}$$

Sandwiching (8) between the radial parts of the electronic wave function leads to the following one electron effective static crystal field Hamiltonian:

$$H_{CF} = -\sum_{i,\ell,m} \langle r_i^\ell \rangle C_\ell^m(\underline{r}_i) \{e\int [\rho(\underline{R}) C_\ell^{m*}(\underline{R})/R^{\ell+1}]\, d^3R\} \tag{9}$$

The subscript i refers to the ith electron on the paramagnetic site. In general, within a given configuration only even values of ℓ contribute, where the maximum value of $\ell \leqslant 2\ell_i$, where ℓ_i is the orbital moment of the i-th paramagnetic electron.

Further, terms in (9) for given ℓ but varying m will either vanish or be related as a consequence of the symmetry of $\rho(\underline{R})$ viewed from the impurity site. Thus, for a cubic site, and d electrons (ℓ=2), only a single term in (9) remains, and we are able to write

$$H_{CF} = A_4{}^0 \langle r^4\rangle[C_4{}^0 + (5/14)^{\frac{1}{2}}(C_4{}^4 + C_4{}^{-4})] \, . \tag{10}$$

For an octahedron of changes, $A_4{}^0 = 14\pi e^2/9R^5$, where R is the distance between the center and the neighboring ligand. For f electrons (ℓ=3) another term is found from (8) in addition to (10):

$$A_6{}^0 \langle r^6\rangle[C_6{}^0 - (7/2)^{\frac{1}{2}}(C_6{}^4 + C_6{}^{-4})] \tag{11}$$

where, for an octahedron of point charges, $A_6{}^0 = 3\pi e^2/13R^7$. In general these coefficients are poorly represented in magnitude by the point charge approximation. For d electrons (transition metals), $A_4{}^0$ is typically found to be of the order of an electron volt ($\sim$ 10,000 cm^{-1}), whereas for f electrons (rare earths) both $A_4{}^0$ and $A_6{}^0$ are an order of magnitude smaller. It is generally accepted that the primary contribution to $A_4{}^0$ for transition metals arises from electron transfer from the ligands (covalency), while for rare earths $A_4{}^0$ and $A_6{}^0$ derive mainly from overlap (13). It is probably fair to state that the calculation of $A_4{}^0$ and $A_6{}^0$ is of sufficient complexity to deter all but the most intrepid. Very recent cluster calculations by Ellis and Freeman (unpublished) show promise, but much remains to be done. For this reason, it is usual to regard the coefficients as adjustable parameters, obtainable from a fit to optical spectra. The forms of H_{CF} in (10) and (11) are valid regardless of the source of the coefficients, arising only from the local site symmetry.

2. Symmetry of Phonon Interaction Hamiltonian. Here, we follow the phenomenology of the first part of this section, but allow the ligands to move. This will give rise to the H_{el-vib} which we shall use to effect relaxation transitions. The simplest way to see the effect of the vibrations is to assume that only the near-neighbor ligands in the vicinity of the paramagnetic defect contribute to the dynamic crystalline potential. For an octahedron of charges, there are 21 degrees of freedom, of which 3 are pure translation and 3 pure rotation. For non-radiative transitions (but not for spin-lattice relaxation!) these modes are ineffective. The remaining 15 can be grouped into the cubic representations:

$$\Gamma_{1g}(1),\ \Gamma_{3g}(2),\ \Gamma_{5g}(3),\ 2\Gamma_{4u}(3),\ \text{and}\ \Gamma_{5u}(3).$$

The number in parenthesis gives the degeneracy. The odd (subscript u) vibrations have no matrix elements within a configuration. Γ_{1g} only shifts the levels up and down, but has no matrix elements between states. It is referred to as the diagonal coupling, and will be important for vibrational quenching and for energy transport. The remaining two (off-diagonal) modes, Γ_{3g} and Γ_{5g}, have their rows labelled for m (θ,e for Γ_{3g}; 0,±1 for Γ_{5g}). We write, in complete analogy to the static terms (10) and (11).

$$H_{el\text{-}vib} = \Sigma_{m,\ell} V(\Gamma_{3g}\ell) C(\Gamma_{3g}\ell,m) Q(\Gamma_{3g},m)$$

$$+ \Sigma_{m,\ell} V(\Gamma_{5g}\ell) C(\Gamma_{5g}\ell,m) Q(\Gamma_{5g},-m)(-1)^m \quad . \qquad (12)$$

The $C(\Gamma_{ig}\ell,m)$ are linear combinations of the Racah normalized spherical harmonics, summed over all electrons, that transform as the mth subvector of the ith irreducible representation. They are listed in ref. (12) and in standard texts (14). The $Q(\Gamma_{ig},m)$ are the normal modes of the octahedral complex which transform as the mth subvector of the ith irreducible representation. They are displayed in ref. (12) following the original development of Van Vleck (15). The coefficients $V(\Gamma_{ig}\ell)$ are the dynamic equivalent of the static $A_\ell^0\langle r^\ell\rangle$ in the crystal field expressions (10) and (11). Thus, for d electrons there are four dynamic coefficients ($V(\Gamma_{3g}2)$; $V(\Gamma_{3g}4)$; $V(\Gamma_{5g}2)$; and $V(\Gamma_{5g}4)$) while for f electrons there are six ($V\Gamma_{3g}6$) and $V(\Gamma_{5g}6)$ in addition). To our knowledge, covalent calculations have yet to be performed. On a point charge model for an octahedral cluster

$$V(\Gamma_{3g}2) = 6e^2\langle r^2\rangle/R^3$$

$$V(\Gamma_{5g}2) = 4e^2\langle r^2\rangle/R^3$$

$$V(\Gamma_{3g}4) = -(5/3)\ 15\ e^2\langle r^4\rangle/R^5$$

$$V(\Gamma_{5g}4) = -(2/3)\ 15\ e^2\langle r^4\rangle/R^5 \ .$$

A comparison with the static point charge values demonstrates that $A_4^0\langle r^4\rangle \sim V(\Gamma_{ig}4)$. This is a general feature of the dynamic coefficients. In general, they are comparable to their static counterparts, an approximation which has proven very useful (16).

Finally, we make the connection to running waves by noting the cluster normal modes at position $\underline{R}_\alpha$ can be expanded as (17):

$$Q_\alpha(\Gamma_{ig},m) = \Sigma_{\underline{K},j}(i/a)[\hbar/2M\omega_j(\underline{K})]^{\frac{1}{2}}$$

$$\times[b_{\underline{K},j}(\exp i\underline{K}\cdot\underline{R}_\alpha) - b^*_{\underline{K},j}\exp(-i\underline{K}\cdot\underline{R}_\alpha)]R_{\underline{K},j}(\Gamma_{ig},m) \ . \qquad (13)$$

Here, as examples

$$
\begin{aligned}
R_{\underline{K},j}(\Gamma_{3g},\theta) &= \left[\frac{1}{2}\, 2e^{z}(\underline{K}j)(\sin K_z a)\right. \\
&\quad \left. - e^{x}(\underline{K}j)(\sin K_x a) - e^{y}(\underline{K}j)(\sin K_y a)\right] \quad ; \\
R_{\underline{K},j}(\Gamma_{3g},e) &= \frac{1}{2}\sqrt{3}\left[e^{x}(\underline{K}j)(\sin K_x a)\right. \\
&\quad \left. - e^{y}(\underline{K}j)(\sin K_y a)\right] \quad ;
\end{aligned} \tag{14}
$$

with similar expressions for the Γ_{5g} expansion. In (14), $e^{\beta}(\underline{K}j)$ is the βth component of the unit polarization vector for the phonon of wave vector $\underline{K}$ and polarization index j. The $b_{\underline{K},j}$ are the destruction operators for the same phonon, M is the mass of the entire crystal, and a the lattice constant. For simplicity we have assumed one atom per unit cell. Eq. (14) is valid at arbitrary $\underline{K}$. The long wave limit, which is probably not very useful for large phonon energies, has the attractive form

$$
\begin{aligned}
R_{\underline{K}j}(\Gamma_{3g},\theta) &= \frac{1}{2}a(2e^{z}K_z - e^{x}K_x - e^{y}K_y) \quad ; \\
R_{\underline{K},j}(\Gamma_{3g},e) &= \frac{1}{2}a\sqrt{3}(e^{x}K_x - e^{y}K_y) \quad .
\end{aligned} \tag{15}
$$

This form is often found in the literature, as it is mathematically tractable and useful for spin lattice relaxation time (T_1) calculations. For our purposes, especially when calculating virtual phonon exchange coupling, it will be necessary to use the full expression (14).

3. Magnitude of the Interaction Constants. The generality of the expressions (12) and (13) enables one to use a variety of techniques to obtain values for the parameters $V(\Gamma_{ig}\ell)$. Examples are static strain splittings or shifts of magnetic resonance or optical spectra. The important point is that (12) is the sum of one electron operators. Therefore, if the coefficients $V(\Gamma_{ig}\ell)$ are determined for a given configuration, (12) can be evaluated for any Russell-Saunders multiplet using these coefficients (see ref. 14).

For Mn^{2+} in MgO, Blume and Orbach (18), and Sharma et al. (19) have calculated the $V(\Gamma_{ig}\ell)$ and obtained good agreement with the static stress experiments of Watkins and Feher (20). To obtain some feeling for the transition metals we quote their results:

$$V(\Gamma_{3g}2) = -11 \times 10^4 \ \text{cm}^{-1} \quad ;$$

$$V(\Gamma_{3g}4) = 3.85 \times 10^4 \ \text{cm}^{-1} \quad ;$$

$$V(\Gamma_{5g}2) = -5.32 \times 10^4 \ \text{cm}^{-1} \ ;$$

$$V(\Gamma_{5g}4) = 1.08 \times 10^4 \ \text{cm}^{-1} \quad .$$

In principle, these values could now be inserted in (12) and non-radiative (one phonon) rates computed for all the Mn^{2+} energy levels (within the $3d^5$ configuration). To get a feeling for the rare earth coupling constants, we use the T_1 data of Sabisky and Anderson (21), taking note of a large over-estimate caused by omission of the rotational relaxation mechanism of Abragam et al. (22). They find for the rare earth Tm^{2+} in CaF_2, SrF_2 and BaF_2, that

$$V(\Gamma_{3g}2) \sim 170 \ \text{cm}^{-1}$$

for all the hosts (with some caution with respect to CaF_2, where the rotation mechanism appears dominant). Thus, we see the enormous differences in coupling constant strengths between the iron group (3d) and the rare earths. This reflects itself in very different spectral characteristics (broad band vs. discrete phonon sidebands (5,23)), and of course in phonon-assisted transfer rates.

II.B. Strong Coupling Limit

1. Reduction of Operator Matrix Elements. As shown in II.A.3, the strength of H_{31-vib} is much larger for transition metal ions than for rare earths. As a result, the electron-vibrational interaction often cannot be treated as a small perturbation. Instead, one must include vibrational states with the electronic states when diagonalizing the Hamiltonian including H_{el-vib}. This leads to coupled electronic-vibrational (vibronic) states. When transition matrix elements are to be calculated, the full coupled vibronic state must be used, even for pure electronic perturbations. Under these conditions, one can apply an "instantaneous" Born-Oppenheimer approximation, in the sense that the electronic

motion will occur about a nuclear position dependent on the electronic level occupied. Under extreme conditions, this can mean a shift of origin for the vibrations (by virtue of the linear electron-vibration interaction) of such a magnitude that matrix elements of pure electronic operators may be quenched, the displaced lattice oscillators having negligible overlap. This well known result was first applied to energy transfer by Soules and Duke (6), and in the very recent work of Kohli and Huang Liu (7). Basically, one finds that the rates we shall calculate in Section III will be reduced by a factor (see ref. (7)),

$$\ell^{-\Phi(T)} \tag{16}$$

where

$$\Phi(T) = \Sigma_{\underline{K},j} \Big| \langle \Sigma_{\ell,m} V(\Gamma_{1g}\ell) C(\Gamma_{1g}\ell,m) Q(\Gamma_{1g},m) \rangle_{ex} - \langle \Sigma_{\ell,m} V(\Gamma_{1g}\ell) C(\Gamma_{1g}\ell,m) Q(\Gamma_{1g},m) \rangle_{gd} \Big|^2 \left(\frac{1}{\hbar^2 \omega_j^2(\underline{K})} \right) \times \coth(\hbar\omega_j(\underline{K})/2k_B T) \tag{17}$$

where the subscripts ex and gd mean that the matrix element of the operators are taken in the excited or ground electronic state , respectively. If either is degenerate, it is then possible that non-diagonal (e.g., Γ_{3g}, Γ_{5g}) vibrations could also contribute to (17) in addition to the totally symmetric vibration Γ_{1g}. Explicit appearance of the coth means that the phonon creation and annihilation operators in Q have already been taken into account. The Debye-Waller-like factor (17) is temperature independent at low temperatures, but the exponent increases linearly with T at high temperatures, leading to a strong temperature dependence for the decrease of electronic transition probability. We shall see that this effect can have strong impact on energy transfer between states where appreciable differences in diagonal phonon couplings exist.

2. Numerical Estimates. An example which has received considerable experimental (24) and theoretical (7) attention is $KMgF_3:V^{2+}$. The coupling constants have been measured (24), along with the appropriate velocities of sound. Kohli and Huang-Liu, using a Debye-like model, and $V(\Gamma_{3g}) = \sqrt{3}\, V(\Gamma_{5g}) = \frac{1}{2}\sqrt{3}\, V$ where $V = 0.85\times10^4 cm^{-1}$ (compare with Mn in MgO, where the coupling is a factor of 4-10 times larger), find at 77K for the $^4T_2 \rightarrow {}^4A_2$ transition:

$$\ell^{-\Phi(T)} \cong 0.2 \quad . \tag{18}$$

This result is weakly range dependent for energy transfer considerations (see Section III). The exponential dependence on V^2 is the reason why, within a 4f configuration, (16) is essentially unity, while for 3d impurities it can almost vanish. For an effective V twice that found for $KMgF_3 : V^{2+}$, the reduction factor would equal 10^{-3} instead of 0.2. One must calculate the matrix element difference for each electronic state in question, before one knows the quantitative importance of such a term.

II.C. The One-Phonon Processes

The non-radiative transition we are considering is to take place between levels separated in energy by ΔE. For such a process, the transition probability per unit time for an atom in its excited state to radiate a phonon and end up in its ground state is given by the Golden Rule (see Fig. 1):

$$W_{ex\to gd} = \Sigma_{\underline{K},j} (2\pi/\hbar^2) |\langle ex|\ H^{\underline{K},j}_{el-vib} |gd\rangle|^2$$

$$\times\ \delta[\omega_j(\underline{K}) - (\Delta E/\hbar)] \qquad (18)$$

where the superscript $\underline{K},j$ designates a particular term in the sum (13). We see from (18) that phonon states must be available with energies $\Delta E = \hbar\omega_j(\underline{K})$ in order that the one phonon emission process can be effective. If this is the case, then (18) is a powerful equilibrating rate. Thus, it is found, in general, that optical levels within $\sim \hbar\omega_D$ of one another are in thermal equilibrium and exhibit the same fluorescent decay rate regardless of which level is pumped.

The thermal behavior of (18) is easily obtained, but the energy dependence is a function of the phonon density of states. For emission,

$$W_{ex\to gd} \sim [n(\underline{K}j)+1] = \{[\exp(\Delta E/k_B T)-1]^{-1} + 1\ .$$

For absorption (the reverse process) n+1 is replaced by n. The temperature dependence of the former is weak: independent of T at low T, and varying as T for high T. That of the latter is very strong at low T (exponentially small), and varies as T at high temperatures.

The frequency dependence of (18) can be easily obtained at small $\Delta E (\ll \hbar\omega_D)$ where the long wave approximation can be used. The sum over $\underline{K}$ is replaced by an integral over $\omega_j(\underline{K})$, yielding two powers of ω. The square of Q yields a term linear in $\omega(\omega^{-1}$ from the square root in (13), and ω^2 from the long wave expression for $R_{K,j}$ (15)), so that a low temperature $\omega^3 \propto (\Delta E)^3$ dependence results. At high temperatures($kT >> \Delta E$) this is replaced by $(kT/\hbar)\omega^2 \propto kT(\Delta E)^2$.

For electronic transitions at larger energy, the transition rate is proportional to $\omega^{-1}N_{ph}(\omega) \propto (\Delta E)^{-1}N_{ph}(\Delta E)$, where $N_{ph}(\omega)$ is the phonon density of states. The prefactor can lead to a diminution in effectiveness of the higher energy end of the phonon spectrum.

It is not very significant to estimate the general strength of the one phonon interaction, for it depends on the details of the phonon spectrum, and the electronic states. We can choose the example of Al_2O_3 doped with $3d^3$ impurities (V^{2+}, Cr^{3+}, Mn^{4+}) to bring out some interesting features, however. The electron-vibration interaction parameters can be obtained from static stress experiments for each of the impurities (26), and the non-radiative relaxation rate determined from T_1 measurements (27) within the excited $\bar{E}$ (doublet) level. Relaxation occurs via transitions to the excited $2\bar{A}$ level (12.3, 29, and 80 cm^{-1} higher in energy, respectively). It is found that the $(\Delta E)^3$ dependence holds, after a slight correction for phonon dispersion for Mn^{4+}. Experimentally, Imbusch et al. (27) find (28):

Impurity in Al_2O^3	ΔE	$W_{ex \to gd}$
V^{2+}	12.3 cm^{-1}	3.5×10^{-9} sec
Cr^{3+}	29 cm^{-1}	2.5×10^{-10} sec (19)
Mn^{4+}	80 cm^{-1}	1.07×10^{-11} sec

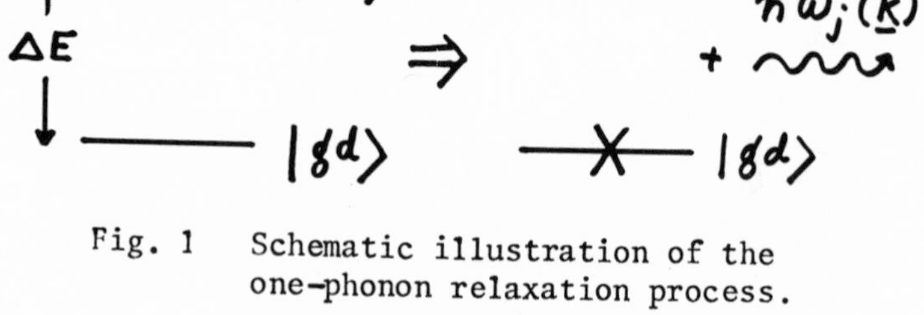

Fig. 1 Schematic illustration of the one-phonon relaxation process.

The departure from a strict $(\Delta E)^3$ dependence is a result of variations in the matrix elements. When these are taken into account, the theoretical estimates track the experimental results well.

II.D. Multiphonon Processes

1. Two-Phonon Processes. As soon as one leaves the one-phonon regime a plethora of quantitative difficulties arises. For a start, the Hamiltonian (12) is insufficient in that only terms linear in Q are present. When multiphonon interactions are important, (12) must be expanded to include terms in Q^2, Q^3, etc. These must be symmetry-grouped, and the minimum number of coefficients determined. Clearly, this presents a formidable problem (though Van Vleck carried it through to order Q^2 in ref. (15). We shall only sketch how the treatment goes.

First, it is possible to use (12) in higher order perturbation theory to effect multiphonon transitions. For example, Fig. 2 displays such a process where, depending on the temperature, a variety of thermal dependences can result. To see this more quantitatively, we write

$$W_{ex\to gd} = \Sigma_{\underline{K},j \atop \underline{K}',j'} (2\pi/\hbar) \quad \times \tag{20}$$

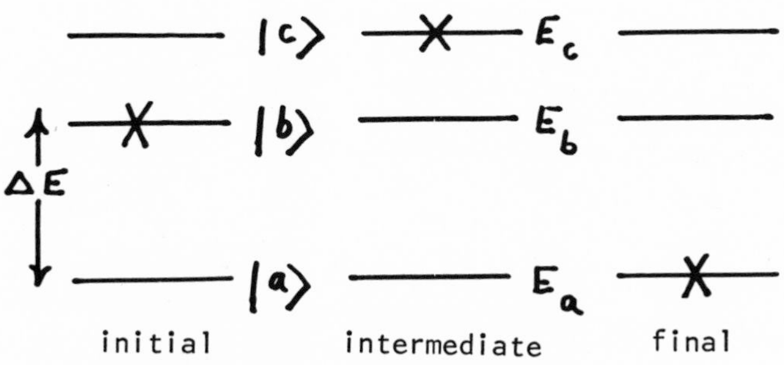

Fig. 2 Schematic illustration of the two-phonon relaxation process.

$$\times \left| \Sigma_{int} \frac{\langle ex|H^{\underline{K}',j'}_{el-vib}|int\rangle\langle int|H^{\underline{K},j}_{el-vib}|gd\rangle}{E_{gd}-E_{int}} \right|^2 \delta(E_{exp}-E_{gd})$$

where |int> means an intermediate electron-phonon state, and all energies are the sum of electronic and lattice energies. Using Fig. 2, we choose the initial, intermediate, and final electronic energies equal to E_b, E_c, and E_a, respectively. Then, for $E_b > E_a$, two choices of phonon emission and absorption are possible as a result of the delta function in (20):

$$\hbar\omega_j(\underline{K}) + \hbar\omega_{j'}(\underline{K}') = \Delta E \quad ; \tag{21a}$$

or

$$\hbar\omega_j(\underline{K}) - \hbar\omega_{j'}(\underline{K}') = \Delta E \quad . \tag{21b}$$

In the former case, two phonons are emitted; in the latter, a phonon is absorbed and another emitted with ΔE larger energy. Which of the two will dominate is determined by the ratio $\Delta E/k_BT$. If:

$\Delta E/k_BT >> 1$ — 2 phonons emitted;

$\Delta E/k_BT << 1$ — 1 phonon absorbed, one phonon emitted.

In general, the large powers of phonon frequency in the integrand of (20) will tend to favor the latter process to temperatures five to six times smaller than $\Delta E/k_B$.

If |b> and |a> are not time-reversed doublets of half-integral spin, (20) will result in a temperature independent process for (21a), and with a temperature dependence for (21b) of:

$$(k_BT/\hbar)^7 \int_0^{\theta_D/T} dx[x^6e^x/(e^x-1)^2]$$

which, at low temperatures (relative to θ_D) varies as T^7, and at high temperatures as T^2.

The energy dependence can also be extracted. In the case (21a), we have two-phonon emission. Since the sum of the phonon frequencies equals ΔE, the condition $\Delta E >> k_BT$ also implies $\hbar\omega_j(\underline{K})$, $\hbar\omega_j(\underline{K}') >> k_BT$ (the two frequencies being comparable for maximum integrand in (20)). Then the Bose factors $[n(\underline{K}j) + 1][n(\underline{K}'j') + 1]$ are essentially unity, and for long waves, we find a $(\Delta E)^6$ dependence. Again, as the temperature is raised, the Bose factors replace this dependence by $\sim(k_BT)^2\ (\Delta E)^4$. The case (21b) will not be important, relative to (21a) (see above) when $T \gtrsim \Delta E/6k_B$, and essentially no energy dependence will be found (the phonon energy in the denominator of (20) exceeding the electronic energy.

There is another interesting feature of the two-phonon process. When J is a good quantum number, the <u>one-phonon</u> transition (18) is forbidden between J=1 and J=0, and between J=0 and J=0. This is a direct consequence of the triangle rule, and will be the case for Sm^{2+} in CaF_2 (29), where the $^7F_1$ level is the terminal laser level of a four level type (30). No accumulation was found for this level, though it was forbidden to relax via one-phonon emission to the ground $^7F_0$ level, some 263 cm^{-1} below it. Orbach showed (29) that the two-phonon process, (21a), was responsible, where $|a\rangle = |^7F_0\rangle$, $|b\rangle = |^7F_1\rangle$, and $|c\rangle = |^7F_2\rangle$. Thus, two-phonon emission is present in this system, even though the intermediate electronic state is higher in energy than the initial excited level. A number of similar situations in excited levels have since been identified (e.g. in $LaCl_3{:}Pr^{3+}$ by German and Kiel (31). They invoke a three-phonon process because of the large energy difference between $^3P_1$ and $^3P_0$ but the result, their eq. (17), is similar to ours).

A final comment is in order regarding equilibrium between states $|a\rangle$ and $|b\rangle$ of Fig. 2. For $k_BT << \Delta E$, selective pumping into $|a\rangle$ can result in weak fluorescence from $|b\rangle$ if the rate $W_{gd\to ex}$ exceeds the radiative and non-radiative relaxation rates of $|a\rangle$. A two-phonon process,may be necessary because $\Delta E > h\omega_D$ (for energy conservation), or because of matrix element considerations (see above, where one-phonon or direct transitions are forbidden). In such a case two-phonon absorption follows from (20), and the temperature dependence is governed by the Bose factors $n(\underline{K},j)\ n(\underline{K}',j')$. The largest contributions to (20) arise from $\hbar\omega_j(\underline{K}) \sim \hbar\omega_{j'}(\underline{K}') \sim \Delta E/2$ (though the integrand of (20) vanishes for exactly equal phonon energies - see ref. (29)). Whence, if $k_BT << \Delta E$,

$$n(\underline{K},j)n(\underline{K}',j') = \exp\{-\hbar\omega_j(\underline{K})/k_BT\}\exp\{-\hbar\omega_j{}'(\underline{K}')/k_BT\}$$

$$\sim \exp\{-\Delta E/k_BT\} \quad .$$

Thus, an activation energy form results with the full optical energy gap regardless of the number of phonons, provided only that each $\hbar\omega_j(\underline{K}) > k_BT$.

The matrix elements in (20) are algebraically complicated, and no purpose will be served by writing them out here. Examples can be found in ref. (12). In addition to using the one-phonon interaction twice, it is also possible to use a two-phonon interaction Hamiltonian (e.g., see ref. (15)) once. Because the one- and two-phonon coefficients have roughly equal magnitudes (32), no significant difference exists between their contributions to the non-radiative relaxation rate. When one considers even higher order processes, the situation is the same.

2. Many-Phonon Processes. For most non-radiative transitions of interest, many more than two phonons are required for energy conservation. As seen in the previous part, a quantitative treatment is hopeless. Even if the one-phonon coupling constants could be determined from static strain measurements, the high order of perturbation theory required would be algebraically prohibitive. If instead one uses an interaction Hamiltonian with many-phonon operators in low order perturbation theory, there is no method available to evaluate the coefficients, not to mention the complex phonon sums.

What we shall do is to use our experience with low order processes to obtain the temperature and energy dependence of the higher order non-radiative rate. The results will enable one to estimate the number of phonons involved, and to extract the dependence of the non-radiative rate on the energy gap ΔE.

We begin with the systematics. Taking an excited level at energy $\Delta E \gg k_B T, k_B\theta_D$; emission to the ground level is proportional to

$$|\langle (VCQ)^N \rangle|^2 \tag{22}$$

where we have used the symbols of (12) in an obvious notation. For phonon emission, the phonon matrix element $|\langle Q^N \rangle|^2$ is proportional to

$$[n(\underline{K},j)+1]^N = \left[\frac{\exp\{\hbar\omega_j(\underline{K})/k_B T\}}{[\exp\{\hbar\omega_j(\underline{K})/k_B T\} - 1]}\right]^N . \tag{23}$$

As first pointed out by Kisliuk and Moore (33), (23) has a peculiar temperature dependence. It is essentially temperature independent at low T, and until $k_B T$ approaches $\hbar\omega_j(\underline{K})/N$. Then, a very sharp increase occurs in (23), giving rise to a precipitate increase in $W_{ex\to gd}$; the larger N, the steeper the rise. Usually one assumes an average phonon energy $\hbar\bar{\omega} = \Delta E/N$, so that the temperature slope of the rate (i.e. $W_{ex\to gd} \propto [k_B T/\hbar\bar{\omega}]^N$) gives a direct measure of N, the number of phonons involved in the transition.

There are other aspects of this decay mechanism which are interesting. The successive values of (22) are smaller as N increases. This is because the expansion parameter is the strain, ε, and as Kiel (34) has shown, it has a value ~ 0.05 for phonons of large energy at low temperatures. Thus, the transition rate can be written as (10)

$$W_{ex\to gd} = A\varepsilon^N = A\,\varepsilon^{E/\hbar\bar{\omega}} = Ae^{\ell n\varepsilon^{\Delta E/\hbar\bar{\omega}}} \tag{24}$$

$$= Ae^{(\Delta E/\hbar\bar{\omega})\ell n\,\varepsilon} = Ae^{-\Delta E(|\ell n\varepsilon|/\hbar\bar{\omega})} \quad .$$

This is the celebrated exponential dependence of the transition rate on the energy gap. It has been observed in an amazingly large number of cases, with Fig. 7 of ref. (10) particularly instructive.

It should be clear that (24) is valid only if the coupling constant and phonon energies are the same for each step of the process (though the former appears only logarithmically). Further, (24) is often used for different transitions with a given impurity energy level diagram, or even worse, for different impurities and different transitions. One's instincts are to reject this approach because of the expected wide variation in non-radiative coupling constants. Amazingly, Riseberg and Moos (10) show that if one plots the log of the the multi-phonon rate against $\Delta E/\hbar\bar{\omega}$, a linear result holds for every measured rare earth impurity transition, each falling on the same line for both $LaCl_3$ and $LaBr_3$, the only difference being the use of properly scaled $\hbar\bar{\omega}$ for the respective host (260 cm^{-1} for $LaCl_3$ and 175 cm^{-1} for $LaBr_3$). N varies from 4 to 6. When a host with different (larger) crystal field strength is used (e.g., LaF_3) a linear dependence is again found. The absolute rate is larger (expected on the basis of the scaling of the dynamic with the static crystalline field magnitude), and the convergence of the transition rate slower (ε larger). Both are found. Empirically Riseberg and Moos (10) give evidence that $N = \Delta E/\hbar\omega_{max}$ where ω_{max} is the maximum phonon frequency in the host. Thus, as a rule of thumb, this means that the lowest order allowed phonon emission process dominates the non-radiative transition state. This last conclusion cannot be true in general, for it depends sensitively on the coupling between the phonon and the paramagnetic defect. Thus, in a host with waters of hydration, or tightly coupled complexes (e.g., vanadates or sulfates), the O-H or intra-complex vibration frequency may be very high but may only couple weakly to the fluorescence center. In such a case, phonons of lower energy, with stronger coupling strengths, would be effective in (24). The phonon sidebands or the optical lines would be a better source for choosing $\hbar\bar{\omega}$: the highest phonon energy with large side band intensity (hence, relatively large coupling constant, should be used in (24)).

Finally, the work of Miyakawa and Dexter (5) treats the multi-phonon emission problem with careful attention paid to the phonon statistics (because of the equivalence of the different orders in which the phonons are emitted). They assume a constant coupling coefficient, and for a narrow electron phonon spectral density (i.e., single phonon energy) they find

$$W_{ex\to gd} = (2\pi/\hbar^2)R^2(1-N/g)^2\{g^N(n+1)^N/N!\} \times \delta[(\Delta E/\hbar) - N\bar{\omega}] \quad . \tag{25}$$

Here g is a dimensionless coupling constant ($\lesssim 5$ for weak coupling) and R a measure of the strength of the nonadiabatic coupling at zero time. The term in curly brackets in (25) can be written as (using Stirling's formula for N!)

$$\exp\{-\Delta E[\ell n\frac{N}{g(n+1)} - 1]/\hbar\bar{\omega}\} \tag{26}$$

which is identical in form to (24) should we identify $|\ell n\varepsilon|$ with $\ell n[N/g(n+1) - 1]$. Again, the exponential dependence of $W_{ex\to gd}$ on $\Delta E/\hbar\bar{\omega}$ is present. The curious prefactor $(1-N/g)^2$ enhances the non-radiative rate over the phonon side band intensity at the same phonon energy as g becomes smaller. It would be interesting to look for this effect, though with all the other approximations in the model (constant coupling constant, single phonon frequency) it may be difficult to isolate.

We have now finished our treatment of single-ion non-radiative transitions. We concentrate next on inter-ion transitions, both energy-matched (resonance), and mismatched (phonon-assisted).

III. ENERGY TRANSFER

We shall divide this section into two main parts: resonant and non-resonant energy transfer. The former will be treated using the basic formalism of Dexter (1), the latter that of Orbach (3), Birgeneau (4), Miyakawa and Dexter (5), Soules and Duke (6), and Kohli and Huang Liu (7). We shall also make use of Imbusch's work on ruby (35), as it represents one of the most interesting and complete pieces of work available.

III.A. Spatial Transfer (Resonant)

We mean by resonant transfer that energy is exchanged between spatially separated ions with transitions occurring within the homogeneously broadened optical line. This corresponds to the situation treated by Dexter (2) who derived the transfer rate:

$$W = (2\pi/\hbar^2)\,|\langle 1,2^*|H_{int.}|1^*,2\rangle|^2 \times \int f_1(\omega)f_2(\omega)d\omega \quad . \qquad (27)$$

Here, $|1,2^*\rangle = |1\rangle|2^*\rangle$ where atoms are located at $\underline{r}_1$ and $\underline{r}_2$, and we denote their excitation by an asterisk (see Fig. 3 for a schematic). According to (27), the interaction Hamiltonian, $H_{int.}$, must coherently connect states $|1\rangle$, $|1^*\rangle$ and $|2^*\rangle$, $|2\rangle$, over a range $\underline{R}_{12} = \underline{r}_1 - \underline{r}_2$. The integral in (27) represents a convolution of the line shapes, and should be used only for the homogeneous portion of an inhomogeneously broadened line. At once a possible difficulty presents itself: let the optical line be severely broadened inhomogeneously, and let H_{int} be of short range. Then it may not be possible to find a site on "speaking terms" with the excited center (within the homogeneous width) in a distance sufficiently short that $H_{int.}$ has significant amplitude. If this is the case, no energy transfer occurs, and we have what has popularly become

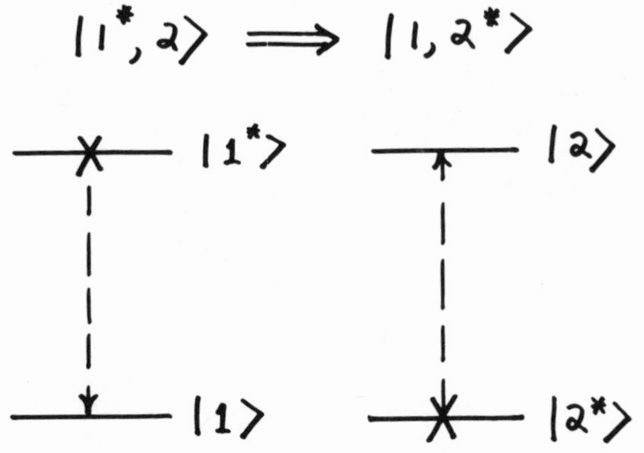

Fig. 3 Schematic of transfer of energy between centers 1 and 2. The kets above the drawing represent product functions of the single ion states in an obvious notation.

known as Anderson localization (36). It is a clear example, in the conduction limit, of variable range hopping (37). We shall postpone this question until part 4 of this section, where we will use Ziman's very nice illustrative calculation (38) to demonstrate this point. We shall also relate it to energy transfer experiments in ruby.

Before we can estimate (27), we must decide on the interaction Hamiltonian capable of effecting the transition exhibited in Fig. 3.

1. Inter-Ion Coupling Mechanisms. As just discussed, an interaction which can induce transfer must act coherently at the two sites 1 and 2, and must possess matrix elements between $|1\rangle$ and $|1^*\rangle$, and between $|2^*\rangle$ and $|2\rangle$. There are a number of possibilities:

1. Radiation of photons (radiative transfer)
2. Electric Multipole Coupling (EMI)
3. Magnetic Dipole Coupling
4. Exchange Coupling
5. Virtual Phonon Exchange

and probably more. The first possibility is essentially that of radiation trapping. A photon is emitted by atom 1, and absorbed by atom 2. This process is known to be important in materials with large oscillator strengths at high concentrations. It can be detected by observing whether radiative trapping is present in the sample. In almost all the cases known to this author, the weakness of the oscillator strength of paramagnetic impurities in solids cause this process to be ineffective. Likewise, the weakness of magnetic dipole-dipole coupling, as compared to electric multipole coupling, means that we can also ignore it except, perhaps, in exceptional circumstances (e.g. when EMI transitions are parity and triangle rule forbidden, as between J=0 and J=1 levels within a configuration).

We are left with three mechanisms which we examine in order. The first was extensively developed by Dexter (2). The electric multipole interaction has been expanded in spherical harmonics by Carlson and Rushbrooke (39). The first term which would yield the matrix element in (27) is dipole-dipole. The square of the matrix element is (these expressions are conveniently found in ref. (35)):

$$|\langle 1,2^*|H_{int}|1^*,2\rangle|^2_{\substack{dip\\dip}} = \frac{2e}{3n^4R_{12}^6}\,|\langle 1|\underline{r}|1^*\rangle\langle 2^*|\underline{r}|2\rangle|^2 \tag{28}$$

where n is the refractive index. For the case of dipole-quadrupole (this will be important for non-resonant transfer, when the donor has large oscillator strength but the acceptor has weak strength):

$$|\langle 1,2^*|H_{int}|1^*,2\rangle|^2_{\substack{dip\\quad}} = \frac{9e}{4n^4R_{12}^8}|\langle 1|\underline{r}|1^*\rangle \times \langle 2^*|\ \Sigma_{j,k}\ x_jx_k|2\rangle|^2 \tag{29}$$

where the subscripts j and k refer to the component of the electronic atomic coordinate. In a different, but more useful form, the quadrupole-quadrupole matrix element

$$|\langle 1,2^*|H_{int}|1^*,2\rangle|^2_{\substack{quad\\quad}} = \frac{16\pi^2e^4\langle r_1^2\rangle^2\langle r_2{}^2\rangle^2}{25n^4R_{12}^{10}} \tag{30}$$

$$\times\ |\langle 1,2^*|6Y_2{}^0(1)Y_2{}^0(2) + 4[Y_2{}^1(1)Y_2{}^{-1}(2) + Y_2{}^{-1}(1)Y_2{}^1(2)] + [Y_2{}^2(1)Y_2{}^{-2}(2) + Y_2{}^{-2}(1)Y_2{}^2(2)]|1^*,2\rangle|^2$$

where $\langle r^2\rangle$ is the mean squared atomic radius and $|1^*,2\rangle \equiv |1^*\rangle|2\rangle$. We shall find that for most iron group and rare earth impurities (30) will dominate (28) and (29), and this is why we have gone to the trouble of expressing (30) in a convenient but lengthy form.

At first sight this is a surprising result. Taking all matrix elements $\sim$ the same, then on the surface it appears (28) is larger than (29) is larger than (30) by the ratio $R^2\langle r\rangle^2/\langle r^2\rangle^2$. This is certainly a large number, and is of course the dominant factor at long distances. However, at the ranges commonly important in physical systems (10-20 Å), the following considerations reverse the inequality. For electric dipole matrix elements to be finite, it is necessary that $\underline{r}$ be sandwiched between odd parity states. Within a (f^n or d^n) configuration, this is impossible and $\langle 1|\underline{r}|1^*\rangle$ vanishes. However, if the site of the paramagnetic impurity lacks inversion symmetry, states of opposite parity can be mixed into $|1\rangle$ and $|1^*\rangle$ by the hemihedral field. This mixing matrix element (u=odd, g=even)

$$V_{hemihedral}/[E(\Gamma_u) - E(\Gamma_g)]$$

squared is of the order of 10^{-3} for transition metal ions, or 10^{-4}

for rare earths. The reduction of $<\underline{r}>$ can also be related to the observed "forced electric dipole" oscillator strength. For example, Imbusch (35) finds

$$f_{el\ dip} = (2m\omega/3\hbar)\,|<1|\underline{r}|1^*|>|^2 \ .$$

Thus, for ruby, the value of $|<1|\underline{r}|1^*>|^2$ can be estimated using $f = 3 \times 10^{-7}$, $\omega = 3 \times 10^{15}$. One finds $|<1|\underline{r}|1^*>| \sim 10^{-3} <r_{at}>$. Noting 10^{-2} is caused by spin selection rules the ratio of (28) to (29) becomes approximately

$$10^{-2} \ \frac{R^2}{<r^2>} \ .$$

Taking $<r^2> \sim 0.5 \times 10^{-16} cm^2$, this means the inequalities are reversed and quadrupole-quadrupole dominates dipole-quadrupole (which in turn dominates dipole-dipole) transfer to distances of ~10-20Å separation. Clearly this does not hold for the next higher order (quadrupole-hexadecapole) for it is only the parity—forbidden character of the $\underline{r}$ matrix element which causes this progression.

Also, if one ion has a large electric dipole oscillator strength (e.g., an allowed s to p transition) but the other does not (e.g., a rare earth), then electric dipole-quadrupole will dominate.

In some later calculations, we shall center our attention on transition metal ions, and thus make extensive use of (30).

Another mechanism is exchange. As Pryce first pointed out (40), and independently Birgeneau (4), the exchange operator may be very potent for transitions of the sort exhibited in Fig. 3. To get a feeling for why, consider the energy level diagram of ruby exhibited in Fig. 4. Then, at first sight, the exchange operator, if as usual, is written as $JS_1 \cdot S_2$ (bilinear in the total Russell-Saunders spin). The ground $^4A_2$ level can connect only with excited A_2 quartets. The ^2E "R" lines need spin-orbit admixture from the $^4T_2$, as does the $^4A_2$ (see Fig. 4) to be connected by exchange. This would reduce the exchange matrix element on each site by $\lambda^2/\Delta E^2$, and would certainly quench the effect. However, this is totally incorrect as one notices immediately if the exchange Hamiltonian is written in its fundamental form:

$$H_{ex} = \Sigma_{\substack{i,j,\\ i',j'}} J(m_{i'}m_{j'}, m_i m_j)\underline{s}_i^{(1)} \cdot \underline{s}_j^{(2)} \qquad (31)$$

Here, i and j label the various orbital levels on sites 1 and 2, respectively. Only for half-filled shells (with maximum spin,) or

$J(m_i m_j, m_i m_j)$ independent of i and j, can the isotropic full spin form for H_{ex} obtain. In the former case (we keep within the same configuration so we drop the primes in (31));

$$H_{ex} = \Sigma_{i,j}\, J(m_i m_j, m_i m_j)\, \frac{\underline{S}_1}{n} \cdot \frac{\underline{S}_2}{n}$$

$$= [\frac{1}{n^2}\, \Sigma_{i,j}\, J(m_i m_j, m_i m_j)]\; \underline{S}_1 \cdot \underline{S}_2 = J_{eff}\, \underline{S}_1 \cdot \underline{S}_2$$

where n is the number of magnetic electrons in the half filled shell (3 for ruby), and we have assumed identical ions. In the latter, where all the J's in (31) are equal

$$H_{ex} = J\, \Sigma_{i,j}\, \underline{s}_i^{(1)} \cdot \underline{s}_j^{(2)} = J\underline{S}_1 \cdot \underline{S}_2 \;.$$

For the case of ruby, the former is the case. Measurement of exchange between Cr^{3+} ions in the ground state will yield J_{eff}. This is useful for determining the scale of the exchange integral, though

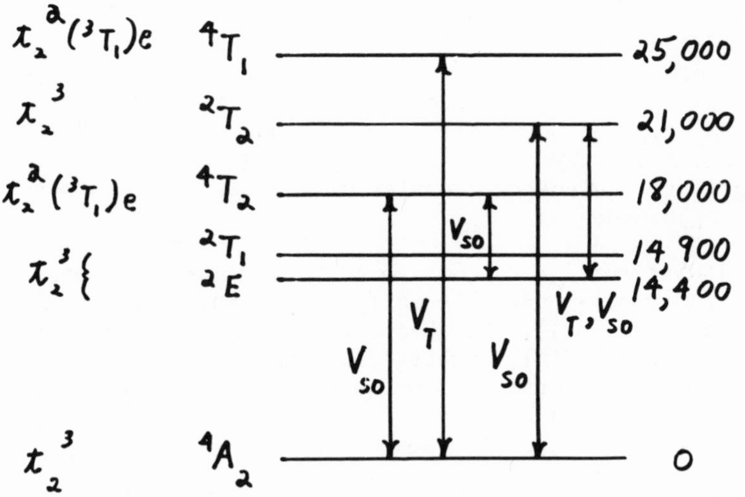

Fig. 4 The energy level diagram of Al_2O_3:Cr (ruby). The arrows labelled V_{so} and V_T display the allowed spin orbit and trigonal field transition matrix elements, respectively.

detailed calculations must be carried out to obtain the relative values of $J(m_i m_j, m_i m_j)$ for all m_i, m_j. Huang (41) has done this for V^{2+} ions (also $3d^3$) in $KMgF_3$, and Birgeneau (4) has made estimates for ruby. To get some quantitative feeling, we use Huang's detailed calculations. First, for the pair of ions, 1 and 2, (31) commutes with the total pair spin S.

For S = 2, Huang shows

$$\langle {}^2E_\Theta, {}^4A_2 | H_{ex} | {}^4A_2, {}^2E_\Theta \rangle = -\frac{9}{2} J_{eff}$$

and

$$\langle {}^2E_e, {}^4A_2 | H_{ex} | {}^4A_2, {}^2E_e \rangle = -\frac{3}{2} J_{eff} \ .$$

Hence, exchange transfer matrix elements are actually larger than the effective ground-state-only exchange. For the example of ruby, the situation is much more complex, though Davydov splittings have enabled Van der Ziel to estimate the larger of the two electron exchange integrals (41). As a function of distance, the exchange falls off as S^2, where S is the ligand-cation, or ligand-ligand overlap integral. This is exponential in distance:

$$J(L) = J_o \, \ell^{-2L|\ell nS|} , \tag{32}$$

where J_o is the closest approach value of the exchange, and L is the additional number of ligand links. The large value of J_o, and for a complex lattice (e.g., Al_2O_3) the rather large distances for relatively modest L (e.g., L = 0 for first through fourth nearest neighbors in ruby) lead to an enchanced effectiveness for exchange over the EMI type of interaction. It is now accepted as the dominant mechanism in ruby. The "shortish" range of J will, nevertheless, aggravate the energy transfer difficulty in the presence of inhomogenous broadening.

It is instructive to compare exchange matrix elements with electric quadrupole-quadrupole matrix elements for energy transfer (e.g., (31a) with (30)). We choose ruby as a representative system. Then, according to (31a), exchange matrix elements are directly allowed for energy transfer. For quadrupolar coupling, however, spin orbit coupling is necessary to break the spin selection rule. The process goes via the ${}^2T_2$ (see Fig. 4):

$$ {}^2E \xrightarrow{H_{quad\ quad}} {}^2T_2 \xrightarrow{H_{so}} {}^4A_2 $$

(where V_T is equivalent to the quadrupolar operators in (30)) so that the transfer matrix element is reduced by $\zeta/[E(^2T_2) - E(^4A_2)]$. We thus must compare

$$<^2T_2, ^2E|H_{int}|^2E, ^2T_2>|_{\substack{quad\\quad}} \times \{\zeta/[E(^2T_2) - E(^4A_2)]\}^2$$

with

$$<^2E, ^4A_2|H_{ex}|^4A_2, ^2E>.$$

An upper limit for the matrix element of H_{int} for a 13 Å separation is $8\times10^{-3}cm^{-1}$ (see ref. (4)), while ζ=170 cm^{-1}, $E(^2T_2)-E(^4A_2)$=21,000 cm^{-1}. Hence, the quadrupole-quadrupole coupling matrix element for energy transfer is no larger than $6\times10^{-7}cm^{-1}$. For exchange, J (4th near neighbor but one oxygen link) $\sim$10 cm^{-1}. At 13 Å, one finds three oxygen links, whence the energy transfer exchange matrix element is approximately

$$10\ cm^{-1} \times S^2 \times S^2 \sim 10^{-3}\ cm^{-1}\ .$$

This is three orders of magnitude larger than quadrupolar-quadrupolar resonant energy transfer, and quite comparable to that value required to yield Imbusch's (35) single ion-single ion transfer rate (1 μsec) in 0.2% ruby.

Finally, there is an additional energy transfer mechanism intimately connected with the electron-vibration Hamiltonian (12). This is virtual phonon exchange quite analogous to the electric coulomb interaction viewed as virtual photon exchange (42). It was first worked out by Sugihara (43), and then developed by a number of authors (17), (44). The algebra is very involved but the basic idea is simple. An ion at site 1 emits a phonon according to (12), and another ion at site 2 absorbs the same phonon according to (12). The inverse also occurs, and absorption and emission is also inverted in two more processes. The result of all this is a Hamiltonian coupling sites 1 and 2, with spin operators of the type $C(\Gamma_{ig}\ell,m)$ on each site. More specifically,

$$H_{int} = \sum_{j,\underline{K}} \left[\frac{<1,2^*|H_{el\ vib}|1,2><1,2|H_{el\ vib}|1^*,2>}{\Delta E \mp \hbar\omega_j(\underline{K})} + \frac{<1,2^*|H_{el\ vib}|1^*,2^*><1^*,2^*|H_{el\ vib}|1^*2>}{-\Delta E \mp \hbar\omega_j(\underline{K})} \right] . \qquad (33)$$

For a particular m subvector of the i-th irreducible representation in (12), this leads to

$$H_{int}^{i,m} = \Sigma_{\ell,\ell'} V_1(\Gamma_{ig}\ell) V_2(\Gamma_{ig}\ell') <1,2^*|C_2(\Gamma_{ig},-m)|1,2>$$

$$\times <1,2|C_1(\Gamma_{ig},m)|1^*,2> \times \{\Sigma_{j,\underline{K}} \frac{\hbar}{2M\omega_j(\underline{K})a^2} \frac{2\hbar\omega_j(\underline{K})}{(\Delta E)^2-[\hbar\omega_j(\underline{K})]^2}$$

$$\times \cos(\underline{K}\cdot\underline{R}_{12}) R_{\underline{K},j}(\Gamma_{ig,m}) R_{\underline{K},j}(\Gamma_{ig,-m})\} . \qquad (34)$$

The subscripts on $C(\Gamma_{ig,m})$ refer to the site upon which the matrix element is to be taken; and a is the lattice constant.

Explicit evaluation of (34) is a mess. It is disastrous to use a Debye approximation, as it leads to a nonsensical range dependence. Orbach and Tachiki (17) were able to evaluate (34) for a simple cubic lattice (s.c.). It is very important to extend their treatment to more complicated lattices involving optical phonon branches, but this has yet to be done. For the s.c., there were two limits for which (34) was evaluated

$\Delta E >> \hbar\omega_D$ extreme retardation

$\Delta E < \hbar\omega_D$ in-band limit .

The former exhibits short range interaction, while the latter can be very long range indeed. For example, for

$i = 3,\ m = \Theta,\ j = \text{longitudinal},\ \Delta E >> \hbar\omega_D$

the curly bracket in (34) becomes

$$\frac{-\hbar^2}{4\rho a^5(\Delta E)^2} (-1)^{(R_{12}/\alpha)} (\hbar\omega_{LA}/2\Delta E)^{[(2R_{12}/\alpha)-4]} . \qquad (35)$$

where ρ is the mass density and ω_{LA} the longitudinal acoustic branch energy at the Buillouin zone surface. The inband results are algebraically more lengthy, and we shall not display them here. They can be found in ref. (17). A representative range dependence is

$$i = 5,\ m = \pm 1,\ j = \text{transverse},\ \Delta E < \hbar\omega_D$$
$$[k_{\Delta E}^{(t)}]^2 \{\cos[k_{\Delta E}^{(t)} R_{12}]/R_{12}\} \tag{36}$$

where t signifies transverse and

$k_{\Delta E}^{(t)} = \Delta E/\hbar v_t$, v_t being the transverse velocity of sound. This is a very long range indeed(falling off as $1/R_{12}$,)but two caviats should be registered. First, (36) fails for distances greater than a phonon mean free path; and second, the coefficient of (36) is relatively small when compared to even magnetic dipole interactions.

One can compare the extreme retardation limit with the quadrupole-quadrupole coupling mechanism considered earlier. For representative terms, dividing the square root of (30) into (34), one finds

$$H_{int}^{5,\pm 1} / H_{quad\ quad} \cong (0.1)(R_{12}/a)^5 \times [\hbar\omega_D/2\Delta E]^{2[(R_{12}/a)-2]} \tag{37}$$

The factor (0.1) in (37),relating the strength of the phonon-induced coupling to the quadrupole-quadrupole coupling strength, is offset by the range of the former as compared to the latter for nearby ions. Eventually, however, the latter dominates because of the former's exponential fall off.

For in-band transitions, McMahon and Silsbee (44) demonstrate that the magnetic dipole-dipole coupling is comparable in magnitude to the phonon-induced coupling; the latter dominating only at distances greater than $1/k_{\Delta E}$.

2. Range Dependence. Transition Energies Less than the Debye Energy. Birgeneau (4) has made compelling arguments concerning the danger of extracting the mechanism for energy transfer from the concentration dependence of the transfer rate. Inokuti and Hirayama (46) show that if donor-donor transfer is negligible compared to donor-acceptor transfer, the fall-off of the acceptor luminescence with time is not exponential. The explicit time dependence allows one to extract the range dependence of the donor-acceptor transfer Hamiltonian, and hence identify the transfer mechanism. However, donor-donor transfer is often resonant, while donor-acceptor transfer is phonon-assisted non-resonant (hence much slower) so that the former can be appreciable at surprisingly low concentrations, whereupon pure exponential decay occurs. In such a case it is only

the concentration of the acceptors which is important, though the non-linearity of the transfer rate dependence on distance requires a statistical sum over individual transfer rates to be made before one can positively identify the transfer mechanism. Both of these limits have been observed for rare earth-rare earth transfer by Gandrud and Moos (46), and Weber (47).

We list below the range dependence of the square of the transfer Hamiltonian matrix element appearing in the transfer rate (27) for comparison purposes:

EMI:

dipole-dipole $(1/R_{12})^6$

dipole-quadrupole $(1/R_{12})^8$

quadrupole-quadrupole $(1/R_{12})^{10}$

Magnetic Dipole $(1/R_{12})^6$

Exchange Coupling $\exp\{-2L|\ln S|\}$

(L is the number of ligand-cation links beyond near neighbor, and S is the cation-ligand or ligand-ligand overlap integral.)

Virtual Phonon Exchange $(R_{12}/a)^5 \exp\{-2[(R_{12}/a)-2]\ln(2\Delta E/\hbar\omega_D)\}$.

Again, one must be very careful in complex lattices when comparing strengths of the various processes. Though the ranges of the last two mechanisms appear to be short, the large magnitudes of the coefficients are sufficiently great as to result in a possible dominance at concentration ranges in the vicinity of 0.5%.

3. Range Dependence. Transition Energies Greater than the Debye Energy. Virtual Phonon Exchange is the only term affected. The leading (longest) range dependence is

$[k_{\Delta E}^{(j)}]^2 \{\cos[k_{\Delta E}^{(j)} R_{12}]/R_{12}\}$ ($k_{\Delta E}^{(j)} = \Delta E/\hbar v_j$, where j labels the phonon polarization branch, and v_j is the respective speed of sound.

4. Inhomogeneous Broadening. Lack of Transport for Short-Range Interactions. We have already dealt with this question in qualitative detail above. The (over-simplified) physical picture is exhibited in Fig. 5. For short range interactions, it may not be possible to find another site in the lattice whose energy is within $\langle H_{int} \rangle$ of ΔE except at such large ranges that H_{int} is negligible. This was first recognized by Anderson (8), and applied to energy transfer in ruby by Lyo (9).

To get some feeling for the mathematical condition for localization, we follow the simplified approach of Ziman (38). We take the limiting case of a near neighbor H_{int} only, which we denote as V. Then, for a simple cubic tight-binding lattice, the bandwidth equals 12V, centered about E=0 (that is, we re-define the center of the optical line to lie at E=0). If every site in the lattice possessed the same energy, the equation for the time dependence of the amplitude and phase of the ℓth site would be given by the (band) solution of

$$i\,\dot{a}_\ell = \Sigma_h V_{\ell,\ell+h}\, a_{\ell+h} = V\Sigma_h a_{\ell+h} \qquad (38)$$

where h labels the near neighbors of a site, and we have taken all the transfer integrals equal. Thus, we are taking the shortest range possible for H_{int}, and we are working with a concentrated system.

We now introduce disorder into the system by assigning (randomly) to each site an energy W_ℓ with a probability distribution for W centered at zero. Then the equation for $\dot{a}_\ell$ becomes,

$$i\,\dot{a}_\ell = W_\ell a_\ell + V\sum_h a_{\ell+h}\,. \qquad (39)$$

For simplicity, we shall assume the W_ℓ to be uniformly distributed over a width W. One can construct the site-site Green function from (39). Its equation of motion is

$$(E-E_\ell)G_{\ell\ell'}(E) - \Sigma_{\ell''} V_{\ell\ell''}G_{\ell''\ell'}(E) = \delta_{\ell,\ell'} \qquad (40)$$

where $V_{\ell\ell''} = V_{\ell,\ell+h}\,\delta_{\ell'',\ell+h}$. The causal propagator from site ℓ is

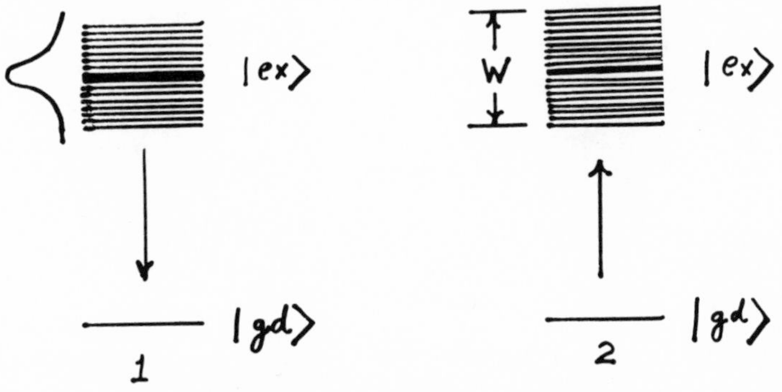

Fig. 5. Inhomogenously broadened optical line at sites 1 and 2. Each line is supposed to represent a site with a distinct energy separation between |ex> and |gd> of ΔE_ℓ. The envelope of the distributions W_ℓ is sketched to the left. W is defined in the text.

found by letting $E \to E + i\delta$, where δ is a positive infinitesimal. We wish to look for singularities of the G matrix. Suppose we found a diagonal element

$$G_{\ell\ell}(E) = \frac{A}{E - E_\ell(E)} \quad . \tag{41}$$

This would be evidence for an eigenstate E_ℓ concentrated spatially mainly at ℓ. Unless E_ℓ is real, the state would decay exponentially in time. Thus the definition of Anderson localization: $\mathrm{Im}E_\ell$ vanishes in the limit $\delta \to +0$.

The next step is to calculate E_ℓ by successive approximations. Conventional perturbation theory yields

$$E_\ell = W_\ell + \Sigma_{\ell'} V_{\ell\ell'} \frac{1}{E - W_{\ell'}} V_{\ell'\ell} + \Sigma_{\ell',\ell''} V_{\ell\ell'} \frac{1}{E - W_{\ell'}} V_{\ell'\ell''} \frac{1}{E - W_{\ell''}} V_{\ell''\ell} + \ldots\ldots \tag{42}$$

Suppose the series converges for E near the real axis (except perhaps at E_ℓ itself). Then $\delta \to +0$ leads to pure real terms converging on a value of E_ℓ without an imaginary part. Our criterion for localization thus boils down to the convergence of (42) for E near the real axis.

To determine convergence, we perform an elementary statistical analysis. Consider the contribution to E_ℓ of the term in (42) with V to order L+1 (i.e., L+1th term). Each site has z neighbors and there will be z^L contributions to this term (this can be seen by analyzing the last term in (42). Let ℓ' have a value. Then since ℓ'' is restricted to the near neighbors of ℓ', there will be z values contributing. But there are z values for ℓ'; hence there are z^2 terms. Noting this is the L=2 term, our result follows. Generalization to arbitrary L is immediate).

We next make a very strong approximation (corresponding to never coming back to the same point): all sub-terms in (42) are independent of each other. There is discussion of this point in Abou-Chaira et al. (36). Then in the Lth term, we have L factors of

$$\frac{V}{E - W_{\ell'}} \equiv T_{\ell'}$$

whence the term becomes $\Sigma V T_{\ell'} T_{\ell''} \ldots T_{\ell'''}$. The independence of the T_ℓ enables us to estimate the statistical average of the log of the product:

$$<\ell n|T_{\ell'} T_{\ell''} \ldots T_{\ell'''}|> = L<\ell n|T|> \quad . \tag{43}$$

Now, the series for E_ℓ (42) would converge absolutely if the series could be cast into a geometric progression, with each successive term smaller than the one preceding it. Thus, the Lth term is, exponentiating (43),

$$<|T_{\ell'} T_{\ell''} \ldots T_{\ell'''}|> = Ce^{<L\ell n|T|>} \tag{44}$$

where C is a number. The next term in the series (43) is then equal to

$$<T_{\ell'} T_{\ell''} \ldots T_{\ell''' +1}> = Cze^{(L+1)<\ell n|T|>} \quad . \tag{45}$$

For convergence, we want the ratio of (45) to (44) less than unity (the C is the same in both equations):

$$ze^{<\ell n|T|>} < 1 \tag{46}$$

We must now evaluate $<\ell n|T|>$. Let the distribution function for W_ℓ have the simple shape

$$P(W_\ell) = \frac{1}{W} \quad |E| \leq W/2$$

$$= 0 \quad |E| > W/2 \quad .$$

This square shaped distribution enables us to evaluate (46) simply and does not affect the essence of our conclusion. Then

$$<\ell n|T|> = \frac{1}{W}\int_{-W/2}^{W/2} \ell n|T|dW = \frac{1}{W}\int_{-W/2}^{W/2} \ell n\left|\frac{V}{E-W}\right| dW \quad . \tag{47}$$

For simplicity, let us calculate E_ℓ near the center of the band (i.e., E=0). This is the most difficult energy for localization. Then

$$<\ell n|T|> = 1-\ell n(\frac{W}{2V}) \quad . \tag{48}$$

Plugging into the inequality (46), we require for convergence that

$$z\,e^{[1-\ell n \frac{W}{2V}]} < 1$$

or, simply, $z\,e < W/2V$. Noting the bandwidth is 12V, we require:

$$\frac{W}{B} > e, \quad \text{for localization .} \tag{49}$$

This is a very significant result. For inhomogeneous distribution widths larger than the bandwidth as determined by the transfer Hamiltonian, resonant energy migration is forbidden — (27) is irrelevant. This point does not seem to have been sufficiently recognized.

5. Inhomogeneous Broadening. Critical Concentration for Energy Transfer. Anderson (8) and Lyo (9) have examined various transfer Hamiltonians discussed in the previous part of this paper (power laws and exponential, respectively). They show that, for a dilute system, a critical concentration exists below which no energy flows. This is equivalent to our result (49), for B is a function of the average spatial separation, for given V. There is an exception to this result: dipolar ($1/R^3$) or weaker range dependences of the transfer integral always exhibit transfer (de-localization).

One might wonder why (27) goes wrong. It has to do with the spatial averaging process inherent in its use to generate an energy "conductivity". Returning to (42), we see that (42) can develop an imaginary component if some of the energy denominators vanish. If the homogeneous widths in $f(\omega)$ are << W, then only a very few terms in (42) will contribute to $\mathrm{Im}E_\ell$. Put another way, $\mathrm{Im}E_\ell$ as $\delta \to +0$ will have zero value with almost unit probability for most of the configurations of the random impurity levels. This means we shall almost always have localization. However, due to an infinitesimal fraction of sites which possess energies which cause the denominator to vanish, the average will be non-zero, leading to the conclusion that transport is possible. This is an erroneous result, arising from the incorrect assumption that the average is a representative value. To quote Lyo (9): "Put another way, because we are interested only in how often the experiment tells us as $\delta \to +0$ that $\mathrm{Im}E_\ell$ is zero (localization) or non-zero (delocalization), it is meaningless to take the average; the weighted average of zero and non-zero elements is always non-zero, obscuring the fact that for the overwhelming majority of configurations $\mathrm{Im}E_\ell = 0$ as $\delta \to +0$, leading to localization."

6. Inhomogeneous Broadening. Relation to Fluroescence Efficiency. By using exchange, Lyo was able to demonstrate that single ion-single ion (resonant) transfer in ruby vanishes for concentrations less than ~0.3-0.4 at .%. This has significance for fluorescence, as discussed above, for it implies that dilute traps would be ineffective for donor concentrations below this magnitude.

Imbusch (35) demonstrated in ruby that the R_1 fluorescence migrates from isolated Cr^{3+} to isolated Cr^{3+}, until in the vicinity of a (scarce) Cr-Cr pair it transfers its energy to the pair (trap) (non-resonantly, as discussed in the next part). This is pictured in Fig. 6. Here, the single Cr^{3+} ions are the donors, and the pairs the acceptors. Though Imbusch's energy transfer experiments did not cover the low concentration range, the fluorescence efficiency studies of Tolstoi and Liu Shun'-Fu (48) on very thin samples did. Radiation trapping is supposed to be absent under these conditions, and their fluorescent lifetime points exhibit a sharp break near 0.3 at.%Cr. This is precisely the critical concentration calculated by Lyo (9), and indicates that energy flow to traps suddenly takes place at this concentration, confirming Lyo's prediction.

It is interesting to note that the presence or absence of donor-donor transfer sharply affects the yield and decay time of donor luminescence as a function of acceptor concentration (45). Thus, as the donor concentration is increased through the critical value, a sudden change from non-exponential to exponential time decay should occur. This would be a very direct determination of donor energy transfer and therefore of Anderson localization. In addition, sharp line excitation, and monitoring of subsequent fluorescence, could demonstrate the energy dependence of the localization criterion (private communication from J. Klafter). Thus, the fluor-

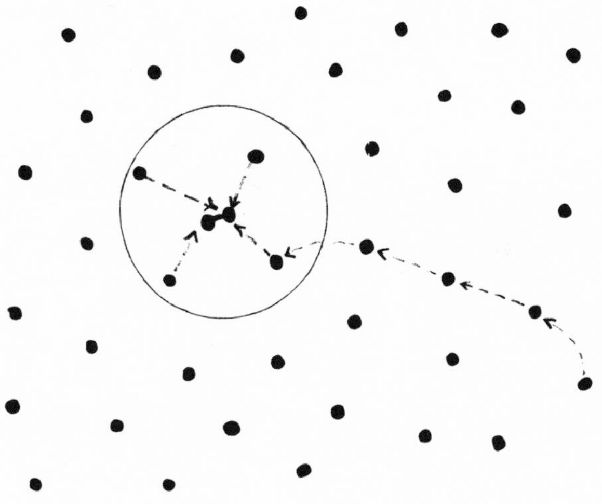

Fig. 6 Schematic of energy migration for single ion Cr^{3+} in Al_2O_3 (donors) and Cr^{3+} pairs (acceptors, traps).

escence decay could be exponential near the center of the line where localization is less likely, and non-exponential in the wings where the localization condition is more easily satisfied.

III.B. Phonon-Assisted Energy Transfer

In this section we examine the problem of donor-acceptor energy transfer when the mismatch in energy greatly exceeds the sum of the homogeneous widths. Under such circumstances, the direct use of Dexter's formula (27) would yield a negligible transfer rate. Though he and Miyakawa (5) claim one can use (27) for large energy mismatch by simply using the full line with sidebands for $f(\omega)$, this is in fact incorrect, as will be shown below, and in refs. (3), (6), and (7).

1. Form of Interaction. The explicit analogue to (27) for large energy mismatch was first formulated by Orbach (3). He noted that phonon emission (or absorption at high temperatures) at either donor or acceptor sites could make up the energy difference required for transfer. His method was expanded upon by Birgeneau (4), and we shall follow the latter derivation here.

To effect transfer, it is necessary to combine the spatial transfer operators of section III.A. with the electron-vibration Hamiltonian of section II. This can be done in the following manner:

$$W = (2\pi/\hbar^2)\ \Sigma_{\underline{K},j} |\langle n(\underline{K},j) + \Delta, 1, 2^*|\ \ H_{eff} | n(\underline{K},j),\ 1^*, 2\rangle|^2$$

$$\times \int g_1(\omega)\ g_2[\omega - \Delta\omega_j(\underline{K})]\ d\omega\ . \tag{50}$$

Here, $\Delta = \pm 1$ corresponding to the creation/destruction of a phonon of energy $\hbar\omega_j(\underline{K})$ in the transfer process. In (50) the effective Hamiltonian is the second order perturbation

$$H_{eff} = \{[H_{int}|\alpha\rangle\langle\alpha|H_{el\text{-}vib}(1) + H_{el\text{-}vib}(2)]/\Delta E\} + c.c. \tag{51}$$

where H_{int} represents any of the transfer mechanisms of section III (e.g., (31)), and $H_{el\text{-}vib}$ is given by (12) for one phonon. The generalization of (51) to multiphonon emission is obvious following the arguments in II.D. We discuss strong coupling effects in Part 5 of this section. Expanding (50) we find

$$W = (2\pi/\hbar^2)|\langle 1,2^*|H_{int}|1^*,2\rangle|^2$$

$$\times \left(\frac{\hbar}{\Delta E_2-\Delta E_1}\right)^2 \Sigma_{\underline{K},j}\Big|\Sigma_{\ell,\Gamma_q,m} M_{11}(\Gamma_q\ell,m)$$

$$\times \langle n(\underline{K},j) + \Delta|Q(\Gamma_q,m)|n(\underline{K},j)\rangle \qquad (52)$$

$$- \Sigma_{\ell',\Gamma_q,m'} M_{22}(\Gamma_{q'},\ell',m')\langle n(\underline{K}j) + \Delta|Q(\Gamma_{q'},m')_o$$

$$|n(\underline{K},j)\rangle \exp(i\Delta\underline{K}\cdot R_{12})\Big|^2 \times$$

$$\times \delta[(\Delta E_1-\Delta E_2)/\hbar-\Delta\omega_j(\underline{K})] .$$

Here, the subscript o means evaluation at $\underline{R} = 0$, ΔE_i is the change in electronic energy at site i, and

$$M_{ii}(\Gamma_q,m) = \langle i^*|V(\Gamma_q\ell)\, C(\Gamma_q\ell,m)|i^*\rangle$$

$$- \langle i|V(\Gamma_q\ell)\, C(\Gamma_q\ell,m)|i\rangle \quad .$$

The sum in (52) over $\underline{K},j$ is hard to handle, and Birgeneau (4) makes the long wave approximation (as did Orbach (3)) $2\pi/|\underline{K}| \gg |\underline{R}_{12}|$. The cross terms between various representations then vanish (see ref. 15), and (52) becomes

$$W = (2\pi/\hbar^2)|\langle 1,2^*|H_{int}|1^*,2\rangle|^2$$

$$\times \frac{(\Delta E_1-\Delta E_2)}{4\pi^2\hbar^2\rho}\Sigma_{\Gamma_q,m}\left[\frac{\langle\alpha^2_{\Gamma_q,m}\rangle_l}{v_\ell^5} + \frac{2\langle\alpha_{\Gamma_q},m^2\rangle_t}{v_t^5}\right]$$

$$\times [M_{11}(\Gamma_q,m) - M_{22}(\Gamma_q,m)]^2 \left\{\begin{matrix} n_j[(\Delta E_1-\Delta E_2)/\hbar] +1 \\ n_j[(\Delta E_1-\Delta E_2)/\hbar] \end{matrix}\right\}$$

$$\approx (2\pi/\hbar^2)|\langle 1,2^*|H_{int}|1^*,2\rangle|^2 \times$$

$$\times \left[\frac{\Delta E_1 - \Delta E_2}{4\pi^2 \hbar^2 \rho} \left(\frac{1}{v_\ell^5} + \frac{2}{v_t^5} \right) |M_{11} - M_{22}|^2 \right] \tag{53}$$

$$\times \left\{ \begin{matrix} n_j[(\Delta E_1 - \Delta E_2)/\hbar] + 1 \\ n_j[(\Delta E_1 - \Delta E_2)/\hbar] \end{matrix} \right\} .$$

We have used symmetry considerations to simplify the first part of (53), and $|M_{11}-M_{22}|^2$ in the second part means an effective average. In the first part of (53), the $\langle\alpha_{\Gamma_q,m}^2\rangle$ are phonon strain spatial averages, defined and evaluated for cubic systems by Van Vleck (15).

For the case of ruby, Fig. 7 displays the single ion to pair transitions. The energy mismatch is typically $\sim$100 cm^{-1}, while $\hbar\omega_D \sim$300 cm^{-1}. The long wave approximation may not be too bad, therefore. Birgeneau (4) uses static strain data (25,26) to estimate $|M_{11}-M_{22}| \sim$ 3,000 cm^{-1}. Taking $\Delta E_1 - \Delta E_2$ = 100 cm^{-1},

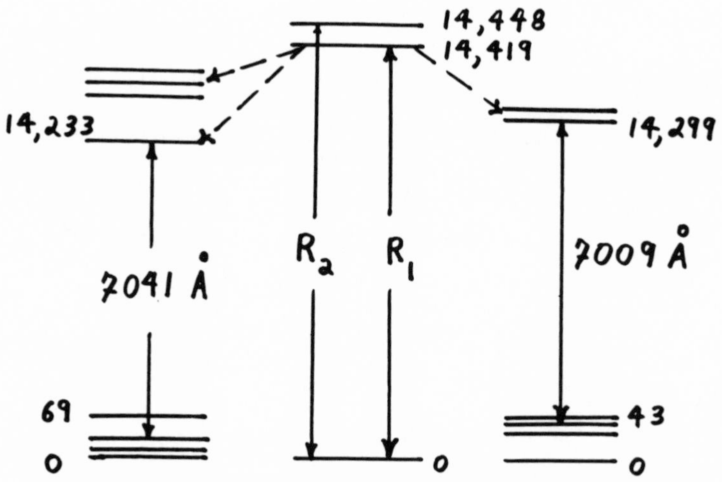

Fig. 7 Energy-level diagram for single chromium ions (center), second-nearest neighbor pairs (left), and fourth-nearest neighbors (right). Fine structure is not shown, and energy values are approximate.

$\rho = 4\text{g/cm}^3$, $v_\ell = 1.1 \times 10^6$ cm/sec, and $v_t = 0.64 \times 10^6$ cm/sec, Birgeneau finds an "effective overlap" of 8×10^{-16} sec, or an energy transfer rate some 10^{-4} times smaller than for resonant transfer. This is not in bad agreement with the experimental ratio of 10^{-3} for non-resonant to resonant energy transfer determined by Imbusch (35).

There are some interesting features of (53) which we shall discuss below. First, however, the long wave approximation is in doubt when $|\Delta E_1 - \Delta E_2| \sim \hbar\omega_D$. In such a case, the $\exp(i\Delta \underline{K}\cdot\underline{R}_{12})$ in (52) cannot be set ~ 1. This results in a mess. The various representations become mixed and one is left with an algebraic horror. For the simplified case of a single representation, one finds that the second factor in (53) is replaced by [(6) and (7)],

$$[M_{11}(\Gamma_g,m) - M_{22}(\Gamma_g,m)]^2 \rightarrow \{[\langle 2^*|VC|2^*\rangle - \langle 2|VC|2\rangle]^2$$
$$+ [\langle 1^*|VC|1^*\rangle - \langle 1|VC|1\rangle]^2 - 2[\langle 2^*|VC|2^*\rangle - \langle 2|VC|2\rangle]$$
$$\cdot[\langle 1^*|VC|1^*\rangle - \langle 1|VC|1\rangle] \times \cos(\underline{K}\cdot\underline{R}_{12})\} \quad . \tag{54}$$

The point of writing (54) out is that in the long wave limit one finds that phonon-assisted energy transfer is ineffective if the electronic matrix elements are the same on both sites (identical transitions). This is untrue for finite $\underline{K}\cdot\underline{R}_{12}$. However, identical transitions are also extremely unlikely in practice because a large energy mismatch must be present. Indeed, the prefactor of (53) vanishes for zero phonon energy (energy matching) transitions. Nevertheless, quantitative corrections could be important, and one should use the long wave result with caution.

Second, the argument used by Miyakawa and Dexter (5) (phonon-assisted transfer can be obtained from resonant transfer theory (12) by substituting the full phonon side band for g (ω)) is incorrect. One sees, even in the long wave limit, that

$$|M_{11}(\Gamma_g,m) - M_{22}(\Gamma_g,m)|^2 = |\langle 2^*|VC|2^*\rangle - \langle 2|VC|2\rangle$$
$$- \langle 1^*|VC|1^*\rangle + \langle 1|VC|1\rangle|^2 \tag{55}$$

contains cross terms which would not be present if one only took overlapping side bands. Writing (55) out, one has,

$$|\langle 2^*|VC|2^*\rangle - \langle 2|VC|2\rangle|^2 + |\langle 1^*|VC|1^*\rangle - \langle 1|VC|1\rangle|^2$$
$$-2|[\langle 2^*|VC|2^*\rangle - \langle 2|VC|2\rangle] \quad [\langle 1^*|VC|1^*\rangle - \langle 1|VC|1\rangle]| \tag{55}$$

The first two terms are what one would obtain from Miyakawa and Dexter's suggestion, but the third, or cross term, would be omitted by their treatment.

We now go on to consider specific aspects of (53).

2. Dependence on Temperature. The result (53) has its temperature dependence in the Bose factors. For phonon emission, we have the same dependence as for phonon non-radiative relaxation:

$$\frac{\exp[(\Delta E_1 - \Delta E_2)/k_B T]}{\{\exp[(\Delta E_1 - \Delta E_2)/k_B T]-1\}} \tag{56}$$

with an obvious expression for phonon absorption.

3. Dependence on Energy. Here, the energy dependence of the phonon density of states plays a role. From (53) one sees the transfer rate is proportional to the energy mismatch. This result is opposite (in sense) to that for resonant energy transfer; it indicates that, within energy differences less than $h\omega_D$, the greater the mismatch the greater the transfer rate. It would be interesting to see if this is true by probing the transfer rate to the various "N" lines in the ruby pair spectrum.

When the energy mismatch exceeds the Debye energy, more than a single phonon is required and one obtains (5) an exponential dependence on transfer energy mismatch of exactly the same form as for multiphonon non-radiative relaxation (see eq. (26)). The same is true for the dependence on temperature (23).

4. Range Dependence. Two elements enter in the range dependence. The first we have already discussed: the factor $\exp(i\Delta \underline{K}\cdot\underline{R}_{12})$ in (52). As the spatial separation increases, the possibility of cancellation between the various matrix elements decreases (actually oscillates —

see eq. (54)), leading to a slight enhancement of the range of phonon-assisted transfer. By far the largest effect comes, however, from the prefactor $|<1,2^*|H_{int}|1^*,2>|^2$ in (52). This is the same term we have considered in section III.A. for resonant transfer, so we shall not consider it in detail. However, one should note that in the present instance, the donor and acceptor ions are most certainly different from one another. It is possible, then, to mix ions with vastly different f numbers to achieve a specific fluorescence "design". Our previous conclusions about the relative importance of various transfer Hamiltonians under such circumstances could then be altered. Also, H_{int} is a new element in the Hamiltonian for finite concentrations. Hence, the Hamiltonian producing phonon sidebands (one ion) may not be the same as that which effects energy transfer (two ion). Thus, there may be no relationship between phonon sideband intensity and phonon-assisted energy transfer (49).

Other features might occur. For example, in ruby where single ion to pair transfer takes place, the strong pair exchange allows the additional processes (see Fig. 4)

$$^4A_2 \overset{H_{ex}}{\rightarrow} {}^2T_2 \xrightarrow{H_{quad\ quad}} {}^2E$$

$$^4A_2 \xrightarrow{H_{quad\ quad}} {}^4T_1 \overset{H_{ex}}{\twoheadrightarrow} {}^2E$$

over what we considered previously. Also, one can modulate exchange on the pairs, an additional interaction process not possible for isolated ions. Every system must be evaluated according to its features before one can draw definitive conclusions.

Finally, one should remark that inhomogeneities in the transition energy will not affect phonon-assisted transfer, though as we have shown they may well quench resonant transfer. This is because the phonons can take up the inhomogeneous shifts in energy, whereas in resonant transfer no energy shifting mechanism is available.

5. Effect of Strong Coupling. As discussed in section II.B., strong coupling can "quench" energy transfer if one of the ions (or both) is strongly coupled to the lattice and a change in occupied electronic orbitals take place. For example, transfer of the R-line excitation in ruby involves no change in electronic orbital occupation, so that strong coupling has no effect. However, for $^4T_2$ to $^4A_2$ excitation transfer in ruby, the change from t_2^2e to t_2^3 causes a large change in lattice coupling, and hence a shift in configurational coordinates. This introduces a factor (refs.(6) and (7))

$$e^{-\Phi(T)}$$

into the transfer rate, where $\Phi(T)$ is obtained from (17) by substitution of (54) for the absolute square in (17). The cross terms here are more important than for the coefficient of the transfer integral because $|\underline{K}|$ is not restricted to a single value because of energy conservation, but can range over all values. For the case of $^4T_2$ to $^4A_2$ transfer, Kohli and Huang (7) find the cross terms in Φ to be such that they reduce the transfer rate by approximately a factor of four from what one would obtain with (17) acting on each site alone. The reduction of the transfer rate caused by vibrational coordinate displacement (16) occurs, of course, for both phonon-assisted (non-resonant) and resonant energy transfer. Also, as the temperature is raised, the exponential dependence on $\coth(h\omega_j/2k_BT)$ can cause rapid diminution in the transfer rate. It would be interesting to search for this thermal quenching in strongly lattice-coupled systems.

IV. SUMMARY AND FUTURE PROSPECTS

We have attempted in this review to delineate the role of the electron-vibrational interaction in the optical spectrum of defects in solids. We have specialized to the case of weak coupling, with only a mention of the procedure to be employed when the coupling is sufficiently strong that vibrational overlap effects significantly reduce the electronic transition rate. When the interaction is this strong, one must shift to the use of configurational coordinates, a topic treated in detail by Professor D. Curie elsewhere in this volume.

We began our treatment with an exposition of the origin of the electron-vibration interaction using simple group theory to simplify the formulation to make it algebraically tractable. Estimates of the strength of the interaction were given for transition metal and rare earth impurities. Then, non-radiative electronic transition rates were calculated for one-,two-, and many-phonon processes. The frequency (energy) and temperature dependence of these processes were explored.

We then examined in some detail the question of energy transfer, dividing our treatment into resonant and non-resonant (phonon-assisted) behavior. In the former case we showed that inhomogeneous broadening could quench energy transfer, depending on the magnitude of the broadening and on the range dependence of the coupling interaction. We discussed a method for establishing the presence of excitation localization (quenched transfer), showing that both the concentration and energy dependence of the localization criterion could be experimentally obtained. Finally, we obtained the energy and temperature dependence of phonon-assisted energy transfer, exhibiting explicitly the electron-vibration selection rules. We also discussed the effect of vibrational overlap quenching of both resonant and non-resonant energy transfer, showing that for finite interaction strength the transfer rate was exponentially quenched, the exponent increasing with the square of the electron-vibration interaction, and with temperature when $k_B T$ exceeds the predominant vibrational frequency.

Many of the results of this cahpter have already been verified experimentally. The related article by Dr. R.K. Watts presents numerous references, and displays methods for treating complex cases involving a number of optical levels and transfer interactions. He also develops the impact of diffusion-limited donor energy transport upon donor fluorescence in mixed (donor-acceptor) systems.

In this author's opinion, future investigations in this field should focus on the dynamical properties of ions in solids. One interesting approach would be to combine optical and magnetic resonance static stress measurements, so as to obtain the individual orbit-lattice coefficients $V(\Gamma_{ig}\ell)$ in full. Acoustic propagation methods could also be utilized. Once in hand, these coefficients would enable the investigator to calculate non-radiative transition rates, as well as phonon-assisted energy transfer rates. This could enable one to "design" advantageous energy transport properties in systems of particular interest.

Another aspect about which we have already spoken is the question of resonant transfer in inhomogeneously broadened systems. We have derived in Section III the condition for excitation localization using Ziman's simplified version of Anderson's derivation. To our knowledge, this effect has not yet been isolated in optical systems, though ruby may exhibit such effects in the "pink" region. We have discussed two methods for determining the presence of excitation localization: the concentration dependence of the fluorescence efficiency, and the time dependence of donor fluorescence. In the former, donor excitation can make small concentrations of acceptors ineffective as traps. As the donor concentration is raised beyond that critical value where localization occurs, "rapid" donor-donor transfer take place, and donor fluorescence behavior will change. This can result in a sharp change (increase) in the effective diffusion constant as the concentration is increased. In principle, it should vary from zero to a finite value. Residual dipolar coupling, or thermal broadening of the lines, can cause some transfer to occur even below the critical concentration. This may have been seen by Krasutsky and Moos (50), but was not discussed by them explicitly. Another feature, in addition to the concentration dependence is the dependence of localization. Near the center of the optical line the effective concentration is larger than in the inhomogeneously broadened wings. At low temperatures, phonons cannot act as a frequency shifting mechanism. In the absence of other mechanisms, this means that sites whose energies fall near the center of the line would be less likely to localize than sites whose frequencies lie in the wings of the line. Thus, by moving a sharp line (dye laser) excitation across the donor transition energy, one could observe a sharp break in donor fluorescence efficiency as a function of excitation position. This would be a direct measure of the energy-dependent localization condition. An analogous argument holds for the donor fluorescence decay properties. In the absence of transfer, it will be non-exponential. In the presence of transfer, it can be diffusion-limited, or very rapid. In the former case, exponential time decay would be seen only at long times. In the latter case, exponential time decay would be seen at all times. Thus, a sharp change in the diffusion constant, or a change from non-exponential to exponential donor fluorescence decay will be seen as the concentration of donors is increased through the critical value.

Another interesting, but as yet untreated, limit is that in which the homogeneous broadening (caused by phonon lifetime effects, for example) begins to be significant (say, of the order of W/10). One can then show that localization in the Anderson sense never occurs --- the wings of the Lorentzian line always have sufficient amplitude to allow transfer to occur, even for concentrations so low that the Anderson criterion would be satisfied in the absence of such broadenings. The amplitude in the wings will be small, however, so that the departure from localization will be small. A sharp increase in transfer rate as the concentration is increased through the critical value would still be expected. However, the value of the critical concentration would presumably be reduced as the homogeneous width increases. It would be extremely interesting to measure this change in critical concentration at different temperatures. Imbusch (49) has already seen signs that such an effect occurs in dilute ruby (the critical concentration is higher at lower temperatures), but a thorough study is necessary to test the theory.

In conclusion, the integration of theory and experiment in the field of optical properties of solids has resulted in a rather satisfactory "static" description. The next decade may well examine the dynamical aspects of this most important subject, and thereby obtain exciting new results and concepts. Schools of the character of this "First Course" at Erice will stimulate and advance this process.

REFERENCES

1. T. Förster, Ann. Physik 2, 55 (1948).
2. D.L. Dexter, J. Chem. Phys. 21, 836 (1953); Phys. Rev. 126, 1962 (1962).
3. R. Orbach, in "Optical Properties of Ions in Crystals", Ed. by H.M. Crosswhite and H.W. Moos (Interscience, New York, 1967), p. 445.
4. R.J. Birgeneau, J. Chem. Phys. 50, 4282 (1969).
5. T. Miyakawa and D.L. Dexter, Phys. Rev. B1, 2961 (1970).
6. T.F. Soules and C.B. Duke, Phys. Rev. B3, 262 (1971).
7. M. Kohli and N.L. Huang Liu, Phys. Rev. B9, 1008 (1974).
8. P.W. Anderson, Phys. Rev. 109, 1492 (1958).
9. S.K. Lyo, Phys. Rev. B3, 3331 (1971).
10. L.A. Riseberg, W.A. Gandrud, and H.W. Moos, Phys. Rev. 159, 262 (1967); L.A. Riseberg and H.W. Moos, ibid, 174, 429 (1968).
11. R. Reisfeld and Y. Eckstein, J. Non-crystalline Solids 11, 261 (1972).
12. "Electron Spin-Lattice Relaxation" by R. Orbach and H.J. Stapleton, "Electron Paramagnetic Resonance", ed. by S. Geschwind (Plenum Press, New York, 1972), page 121.
13. E. Simanek and Z. Sroubek, Phys. Status Solidi 4, 251 (1964); R.E. Watson and A.J. Freeman, Phys. Rev. 134, A1526 (1964); R.E. Watson and A.J. Freeman 156, 251 (1967).
14. A.R. Edmonds, "Angular Momentum in Quantum Mechanics" (Princeton University Press, Princeton, 1957); J.S. Griffith, "The Theory of Transition Metal Ions" (Cambridge University Press, Cambridge, 1961).
15. J.H. Van Vleck, J. Chem. Phys. 7, 72 (1939); Phys. Rev. 57, 426 (1940).
16. R. Orbach, Proc. Roy. Soc. A264, 458 (1961); P.L. Scott and C.D. Jeffries, Phys. Rev. 127, 32 (1962).
17. R. Orbach and M. Tachiki, Phys. Rev. 158, 524 (1967).
18. M. Blume and R. Orbach, Phys. Rev. 127, 1587 (1962).
19. R.R. Sarma, T.P. Das, and R. Orbach, Phys. Rev. 149, 257 (1966); ibid, 155, 338 (1967); ibid, 171, 378 (1968).
20. G. Watkins and E. Feher, Bull. Am. Phys. Soc. 7, 29 (1962); Elsa Rosenwasser Feher, Phys. Rev. 136, A145 (1964).
21. E.S. Sabisky and C.H. Anderson, Phys. Rev. B1, 2028 (1970).
22. A. Abragam, J.F. Jacquinot, M. Chapellier, and M. Goldman, J. Phys. C. (London) 5, 2629 (1972).
23. M. Wagner, J. Chem. Phys. 41, 3939 (1964).
24. H. Rosenberg and J.K. Wigmore, Phys. Letters 24A, 317 (1967).
25. M. Blume, R. Orbach, A. Kiel, and S. Geschwind, Phys. Rev. 139, A314 (1965).
26. A.L. Shawlow, in "Advances in Quantum Electronics", ed. by J.R. Singer (Columbia University Press, New York, 1961), p. 50; M.D. Sturge, quoted in ref. (27).

27. G.F. Imbusch, S.R. Chinn, and S. Geschwind, Phys. Rev. 161, 295 (1967).
28. Actually, what is quoted in (27) is the spin flip non-radiative rate from the 2Ā to the Ē level. We have used the theoretical ratio, (1/60), given in (25) for this rate to the non-spin-flip non-radiative rate, to obtain these values. Work on photon echoes by I.D. Abella, N.A. Kurnit, and S.R. Hartmann, Phys. Rev. 141, 391 (1966) suggest these times may be too short. That is, the factor of 1/60 is too large, and the actual non-spin-flip non-radiative rate is around an order of magnitude slower than given in eq. (19).
29. R. Orbach, Phys. Rev. 133, A34 (1964).
30. P.P. Sorokin and M.J. Stevenson, IBM, J. Res. Development 5, 56 (1961); see also W. Kaiser, C.G.B. Garrett, and D.L. Wood, Phys. Rev. 123, 766 (1961); and P.P. Sorokin, M.J. Stevenson, J.R. Lankard, and G.D. Petit, Phys. Rev. 127, 503 (1962) for a study of $SrF_2:Sm^{2+}$.
31. K.R. German and A. Kiel, Phys. Rev. B8, 1846 (1973).
32. R. Orbach, Proc. Roy. Soc. A264, 458 (1961).
33. R. Kislink and C.A. Moore, Phys. Rev. 160, 307 (1967).
34. A. Kiel, "Quantum Electronics", ed. by P. Grivet and N. Bloembergen (Columbia University Press, New York, 1964), Vol. 1, page 765.
35. G.F. Imbusch, Phys. Rev. 153, 326 (1967).
36. P.W. Anderson, Phys. Rev. 109, 1492 (1958). See also an "improved treatment" which curiously leads to the same conclusions: R. Abou-Chaira, P.W. Anderson, and D.J. Thouless, J. Phys. C. 6, 1734 (1973).
37. N.F. Mott, Phil. Mag. 19, 835 (1969).
38. J.M. Ziman, J. Phys. C. 2, 1230 (1969).
39. B.C. Carlson and G.S. Rushbrooke, Proc. Cambridge Phil. Soc. 46, 626 (1950).
40. P. Kislink, N.C. Chang, P.L. Scott, and M.H.L. Pryce, Phys. Rev. 184, 367 (1969).
41. J.P. van der Ziel, Phys. Rev. B9, 2846 (1974).
42. W. Heitler, "The Quantum Theory of Radiation", (Oxford University Press, New York, 1957), page 231.
43. K. Sugihara, J. Phys. Soc. (Japan) 14, 1231 (1959).
44. L.K. Aminov and B.I. Kochelaev, J. Exptl. Theor. Phys. (U.S.S.R.) 43, 1303 (1962)[Engl transl: Soviet Phys. J.E.T.P., 15, 903 (1962)]; D.H. McMahon and R.H. Silobee, Phys. Rev. 135, A91 (1964). See also many references to Birgeneau's work in ref. (4).
45. M. Inokuti and F. Hirayama, J. Chem. Phys. 43, 1978 (1965). Confirmatory experiments were made by E. Nakazawa and S. Shionoya, J. Chem. Phys. 47, 3211 (1967).

46. W. B. Gandrud and H. W. Moos, J. Chem. Phys. 49, 2170 (1968).
47. M. J. Weber, Phys. Rev. B4, 2933 (1971).
48. N. A. Tolstoi and Liu Shun'-Fu, Opt. i Spectroskopiya 13, 403 (1962) [Opt. Spectry. (USSR) 13, 224 (1962)]. More recent experiments by Imbusch on finely powdered ruby also indicate this behavior.
49. G. F. Imbusch, private communication.
50. N. Krasutsky and H. W. Moos, Phys. Rev. B8, 1010 (1973).

CHARGE TRANSFER SPECTRA

Donald S. McClure

Department of Chemistry, Princeton University

Princeton, New Jersey 08540

ABSTRACT

Charge transfer spectra in highly ionic crystals correspond to electron transfer between neighboring atoms. They can be classified as donor or acceptor, depending upon whether the metal atom donates or accepts an electron. The essential differences between these two kinds of transitions are emphasized, and a simple criterion to establish which type is responsible for observed spectra are stated. The relationship of optical and chemical charge transfer processes are examined. The theories used to explain and correlate charge transfer spectra are finally discussed.

I. INTRODUCTION

The name charge transfer suggests the essential difference between these and localized transitions. They are basically more interesting as they involve the electronic structure of the crystal more than do localized transitions. The amount of detailed information is often less because these spectra are typically broader than localized transitions. Some especially sharp spectra have been found however (2nd and 3rd row transition metal halides in Cs_2ZrCl_6 host crystals). The use of MCD has provided detailed information which has led to band assignments for these systems (1). At this point in time, our understanding of these spectra is becoming quite detailed, but there are many other charge transfer spectra about which we know only the rudimentary facts.

The simplest view of a charge transfer process is that an electron on one atom is transferred to a neighbor. In highly insulating crystals the transfer does not extend very far into the lattice because the wavefunctions are localized. In ZnS, GaP, etc., donor-acceptor pairs display charge transfer transitions, but these are described by band wavefunctions and can occur for widely separated ion pairs. Here we will discuss the localized charge transfer transitions which occur between an anion and cation in ionic crystals.

II. ELEMENTARY ASPECTS OF CHARGE TRANSFER

The elementary idea of charge transfer is something which can easily be tested against observations. If a charge moves by an angstrom or so, there is a large transition moment M = 1 x e, corresponding to a large oscillator strength. Of course there can be selection rules in symmetrical systems which could result in weak transitions. Nevertheless, large absorption strengths have been one of the principal means of identifying these transitions.

The other principal identification method is to use the formula for charge transfer energy:

$$\Delta E = I_D - E_A - e^2/R_{DA} \qquad (1)$$

where I_D is the donor ionization potential, E_A the acceptor electron affinity and R_{DA} is the distance of separation. The ionization potential and electron affinity have to be corrected for crystal field effects in solids. Usually it is not possible to do this completely enough and one corrects for d-electron splittings and inserts an empirical constant which can be determined for a series of transition elements. In this case, the constant also includes the transfer length, since it could be assumed to be constant in a transition series. The charge transfer energy is then

$$E_{CT} = I_D' - E_A' + C \qquad (2)$$

where the primes indicate that crystal field corrections have been made. The identity of a charge transfer process can be established if one compares the spectra of a series of metal atoms acting as electron acceptors with a constant donor atom. For example, the dichlorides of the first row transition metal atoms could be examined. The charge transfer band energy should then vary as the electron affinity of the metal atom.

I will refer to two kinds of charge transfer transitions as donor and acceptor types depending on whether the metal atom donates or accepts an electron from the lattice. This terminology corresponds to that which is used to describe donors and acceptors in semiconductors; in our case the only impurities to be considered are transition metal or rare earth ions.

The crystal field correction for transition metal atoms is illustrated by Fig. 1. We can use the ionization potentials of a transition metal atom and one of the diagrams of Fig. 1 for the initial and final state of the atom. We also need Dq values for each charge state of the atom. These are found from Table 1 where typical Dq values for +2 and +3 ions are given, for octahedral and tetrahedral coordination.

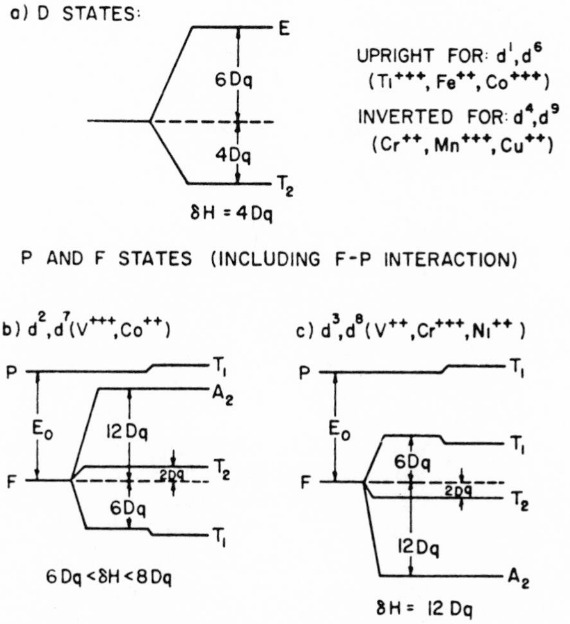

Fig. 1 Splitting of P, D, and F states in a cubic field. These diagrams are appropriate for octahedral coordination as they stand, and when inverted are appropriate for tetrahedral or cubic coordination. The value of δH is the energy of the lowest component below the unperturbed level. The strongest bands of the spectrum are due to transitions from the lowest to the higher levels shown here. Other levels (not shown) give rise to weak bands. The method of deriving Dq from the spectrum is to identify the bands and to divide the observed energy by the corresponding theoretical factor. Part a of the figure serves as a splitting diagram for both a D state and a d orbital. The t orbitals are lowest in an octahedral field, and highest in a tetrahedral field.

For example in the case of Ni^{++}, an acceptor type transition would involve the process $e + Ni^{++} \rightarrow Ni^{+}$, so E_A is $-I_2(Ni)$ + crystal field corrections. The latter are $+ 12\,Dq(Ni^{++}) - 6Dq(Ni^{+})$ and the transition energy is

$$\Delta E_{CT} = I_D + C - I_2 + 12Dq(Ni^{++}) - 6Dq(Ni^{+}) \qquad (3)$$

For a series of oxides, for example, we would take $I_D + C$ as a constant for the entire series, and compare ΔE_{CT} to the variations predicted by the other terms of eq. 3. (The values of Dq for monovalent ions are not known and have to be estimated).

Table 1. Crystal field theory data for transition metal ions.

Number of *d*-electrons	Ion	Free ion ground state	Octahedral field ground state	Tetrahedral field ground state	*Dq* cm^{-1} oct.	*Dq* cm^{-1} tetr.	Stabilization, kcal oct.	Stabilization, kcal tetr.	Oct. site preference energy kcal/mole
1	Ti^{+++}	^2D	$^2T_{2g}$	^2E_g	2030	900	23.1	15.4	7.7
2	V^{+++}	^3F	$^3T_{1g}$	$^3A_{2g}$	1800	840	41.5	28.7	12.8
3	V^{++}	^4F	$^4A_{2g}$	$^4T_{1g}$	1180	520	40.2	8.7	31.5
	Cr^{+++}	^4F	$^4A_{2g}$	$^4T_{1g}$	1760	780	60.0	13.3	46.7
4	Cr^{++}	^5D	^5E_g	$^5T_{2g}$	1400	620	24.0	7.0	17.0
	Mn^{+++}	^5D	^5E_g	$^5T_{2g}$	2100	930	35.9	10.6	25.3
5	Mn^{++}	^6S	$^6A_{1g}$	$^6A_{1g}$	750	330	0	0	0
	Fe^{+++}	^6S	$^6A_{1g}$	$^6A_{1g}$	1400	620	0	0	0
6	Fe^{++}	^5D	$^5T_{2g}$	^5E_g	1000	440	11.4	7.5	3.9
	[a]Co^{+++}	^5D	$^1A_{1g}$	^5E_g		780	45	26	19
7	Co^{++}	^4F	$^4T_{1g}$	$^4A_{2g}$	1000	440	17.1	15.0	2.1
8	Ni^{++}	^3F	$^3A_{2g}$	$^3T_{1g}$	860	380	29.3	6.5	22.8
9	Cu^{++}	^2D	^2E_g	$^2T_{2g}$	1300	580	22.2	6.6	15.6
10	Zn^{++}	^1S	$^1A_{1g}$	$^1A_{1g}$	0	0	0	0	0

The data given are:
1. Number of *d* electrons.
2. Transition metal ions.
3. Free ion Russell–Saunders ground term (spin-orbit coupling is neglected in the term designation).
4. The standard group theory symbol for the ground state of the octahedrally coordinated ion in a solid (using nomenclature of Eyring, Walter, and Kimball, *Quantum Chemistry*).
5. The group theory symbol for the tetrahedrally coordinated ionic ground state.
6. *Dq* values for octahedral hydrates of the ions.
7. *Dq* calculated for tetrahedral coordination.
8, 9. The thermodynamic stabilization in octahedral or tetrahedral fields.
10. The octahedral site preference, or the difference between columns 8 and 9.

[a] The octahedral site stabilization of Co^{+++} was estimated from the heat of hydration increment caused by the crystal field, and the tetrahedral site stabilization was taken to be the same as for Cr^{+++}.

This type of comparison has been done by Tippins for most of the trivalent transition metal ions in Al_2O_3 (2). There is moderately good agreement with eq. 3 for the case that the metal ion is an acceptor. Ni^{+++} was not in line, but we believe that Tippins misinterpreted a Ni^{+4} d → d band as a Ni^{+++} CT band, and that the amount of Ni^{+++} was very small.

The mis-identification of these bands is a serious problem. The chemical analysis usually shows traces of several elements in a sample. Even without impurities due to other elements, the oxidation state is sometimes difficult to establish. Those ions having unfavorable thermodynamic crystal field stabilizations are sometimes difficult to form in the crystal, and another oxidation state forms instead.

One of the best known sets of charge transfer bands are the halogen compounds of transition metals. Fig. 2 shows the $NiX_4^{=}$ halide ion spectra (3). There is a shift to the red of nearly 1 ev in each step of the progression from Cl^- to Br^- to I^-. This must correspond to the change in ionization energy of a halide site (the donor). It is remarkably constant from one series of metal halides to another.

There is now a large and growing body of literature on charge transfer spectra of inorganic complex ions having halide and oxide ligands, much of which has been reviewed by Jorgensen (4). The decrease of spectral energy in going from Cl^- to Br^- to I^- is a familiar feature, and it led Jorgensen to propose the idea of an "optical electronegativity" scale. Thus a corrected absorption band position is given by:

$$\sigma_{corr} = 30{,}000 \text{ cm}^{-1} \left[\chi_{opt}(X) - \chi_{opt}(M)\right] \qquad (4)$$

The correction will be for the spin pairing energy (a few thousand cm^{-1}) and for the crystal field state of the electron. The values χ_{opt} for halides are: F^- 3.9, Cl^- 3.0, Br^- 2.8, I^- 2.5, and they roughly parallel Pauling's electronegativity scale. Values can be assigned to the metals as well, taking account of their coordination number. Eq. 4 gives a good first approximation to the position of the charge transfer bands of a large number of systems.

The reverse process, metal → ligand, is not often observed and for an obvious reason. There are no vacant low-lying orbitals on a halogen, oxide, or other common electronegative ligand ion. There probably are such transitions to CN^- ligands, since these have low-lying π-type molecular orbitals; and other organic ligands probably show this process as well.

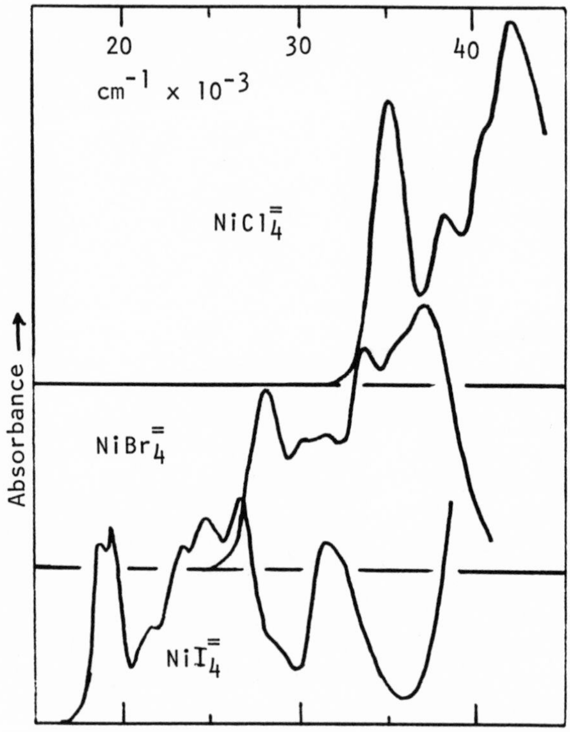

Fig. 2 Charge transfer spectra of the complex ions $NiX_4^{=}$, $X = Cl^-$, Br^-, I^-, showing the characteristic changes in charge transfer energy with halogen. (From ref. 3, redrawn.)

Direct donor-type charge transfer processes in crystals would be electron transitions from the impurity into the conduction band of the host. Instead of these direct processes, it appears that one sees higher internal transitions in rare earth and transition metal ions in crystals. In the rare earth series the transitions observed at energies higher than the 4f → 4f transitions are 4f → 5d (5). These are parity permitted and therefore are quite strong. Recently, Sabatini has observed absorption bands in the vacuum ultraviolet region of transition metal ions in fluoride host crystals, such as Co^{++}in MgF_2 (6). These bands are quite weak. When the entire series is examined and eq. 3 applied, it is found that they act like donor-type charge transfer bands. On comparison with free ion energies, they could also be 3d → 4s bands. The latter interpretation may be closer to the truth since 3d → 4s is parity forbidden, consistent with the weakness of the bands. Thus there are two examples in which internal transitions are found. True donor-type charge transfer processes must have small spectroscopic cross sections, but could be looked for using photoconductivity. So far, no systematic work in this area has been done.

These internal interconfigurational transitions, 4f → 5d in rare earths and 3d → 4s in transition metals, could be regarded as "proto charge transfers," i.e., almost charge transfers, because the upper orbital 5d or 4s is much less localized than the lower 4f or 3d orbitals. The upper orbital is in fact partially on the surrounding atoms. As will be shown in Section IV, the variation of the energy of such transitions with changing atomic member is almost the same as the variation of the ionization energy. This is because the upper orbitals have little binding energy, and hence little sensitivity to the details of the atomic core to which they belong, while the principal part of the variation with atomic member is due to the binding energy of the electron in its initial state.

III. RELATION BETWEEN OPTICAL AND CHEMICAL CHARGE TRANSFER

Optical charge transfer processes are related to chemical charge transfer processes. There is a class of chemical reactions in aqueous solution for which a half-reaction is: $M^{++} \rightarrow M^{+++} + e$. The electron is transferred to an acceptor in the full reaction. In an electrochemical cell, the free energy of the half reaction is measured relative to some standard acceptor process such as $e + H^+ \rightarrow \frac{1}{2} H_2$. The free energy of such a reaction varies according to the third ionization potential of M when one examines the data for Ti^{++} through Cu^{++}; but to observe this trend the crystal field corrections have to be made. The correlation between the half-reaction potential and the third ionization potential is illustrated in Fig. 3.

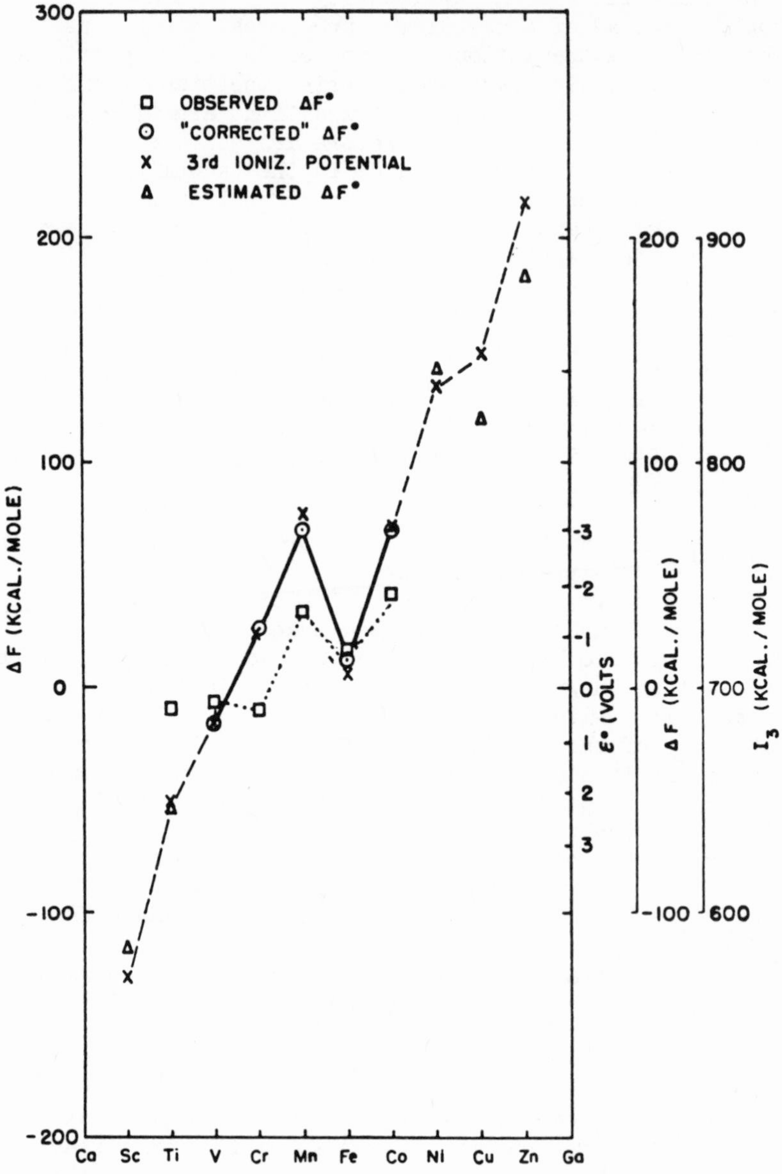

Fig. 3 The $M^{2+} \rightarrow M^{3+}$ oxidation potentials of the first transition series. The observed values, when corrected for ligand field effects, are seen to lie close to the I_2 values, also shown in the figure.

It should also be possible to correlate the charge transfer absorption band energies to this half-reaction oxidation potential. This cannot be done at present because the donor-type transitions have only been observed in fluoride matrices, not in oxide or aqueous media. Nevertheless, there is a linear relationship between the $3d \rightarrow 4s$ transition energies of transition metal ions in fluoride matrices and the oxidation potential of these ions in water. This result confirms the idea that the transitions in the fluoride host crystals are in the direction of electron donation, rather than being in the other direction. The $3d \rightarrow 4s$ process corresponds nearly to a donation, since the 4s electron is in an extended orbit. The theory of these processes will be examined later.

The $+2 \rightarrow +3$ oxidation potential should also be correlated to the charge transfer spectrum of the reverse process. Thus the acceptor-type charge transfer spectra of M^{+3} should vary as $-I_3$. There is evidence that this is true from Tippin's work(2) as well as more fragmentary data on the trivalent halide spectra.

IV. THEORY OF CHARGE TRANSFER SPECTRA

There are two important aspects of the theory of charge transfer transitions which we will discuss. In view of the significance of the variation of the absorption band energy with atomic number, it is important to understand the theory behind this variation. If one knew the ionization potentials in every case, a calculation would not be necessary, but for the rare earth series, these are not known. The Slater-Condon theory of atomic spectra(7) gives a reasonable account of donor type spectra for these atoms.

The other aspect of the theory is the application of molecular orbitals(8). This method has been applied successfully to the detailed interpretation of the multiband structure of acceptor type transitions.

Neither type of theory is complete in itself, and the two will ultimately be replaced by a self-consistent theory(9) or other more sophisticated theory.

IV.A. Change of Energy with Atomic Number

As stated earlier, the main trends in the donor energies with atomic number are explained by the trend in the energy of the ground state with atomic number, since the excited state is either unbound or only weakly bound. Therefore, it is necessary to calculate the energy of the lowest state of a configuration $4f^n$, $3d^n$, etc. The first term to consider is the average energy of the electrons of the l^n configuration as a function of n. This energy depends on the state of ionization of the ions of the series, i.e., on whether it is a +2 or +3 or other series, and on the value of n. One would expect a monotonic increase in this average binding energy with n in a series of constant state of ionization since the atomic number, Z, increases as n increases. Thus there are two terms W(i) and nE(i) to express these two effects, where i represents state of ionization. Both W and E are considered to be empirical. The splitting of the l^n configuration by interelectronic repulsion is considered next. This can be worked out quite easily by methods found in Condon and Shortley's book(7), since all we need is the ground state, usually taken as the lowest state given by Hund's rule. Finally, the additional stabilization of the ground state due to spin-orbit coupling must be added. The final atomic or ionic ground state energy is

$$E(n,i) = W(i) + nE(i) + Q(n,i) + S(n,i) \qquad (5)$$

where Q(n,i) is the electrostatic energy of the ground state measured from the average energy of the l^n configuration and S(n,i) is the spin-orbit energy of the lowest multiplet component, measured from the average of the lowest multiplet levels.

An example of the application of this process to the 4f → 5d energies of the divalent rare earth ions in CaF_2(5) is shown in Fig.4. Small corrections for the variation of the 5d binding energies with atomic number have actually been included in the plotted points in the figure, and these were made in the same way as for the 4f energies. The experimental 4f → 5d energies fit the calculations quite well except for La, Ce, Gd and Tb. For these, however, the transitions are 5d → 4f, and when this is taken into account, the agreement is uniformly good. The details are given in the original article (5).

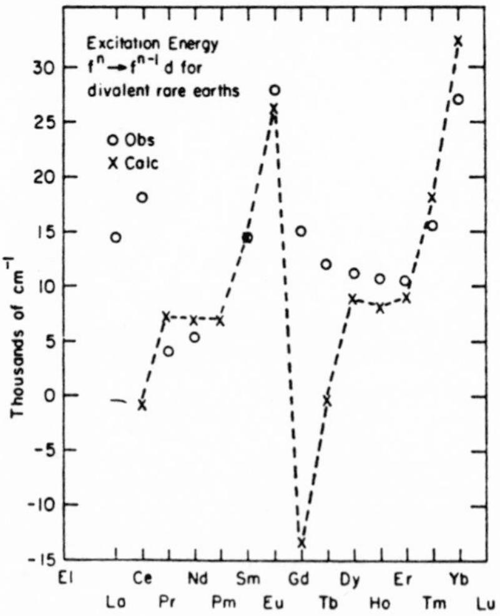

Fig. 4 Comparison of calculated and observed onset of absorption in the 4f → 5d transitions of the divalent rare earths.

Recent work along this same line has been done by Nugent and his collaborators(10), and a great deal of spectroscopic and chemical data on lanthanides and actinides has been correlated, and predictions of unmeasured quantities have been made.

(NOTE: J. Sugar and J. Reader have recently published good estimates of the first four ionization energies of the rare earth atoms, J. Chem. Phys. 59, 2093 (1973).)

IV.B. Application of Molecular Orbitals

Charge transfer transitions of impurity ions in crystals are probably quite localized on the central atom and a few of its near neighbors. Since a charge separation is always involved, the electron and hole are attracted by a Coulomb field, and at least the lowest charge transfer states are bound states for this reason. The simplest and most useful theoretical description of these states is the molecular orbital, ℓ.c.a.o. description (8). It is possible to parametrize this theory and determine the parameters from the spectra. Theoretical estimates of the parameters can also be made.

1. Molecular Orbital Theory. The general method, illustrated for a pair of atoms a and b is to form a wavefunction

$$\psi = C_a\varphi_a + C_b\varphi_b \tag{6}$$

as a linear combination of atomic orbitals φ_a, φ_b. Then using the hamiltonian $\mathcal{H}$, whose form can be specified later we must have

$$\mathcal{H}\psi = e\,\psi \tag{7}$$

where e is an energy eigenvalue. Inserting ψ from eq. 6 and multiplying through by φ_a or φ_b and integrating gives the two equations

$$\begin{aligned} C_a(e_a - e) + C_b(\beta - Se) &= 0 \\ C_a(\beta - Se) + C_b(e_b - e) &= 0 \end{aligned} \tag{8}$$

where $e_a = (\varphi_a|\mathcal{H}|\varphi_a)$, $e_b = (\varphi_b|\mathcal{H}|\varphi_b)$, $\beta = (\varphi_a|\mathcal{H}|\varphi_b)$, $S = (\varphi_a|\varphi_b)$.

These two equations each give a value of C_a/C_b; in order for them to give the same value, e must be restricted to the values given by the determinantal equation

$$\begin{vmatrix} e_a - e & \beta - Se \\ \beta - Se & e_b - e \end{vmatrix} = 0 . \tag{9}$$

If the values of e given by eq. 9 are inserted in eq. 8 and the normalization condition

$$C_a^2 + 2SC_aC_b + C_b^2 = 1 \qquad (10)$$

is used, then the individual values of C_a and C_b can be found. There are two values of e which fit the consistency condition, and therefore two sets of C_a, C_b. These define two molecular orbitals which are mutually orthogonal.

The algebra of M.O. theory becomes somewhat involved as even this simple example would show if the results were fully written out. For applications to transition metal atoms in crystals, however, some simplifications can be made because one of the coefficients will be much smaller than the other, and the overlap integral, S, is small. An added complication, however, is that an entire group of neighboring atoms must be taken into account. For example in an octahedral complex, MX_6^{n-}, the six neighbors are equivalent by symmetry and all have the same overlap with M. The partial ℓ.c.a.o.'s involving only the X-atoms can be grouped into symmetry-adapted combinations and then treated as single orbitals as in eq. 6.

The symmetry-adapted ℓ.c.a.o. M.O.'s for an octahedral array of closed-shell atoms around a transition metal atom are shown in Table 2. The atomic basis functions are the ns and np orbitals of the ligands and the 3d, 4s, 4p orbitals of the first row transition metal atom. Analogous orbital arrays could be constructed for other coordination numbers and types of atoms. The phase conventions for setting up the basic orbitals must be carefully defined. A first

Table 2. Orbitals for octahedral array of anions around a central cation. Col. 1 gives orbital symmetry. Symmetry-adapted basis functions are shown for central atom and outer atoms. These are combined to give molecular orbitals as in eq. 6, and they are symbolized in the last column.

Symmetry	Central atom	Outer atoms (ligands)	Combined orbital
A_{1g}	s	$\frac{1}{\sqrt{6}}(\sigma_x - \sigma_{\bar{x}} + \sigma_y - \sigma_{\bar{y}} + \sigma_z - \sigma_{\bar{z}})$	σa_{1g}, sa_{1g}
E_g	$de = (2x^2 - y^2 - z^2)$,	$\frac{1}{2\sqrt{3}}[2(\sigma_x - \sigma_{\bar{x}}) - (\sigma_y - \sigma_{\bar{y}} + \sigma_z - \sigma_{\bar{z}})]$	de_g, σe_g
	$(y^2 - z^2)$	$\frac{1}{2}[\sigma_y - \sigma_{\bar{y}} - \sigma_z + \sigma_{\bar{z}}]$	
T_{1u}	$p = (x,y,z)$	$\frac{1}{2}(\sigma_x + \sigma_{\bar{x}})$, etc.	pt_{1u}, πt_{1u}, σt_{1u}
		$\frac{1}{2}[(x_z + x_{\bar{z}}) + (x_y + x_{\bar{y}})]$, etc.	
T_{1g}		$\frac{1}{2}[(x_z - x_{\bar{z}}) - (z_x - z_{\bar{x}})]$, etc.	πt_{1g}
T_{2u}		$\frac{1}{2}[(x_z + x_{\bar{z}}) - (x_y + x_{\bar{y}})]$, etc.	πt_{2u}
T_{2g}	$dt = (xz$, etc.)	$\frac{1}{2}[(x_z - x_{\bar{z}}) + (z_x - z_{\bar{x}})]$, etc.	πt_{2g}, dt_{2g}

good guess as to the order of energy of these orbitals can be based on the reasonable assumptions that $E_s < E_p$ and $E_\sigma < E_\pi$ and finally the order of energies within, say, the set of π orbitals can be determined from the node patterns (11). Fig. 5 shows the node patterns of the orbitals of Table 2. The greater the number of nodes, the higher is the orbital energy. The highest energy ligand orbital should be the one labeled πt_{1g} since it is non-bonding with respect to the d-shell and has the maximum number of nodes between ligands. This should be the highest filled orbital and the one most easily excited in a charge transfer transition. Since it has even parity as do the d-orbitals, a $\pi t_{1g} \rightarrow nd$ transition is expected to be weak. This feature has been observed for example in the charge transfer spectrum of $IrCl_6^{=}$. A weak band at the beginning of the spectrum is identified with this forbidden transition (1). An energy level diagram giving a reasonable order of levels in an octahedral complex is shown in Fig. 6. The actual order of levels has been established experimentally for a number of octahedral second and third row transition metal halide complexes, and will be discussed later.

The linear combinations of the d- and ligand-orbitals of Table 2 may be described as bonding or as antibonding. The bonding orbitals are largely localized on the ligands while the antibonding ones are nearly pure d-orbitals. The $T_2\xi$ orbitals for example are

$$\psi(T_2\xi, b) = N_b^{-1/2}[\gamma_\pi \xi_d + \xi_\ell]$$

$$\psi(T_2\xi, a) = N_a^{-1/2}[\xi_d - \lambda_\pi \xi_\ell]$$

where γ_π and λ_π are positive and are small compared to unity. When these two orbitals are made orthogonal to each other one finds the relationship between γ and λ :

$$\lambda = (\gamma + S)\ (1 + \gamma S\ )$$

where S is the overlap integral: $S = \langle \xi_d | \xi_\ell \rangle$. The parameter γ is called the covalency parameter. A detailed discussion of covalency in this form is given in ref. 12, and will not be pursued here.

The complete specification of the number of electrons in each of the occupied molecular orbitals is called a configuration. Sometimes a configuration is a rather good description of a state, but often a mixture of several configurations is necessary for a reasonably accurate description. In the ground state, the ligand M.O.'s are filled, and the configuration of a complex such as NiF_6^{-4} would be, using the combined orbitals of Table 2 and the energy order of Fig. 6:

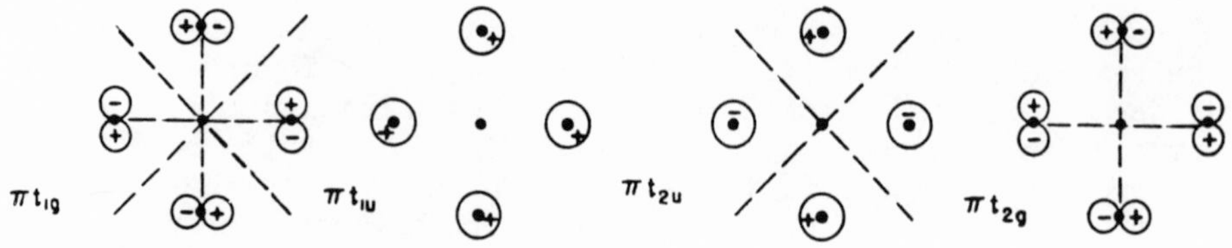

Fig. 5 Node patterns for one partner each of the π-type ligand M.O.s arranged in order from most weakly bound on left to most strongly bond on right, according to results shown in Fig. 7.

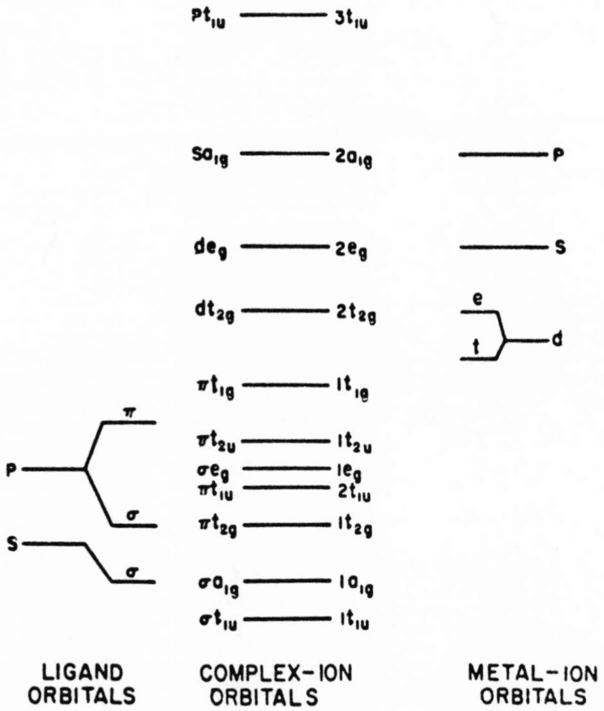

Fig. 6 Schematic M.O. energy diagram for octahedral complex. On left are strongly bound ligand basis orbitals, sσ, pσ, pπ and on right the more weakly bound 3d and unfilled 4 s and 4 p metal orbitals. The combined orbitals describing the complex are in the center. The order of binding energies is approximate. The experimentally determined order in one case can be deduced from Fig. 7.

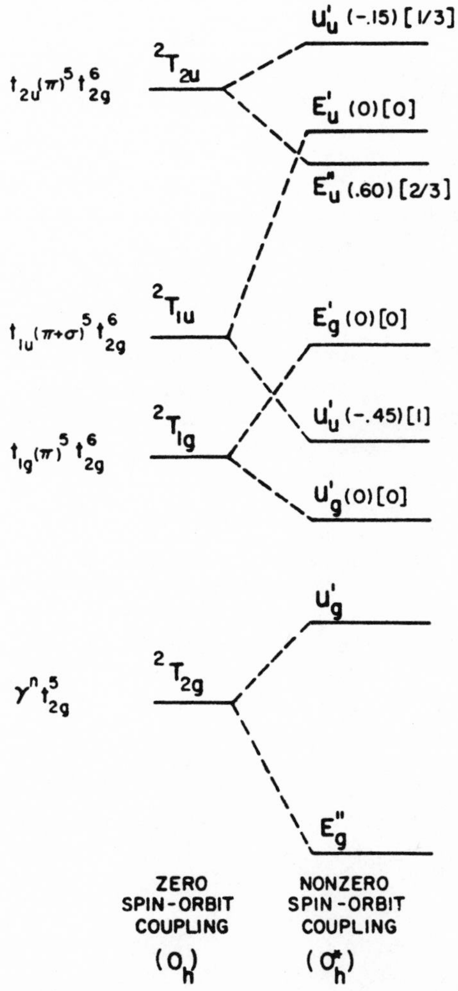

Fig. 7 Excitation energies of states of octahedral complex $IrBr_6^{2-}$ in Cs_2ZrBr_6 crystal (ref. 13), showing spin-orbit splitting.

$$(\sigma a_{1g})^2(\sigma e_g)^4(\sigma t_{1u})^6(\pi t_{2g})^6(\pi t_{1u})^6(\pi t_{2u})^6(\pi t_{1g})^6(dt_{2g})^6(de_g)^2$$

where the orbitals are listed from most tightly bound on the left to most weakly bound on the right. With the spins of electrons in $(de_g)^2$ parallel, this configuration gives the $^3A_2$ ground state of octahedral Ni^{++}. In a charge transfer transition an electron from one of the seven filled orbitals localized mainly on the ligands is promoted to the unfilled de_g orbital of the metal. This process leaves a hole in a ligand orbital and a hole in the de_g orbital.

The simplest examples of charge transfer processes are those in which there is only one open shell in each state of the transition: that is, the metal atom (acceptor) must have a single hole in an otherwise filled shell in the ground state, and in the excited state the hole transfers to the ligand orbitals. This special case will be discussed first.

2. The Case of One Open Shell. The halide complexes of tetravalent iridium have strong absorption bands in the visible region. The details of the spectra of these complexes have been worked out in a series of beautiful papers by Schatz and his coworkers (1).

Ir^{+4} has a $5d^5$ shell, and in a very strong crystal field the lowest term is $^2T_2(t^5)$. Because of the great radial spread of the 5d orbital (relative to the 3d) the ions of the third transition group experience large crystal fields and their ground states are derived from the lowest crystal field configurations, t^5 in this case. The acceptor-type charge transfer transition places an electron in the t^5 shell, filling it and leaving a hole in the ligand orbitals. Thus there is only one open shell in the ground and in the excited state.

As shown in Fig. 6, the most weakly bound of the ligand molecular orbitals is the πt_{1g} orbital whose nodal pattern shows it to be antibonding between all ligands (Fig. 5). The transition $\pi t_{1g} \rightarrow dt_{2g}$ is parity forbidden and therefore weak. The weak band in the spectrum of $IrCl_6^{4-}$ near 17000 cm^{-1} is attributed to this transition (1). The next absorption region is very strong and, according to Schatz et al., corresponds to the $t_{1u} \rightarrow dt_{2g}$ transition of Fig. 6. This transition is out of order on the energy diagram constructed from the node patterns, Fig. 5, yet the assignment based on the use of magneto circular dichroism, seems quite certain. The next two transitions, from the t_{2u} and t_{2g} π-levels, are in the expected order.

An energy level diagram of the $IrBr_6^{=}$ complex (13) showing the spin-orbit splitting is given in Fig. 7. This is a state-diagram, not an orbital diagram as is Fig. 6, and it represents the spectroscopic band positions. The spin-orbit product notation is that of Griffith (14). Next to each symmetry symbol is given the theoretical value of the C-term of M.C.D., and then the dipole strength. The agreement in sign and magnitude between the observed and calculated C-value as well as the agreement with the D-value confirm the spectroscopic assignments. Those states having D=0 are observed to have vibronically induced spectra, showing both the expected temperature dependence and the spectral structure. When the ligands are Cl^- instead of Br^-, the spin orbit splitting of the excited states is about a fourth as great, and the splittings shown in Fig. 7 are reduced accordingly.

3. More Than One Open Shell. The $OsX_6^=$ spectra look remarkably similar to those of $IrX_6^=$ (15). This is surprising because the ground configuration of Os^{+4} is t^4 and the acceptor-type excited states must belong to configurations $(L)^5(dt_{2g})^5$ and therefore there are two unfilled shells. For each ligand hole, say for example $(\pi t_{2u})^5$, there will be many states, as is shown by multiplying out the representations $t_{2u} \times t_{2g}$, giving four states, and when spin orbit coupling is included, there will be 8 states. The selection rules, however, permit only two of these states to have high intensities, the same number as in the corresponding configuration of $IrX_6^=$. This remarkable simplification applies to the other configurations of $OsX_6^=$ and is commonly observed in other complexes (3).

The electrostatic interaction splits the orbital components, but the magnitude of this splitting is not known. It is presumed to be small, since the $5dt_2$ and $np\pi$ orbitals are localized on different atoms, and furthermore and $np\pi$ hole is distributed over six halogen atoms.

In recent work, the magnitude of an electrostatic splitting in $Mo^{4+}(Cl)_6$ has been found (16). For the $t_{1u} \rightarrow t_{2g}$ excitation, the $^3T_{2u}$ - $^5T_{2u}$ splitting is close to 515 cm^{-1}. This is certainly small.

As a result of the work on the charge transfer transitions in these heavy metal systems, the order of orbital levels in octahedral complexes (averaged spin-orbit levels) has been found to be that shown on Fig. 7. This remarkable constancy may extend to other systems, as the level order depends more on ligand-ligand interaction than on the identity of the central metal atom.

V. CONCLUSIONS

We have a fairly good picture of the wavefunctions and states of acceptor-type complexes mainly as a result of the work with the heavy-metal systems. Less is known about high s-states of transition metal ions in crystals. These states must be somewhat delocalized, but it is not known how much. An interesting extension of work in charge transfer is the transfer of electrons between two metal ions; several such systems have been investigated.

REFERENCES

1. S.B. Piepho, J.R.D. Dickenson, J.R. Spencer, and P.N. Schatz, J. Chem. Phys. 57, 982 (1972).

2. M.M. Tippins, Phys. Rev. B1, 126 (1970).

3. B.D. Bird and P. Day, J. Chem. Phys. 49, 392 (1968).

4. C.K. Jorgensen, Electron Transfer Spectra in Progress in Inorganic Chemistry, Vol. 12 (S.J. Lippard, ed.), pp 101-58 Interscience Publishers, New York (1970).

5. D.S. McClure and Z.J. Kiss, J. Chem. Phys. 39, 3251 (1963).

6. J. Sabatini, Ph.D. Thesis, Princeton University, 1973.

7. E.U. Condon and G.H. Shortly, The Theory of Atomic Spectra, Cambridge University Press, 1935.

8. C.J. Ballhausen and H.B. Gray, Molecular Orbital Theory, W.A. Benjamin, Inc., New York, 1964.

9. T.F. Soules, J.W. Richardson and D.M. Vaught, J. Chem. Phys. 3, 2186 (1971).

10. L.J. Nugent, R.D. Baybarz, J.L. Burnett and J.L. Ryan, J. Phys. Chem. 77, 1528 (1973).

11. D.S. McClure, Electronic Spectra of Molecules and Ions in Crystals in Solid State Physics Vol. 9 (F. Seitz and P. Turnbull, eds.) Academic Press, 1959 .

12. S. Sugano, Y. Tanabe and H. Kamimura, Multiplets of Transition Metal Ions in Crystals, Academic Press, New York, 1970.

13. J.R. Dickenson, S.B. Piepho, J.A. Spencer and P.N. Schatz, J. Chem. Phys. 56, 2668 (1972).

14. J.S. Griffith, The Theory of Transition Metal Ions, Cambridge University Press, 1961.

15. B.D. Bird, P. Day and E.A. Grant, J. Chem. Soc. 100-109 (1970).

16. J.C. Collingwood, R.W. Schwartz, P.N. Schatz and H.H. Patterson, J. Chem. Phys., to be published.

THE ROLE OF THE JAHN-TELLER EFFECT IN THE OPTICAL SPECTRA OF IONS IN SOLIDS

Thomas L. Estle

Department of Physics, Rice University

Houston, Texas 77001

ABSTRACT

The Jahn-Teller theorem is stated and then illustrated by a simple two-dimensional model. The model is used to show the Coulomb origin of the effect and to discuss the static and dynamic Jahn-Teller effects. This is followed by a fuller discussion of the Hamiltonian describing a magnetic ion in a crystal and a presentation of the multidimensional configuration coordinate diagrams for actual systems of interest. These diagrams are then used to discuss the broad-band optical spectra which can occur. The vibronic level structure for electronically degenerate states is then discussed employing correlation diagrams and evidence of this structure is cited from selected optical spectra.

I. INTRODUCTION

The subject of this review is the Jahn-Teller effect (JTE) as it occurs for magnetic impurity ions in crystalline solids and the way the JTE influences the optical spectra of these ions. The Jahn-Teller effect occurs for orbitally degenerate electronic states. Similar physical features may result from nearly degenerate electronic states. The latter is sometimes referred to as the pseudo-Jahn-Teller effect. Of the 32 crystallographic point groups representing the possible symmetries of an impurity ion in a crystal, five correspond to cubic symmetries and either 2-fold or 3-fold orbital degeneracies are possible. Of the remaining 27, all of the hexagonal, trigonal, and tetragonal point symmetries (a total of 19) permit 2-fold orbital degeneracies. For ions with an incomplete d-shell in sites for which the symmetry is one of those just mentioned, orbitally degenerate states are comparable in number to nondegenerate states. Even for lower symmetries (orthorhombic, monoclinic and triclinic) near degeneracies often occur. To put this another way, the crystalline field symmetry is often approximately cubic and thus the abundance of orbital degeneracies for cubic symmetry suggests that the JTE or the pseudo-Jahn-Teller effect (the difference may be more semantic than physical) must be considered here as well. Thus we are left with the correct impression that to understand the energy levels and hence the optical spectra of transition-group ions having incomplete d-shells, we must understand the role of the Jahn-Teller effect. Indeed, optical spectroscopy is one of the most powerful methods of studying the JTE (electron paramagnetic resonance is the other principal technique).

The Jahn-Teller effect is not restricted to point defects which are impurity ions with incomplete d-shells. However we will concentrate our attention on such ions and virtually ignore ions with incomplete f-shells, such as the rare earths, because for them the spin-orbit interaction is large and the vibronic coupling is small. Under these circumstances the Jahn-Teller effect should be considerably less important. We will also restrict our considerations to isolated impurity ions.

Several recent reviews of various aspects of the JTE exist and should be consulted for more detail than can be included in this brief discussion (1-3).

II. THE JAHN-TELLER THEOREM

The Jahn-Teller theorem may be stated as follows: An orbitally degenerate state of a nonlinear molecule or crystalline defect is unstable against at least one asymmetric distortion which lowers the energy and removes some of the degeneracy. The Jahn-Teller theorem (4) is an existence theorem derived by group theory; no questions as to the magnitude of the effect or its physical origin are involved. When applied to any of the crystallographic point symmetries allowing orbital degeneracy it simply says that at least one distortion with the required properties must exist. Physically however it is important to know whether any such distortion is strongly coupled to the degenerate state. This vibronic (vibrational-electronic) coupling is normally quite strong because it arises from electrostatic forces between the d electrons and the charges of the crystalline environment. We will not try to understand this coupling in detail but will usually view it from the crystalline field point-of-view, even though more complicated molecular orbital and bonding concepts would clearly be required to obtain a quantitative understanding. In any case the coupling is Coulomb in nature. It is because of the Coulombic nature of the vibronic coupling that we restrict our consideration to orbital degeneracy. Spin degeneracy in the absence of orbital degeneracy will not lead to a Jahn-Teller effect even though the Jahn-Teller theorem has been extended to argue that modes exist to lift all but Kramers' degeneracy (5). The coupling is insufficient for this to occur.

III. A MODEL ILLUSTRATING THE JAHN-TELLER EFFECT

To get a clearer impression of the Jahn-Teller effect and its various characteristics, let us examine a simple two-dimensional model. Consider a binary ionic crystal with square symmetry, as shown in Figure 1 (this is the 2-dimensional analog of the NaCl structure).

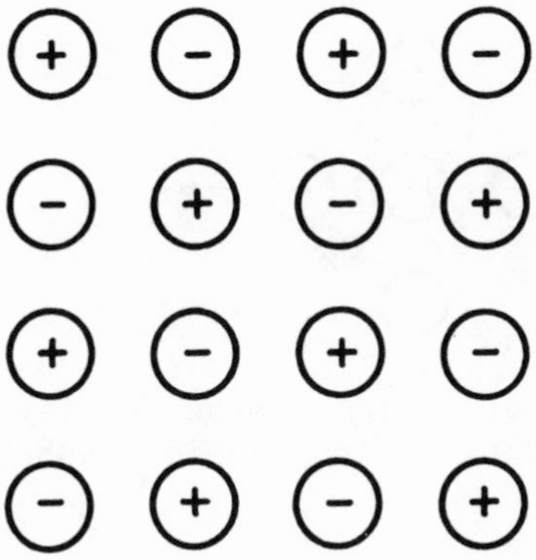

Figure 1 Two-dimensional binary ionic crystal.

Each ion is surrounded by four nearest neighbours or ligands with the opposite charge. For an impurity ion substituting for one of the positive ions of this crystal, consider the complex consisting of the impurity and the 4 negative ions which are its nearest neighbours. Since we are interested in understanding the effect of the motion of the ions on the energy of the impurity in the crystal, we focus our attention only on the closest ions and argue that the other ions play a less important role. The qualitative features are all we seek and they can be obtained from this five ion "molecular" complex.

Take the electronic configuration of the impurity to be one in which all shells are closed except one, which contains a single p electron. A p orbital is doubly degenerate in two dimensions. It simply means that there is one unit of orbital angular momentum for motion in the plane (i.e., about an axis normal to the plane). The rotation can be either clockwise or counter-clockwise, hence the 2-fold degeneracy. We can of course choose standing waves, which are linear combinations of the rotating waves, and obtain eigenfunctions which we can label as p_x and p_y. They are identical charge distributions except that they are rotated by 90°, a symmetry operation of the square symmetry. The orbital p_x has a greater charge density near the x-axis as illustrated in Figure 2.

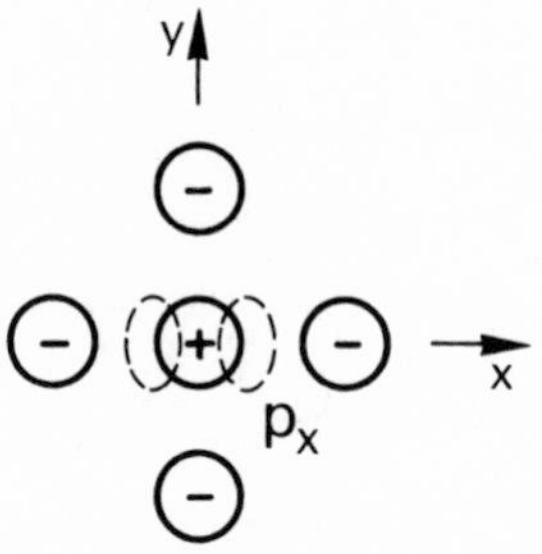

Figure 2 The regions of high charge density for a p_x orbital (indicated by regions inside dashed lines).

We wish now to investigate the effect of displacements of the ions on the energy of the complex. This includes the energy of the p electron plus the elastic energy. If we choose displacements using symmetry arguments then we must examine the effect of the normal modes of the complex. With five ions and two degrees of freedom for each, we would have a total of ten degrees of freedom. Three of these correspond to translation of the complex as a whole (2) and rotation of the complex (1). The seven possible resultant vibrational normal modes include a completely symmetric breathing mode, Q_1, transforming like the Γ_1 irreducible representation of the point group C_{4v}. There are two asymmetric modes, Q_3 and Q_4, transforming like Γ_3 and Γ_4 and the remaining 4 normal modes are a pair of doubly degenerate modes each transforming like Γ_5.

We now consider departures of the Q_i from their equilibrium values of $Q_i = 0$. (For a square crystal and the vast majority of impurities, all Q_i except Q_1 must have zero for their equilibrium values. We can choose the zero of Q_1 to be the equilbrium value since it is a motion which does not destroy the symmetry.) The elastic energy has a quadratic dependence upon Q_i (we will ignore higher powers for the present). In addition we must calculate the change in the Coulomb interaction of the p orbital with the neighbouring ions.

We will assume no overlap and consider the limit of small displacements (i.e., keeping only the contribution linear in Q_i in a power series expansion of the electrostatic energy).

In Figure 3 the normal mode Q_1 is shown and also the effect of $Q_1 \neq 0$ on the energy. Since Q_1 does not destroy the square symmetry, the 2-fold p degeneracy remains.

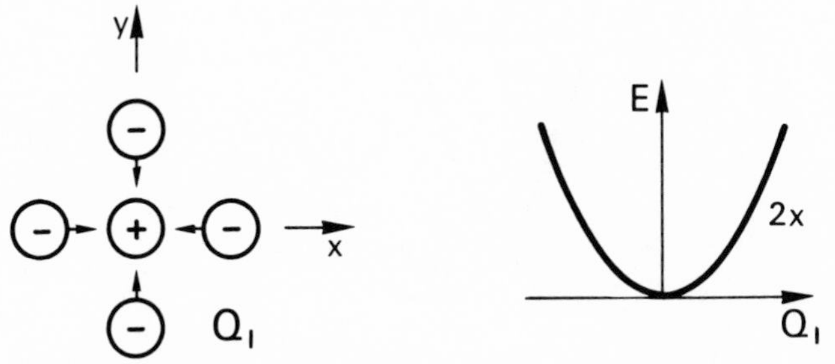

Figure 3 The breathing mode Q_1 and its effect on the energy of the complex (note that the double degeneracy remains).

By contrast if we consider the effect of the normal mode Q_3 the results are quite different. Near $Q_3 = 0$ the elastic energy will vary slowly since it is quadratic in Q_3. However, since this is a symmetry-lowering distortion it need not influence the two p states identically and a term linear in Q_3 arises from the Coulomb energy. This is illustrated in Figure 4. Because the p_x orbital is most sensitive to the motion of the ions along the positive and negative x axis we find that the energy of the p_x orbital increases as Q_3 increases and vice versa for $Q_3 < 0$. The p_y orbital behaves exactly the opposite. The energies change linearly in Q_3, as can be seen by expanding the 1/r factor of the Coulomb interaction in

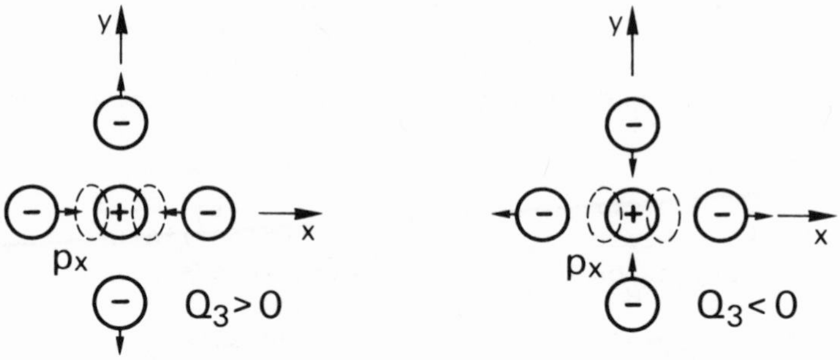

Figure 4 The normal mode Q_3. The energy of the p_x orbital increases for $Q_3 > 0$ and decreases for $Q_3 < 0$.

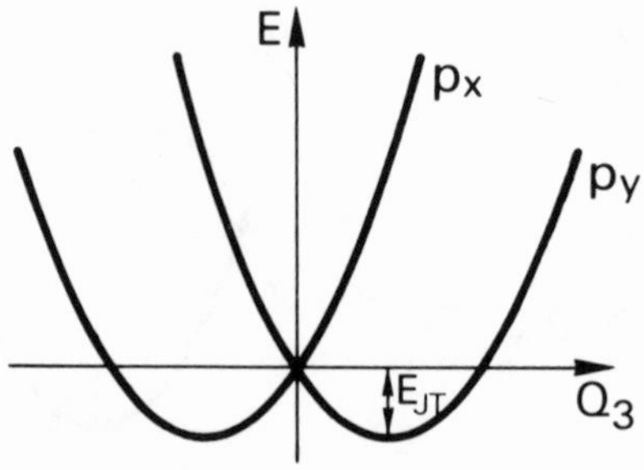

Figure 5 The effect of the normal mode Q_3 on the energy of the doubly-degenerate p state. The two curves are labeled by the appropriate electronic orbitals and the reduction in energy, E_{JT}, resulting from the coupling of the electronic state to the ionic displacements is shown.

a power series in Q_3. The sum of this linear Coulomb energy and the quadratic elastic energy yields a curve qualitatively like that of Figure 5.

Both the curve of Fig.3 and that of Fig. 5 are configuration coordinate diagrams; however the latter has many features which do not occur for orbital singlet states, which generalize to more realistic systems, and which characterize the Jahn-Teller effect. The more important features are:

1. The minimum energy is lower than the energy of the symmetric geometrical configuration by E_{JT}. This can be large since it results from the Coulomb force.
2. The electronic degeneracy is removed, except for no distortion ($Q_3 = 0$).
3. The energies vary linearly with Q_3 for small Q_3 but the mean energy changes little, hence one energy level decreases linearly with Q_3.
4. Two equivalent minima of the energy result, both with the same energy and equilibrium value of $|Q_3|$. They correspond to two different electronic states, p_x and p_y. Both the distortions and the electronic orbitals are related by a 90° rotation; i.e., they are equivalent because of the square symmetry of the problem.

At this point we should examine the degeneracy question more carefully. Have we actually removed the degeneracy because of the Jahn-Teller effect? We may regard the states of the system to be of two types. For one the system vibrates near the minimum in energy for $Q_3 > 0$ and the electronic state is p_y. For the other the vibration is about the other minimum and the electronic state is p_x . For zero-point vibration these two (vibronic) states are degenerate and thus, although we removed the original 2-fold electronic degeneracy, we end up with a 2-fold vibronic degeneracy. Thus the JTE does not actually remove degeneracy but rather it changes the degeneracy from electronic to vibronic. We are enlarging our basis states by adding vibrational degrees of freedom and changing the characteristics of the basis states in this way. However degeneracies will remain.

Of the two normal modes of the complex which we have considered only one led to the Jahn-Teller effect, that is to a splitting of the electronic state linear in the displacement. Of the remaining five normal modes, only one other has this property and it is the symmetry-lowering Q_4 distortion. Before we examine this mode and its effects when combined with Q_3, let us dispense with the Γ_5 modes. In Figure 6 we show two possible pairs of vibrational modes which have the required symmetry properties (i.e., they transform as x and y). Any actual Γ_5 vibration of such a complex would be some linear combination of the two motions pictured in Fig. 6.

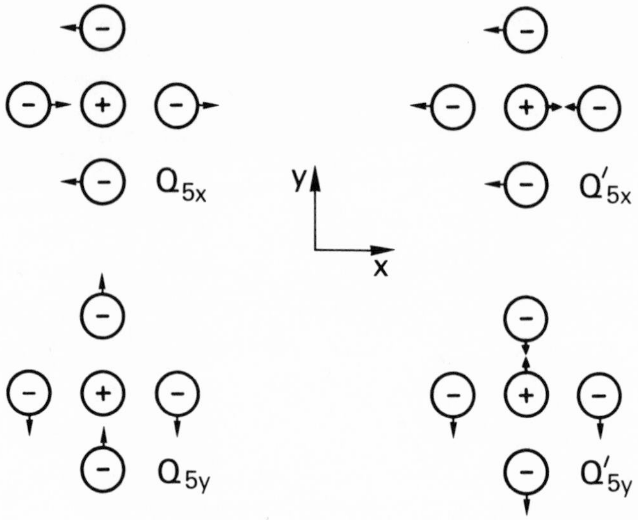

Figure 6 The pairs of Γ_5 normal modes. The two components of the doubly-degenerate modes are indicated by the subscripts x and y and the two different modes by the presence or absence of a prime.

If we now ask what a Γ_5 distortion will do to the energy of our complex, we can see from Fig. 7 that no change in the energy results which is linear in Q_i. Consider specifically the effect of Q_{5x} (as shown in Fig. 7). The effect of the motion of the negative ion lying along the positive x axis is to decrease the energy of the two p orbitals linearly in Q_{5x} (p_x being reduced more than p_y). However the effect of the motion of the negative ion lying along the negative x axis exactly cancels this linear variation and one is left with only a quadratic dependence on Q_{5x}, which for small displacements will be small. The effect of the two ions along the + and -y axes lacks a linear variation, since it is the angle between the + and - ions which changes in first order, not their separation. Thus no linear variation of energy occurs and no Jahn-Teller effect. Similar arguments can be made for Q_{5y}, Q'_{5x} and Q'_{5y}.

Let us now return to the mode Q_4 which is illustrated in Figure 8. In this case the two orthogonal p orbitals whose energies are changed as Q_4 varies are inclined at 45° with respect to the x and y axes. The mode Q_4 is not as effective

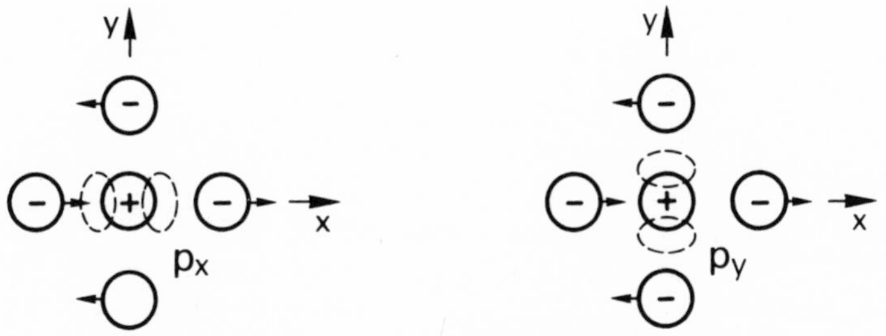

Figure 7 The geometrical relationships of the p_x and p_y orbitals to the Q_{5x} mode of Fig. 6.

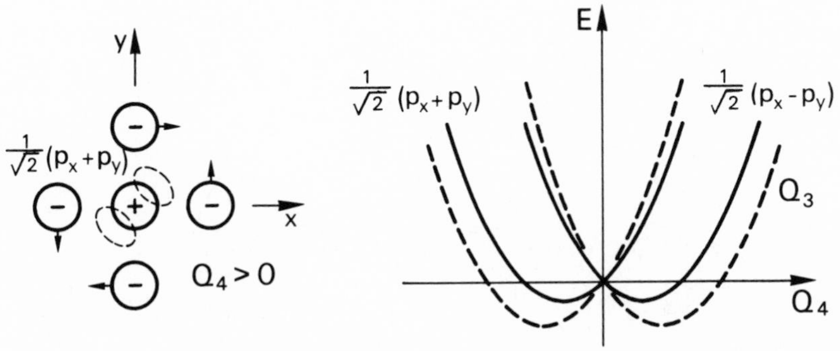

Figure 8 The normal mode Q_4. The energy of the p orbital at an angle of 45° with the x axis, as shown, is increased if $Q_4>0$, whereas the orbital at −45° has its energy decreased. The variation of the energy with Q_4 is shown on the right. The dashed curves are the corresponding results for Q_3 shown in Fig. 5.

in increasing or decreasing the energies of these p orbitals as was Q_3 for the orbitals p_x and p_y. This is simply a consequence of the geometry. Thus the resultant configuration coordinate diagram of E vs. Q_4 is similar to that for Q_3 (see Fig. 5), but the stabilization energy, E_{JT}, is not as great and the equilibrium distortion is smaller (compare the solid and dotted curves on the right side of Fig. 8). The change in energy is however still linear in Q_4, as it was for Q_3, for small distortions. Thus the qualitative features are the same as for Q_3 except the p orbitals are rotated by 45° and, of course, the distortion is different.

Distortions which are a combination of Q_3 and Q_4 may also occur and these will result in energy diagrams qualitatively similar to Figs. 5 and 8. The electronic states associated with the two energies will be appropriate linear combinations of p_x and p_y, the linear combination depending on the relative amount of Q_3 and Q_4 in the distortion. We can thus construct a configuration coordinate diagram for a two-dimensional configuration coordinate space, Q_3 and Q_4 . This is essentially a plot of E versus Q_3 and Q_4 and we will represent some of the properties of this three-dimensional object by drawing contours of constant E in Q_3 - Q_4 space. For simplicity and so as to deal with only the lowest electronic state, we restrict ourselves to E < 0. We show such a sketch in Fig. 9.

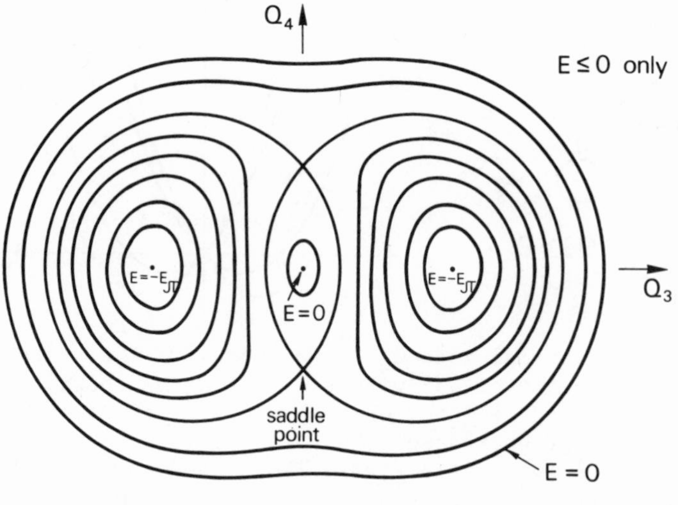

Figure 9 Contours of constant energy for E less than zero, its value for Q_3=Q_4=0. There are two minima along the + and - Q_3 axis, two saddle points along the + and - Q_4 axis, and a maximum(for this range of energies) at the origin. The energy of the saddle points or barriers is simply the negative of E_{JT} for the configuration coordinate Q_4 alone (see Fig. 8).

We must keep in mind that the electronic state for the minimum with $Q_3 > 0$ is the p_y state and that for the minimum with $Q_3 < 0$ is the p_x state. If the two minima are sufficiently deep, then the two lowest (vibronic) states of the system may be regarded a nearly-p_y electronic state times the zero-point vibrational state near the $Q_3 > 0$ minimum and a nearly-p_x electronic state times the zero-point vibrational state near the $Q_3 < 0$ minimum (note that we have explicitly used the fact that what we have heretofore called the energy is actually just the potential energy when the nuclear kinetic energy is added and the Born-Oppenheimer approximation is valid). This is sometimes referred to as the static Jahn-Teller effect. Viewed this way the system has distorted in either of two equivalent ways and the charge distributions of the electrons are different, but equivalent. In one case the charge is primarily along the y axis, in the other it is primarily along x.

A particularly simple version of the (dynamic) Jahn-Teller effect occurs if we make the unphysical assumption that Q_3 and Q_4 are equivalent. In other words their coupling to the p orbitals are equal and the force constants are the same. Thus the configuration coordinate diagrams would be the same for the two and hence also for any linear combination of the two. In this case the two-dimensional configuration coordinate diagram is rotationally invariant about the E axis. Such a diagram is sketched in Fig. 10.

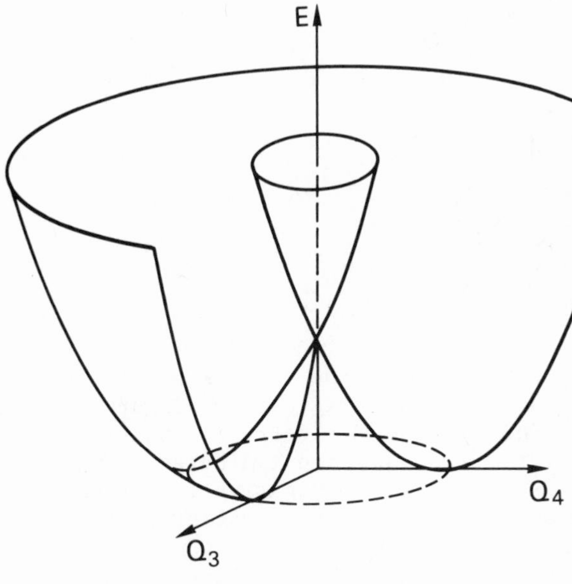

Figure 10 Two-dimensional configuration-coordinate diagram for the case that Q_3 and Q_4 are equivalent.

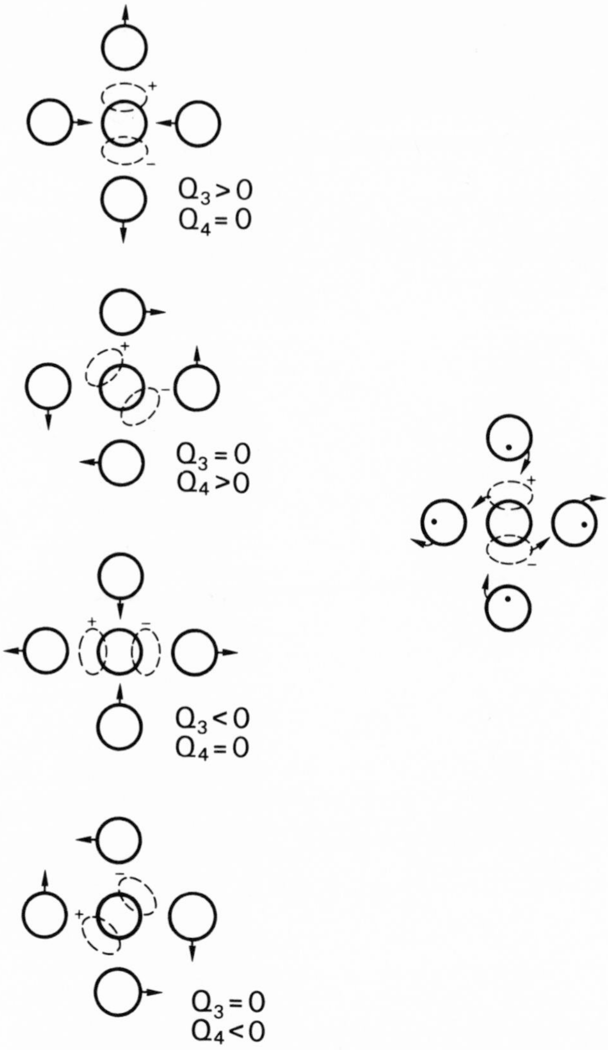

Figure 11 Motion of the ions of the complex resulting from a CCW motion about the origin in configuration- coordinate space. The associated electronic state is shown with the phase of the wave function indicated by the signs. Note that a CW motion of the negative ions results in a CCW motion of the p orbital and that after a 360° rotation the sign of the electronic wave function has changed (i.e., the electronic charge distribution rotates at half the angular velocity that the neighboring ions do).

If we now ask for the vibrational motion in such a potential, we conclude that harmonic-oscillator-like radial vibrations occur but that the angular motion is unhindered. Thus each radial-vibrational state will have associated with it a series of more closely spaced rotational levels. Since rotation can occur in either a CW or CCW sense, each rotational level will be 2-fold degenerate. Radial motion in Q_3-Q_4 space does not change the electronic state but angular motion does. This is illustrated in Figure 11 for the case of a CCW motion about the trough of minimum energy. Because a rotation of 360° about the origin in Q_3-Q_4 space rotates the p wave function by only 180° we find that the electron wave function changes sign, hence so must the "rotational" vibrational function in order that the product vibronic function be unchanged (i.e., single-valued). Thus the rotational states correspond to odd half-integral values of the quantum number analogous to angular momentum.

IV. THE HAMILTONIAN FOR A MAGNETIC ION IN A CRYSTAL

Let us now return to the problem of a magnetic ion in a crystal and examine it less superficially than we have up until now. We shall start by writing an approximate Hamiltonian for the problem. Making the crystal-field approximation, we may write this Hamiltonian as

$$H = H_o + V_c + H_{JT} + V_{VIB} + T_n + H_{SO} + V_c' + H_p \ .$$

The first term, H_o, is the free-ion electrostatic Hamiltonian (to the accuracy of the present analysis it is the free-ion Hamiltonian without spin-orbit interaction; i.e., it is the sum of the kinetic energy of the electrons of the ion plus the Coulomb interaction of the electrons with the nucleus and with each other). The second term, V_c, is the static crystal field or in some cases the largest (cubic) part; the smaller lower-symmetry crystal fields being represented by the quantity V_c'. The third term, H_{JT}, is the electron-vibrational (vibronic) coupling, which led to the linear dependence of E on Q_i in our model. The fourth term, V_{VIB} is the vibrational potential energy or elastic energy and the fifth term, T_n, is the kinetic energy of the nuclei or the vibrational kinetic energy. The sixth term, H_{SO}, is the spin-orbit interaction and the last term, H_p, represents a variety of small perturbations to the earlier terms. Examples of these perturbations are the Zeeman interaction, interactions with internal or applied strains, and the hyperfine interaction. Since our interest is primarily in impurity ions with incomplete d shells, we may start our analysis by considering the first two terms, $H_o + V_c$. The solutions to the resulting eigenvalue problem give us the electronic states in the static crystal field approximation neglecting at present both spin (i.e., H_{SO}) and possible weak crystal fields (i.e., V_c').

If we now consider only one degenerate electronic state at a time, we can add this electronic energy to the vibronic coupling, H_{JT}, and the elastic energy, V_{VIB}, and in this way obtain in general a multidimensional configuration-coordinate diagram (the electronic energy in this case corresponds to the equilibrium nuclear positions for zero vibronic coupling). We will also assume that we need consider only the normal modes of the impurity and its ligands (since these would normally have the largest effect by electrostatic arguments). We will usually ignore all of these normal modes except those giving a splitting of the degenerate levels, i.e., which lead to the JTE.

V. CONFIGURATION-COORDINATE DIAGRAMS

Configuration-coordinate diagrams are frequently employed to explain various aspects of optical spectra. Frequently these involve only a single configuration coordinate and nondegenerate states. Here we describe, but do not derive, the corresponding diagrams for degenerate states coupled to modes which yield the JTE. These diagrams must in general be multidimensional; i.e., 2 or more configuration coordinates are involved.

V.A. E Electronic States in Cubic Symmetry

If we consider first the case of an orbital doublet in cubic symmetry, then its coupling with a doubly-degenerate (E) mode of its ligands will yield the JTE. If we consider only linear vibronic coupling and quadratic elastic energy then the resultant configuration coordinate diagram is shown in Fig. 12. There we also show the two orthogonal components used to describe the E mode for the case of octahedral coordination and the corresponding charge distributions for a doubly degenerate d orbital (recall that a d orbital breaks up into a doublet and a triplet in a cubic environment). The configuration-coordinate diagram so obtained is the same as that of Figure 10 which we obtained by making a simple but unphysical assumption concerning the model presented in section III. In the present case the result is a consequence of symmetry as long as nonlinear coupling and anharmonicity of the elastic energy are both neglected. The properties of this "Mexican Hat" are much like those we obtained in section III. The electronic state depends on θ

$$\theta = \tan^{-1} \frac{Q_\varepsilon}{Q_\theta}$$

but not on Q. The minimum energy again occurs for a fixed value of Q $\left(Q = \sqrt{Q_\theta^2 + Q_\varepsilon^2}\right)$ but any angle θ. Again we have radial vibrations and doubly degenerate rotational states. The rotational states again correspond to half-integral "angular momentum" quantum numbers, although the demonstration in this case would not be as easy as for our simple model.

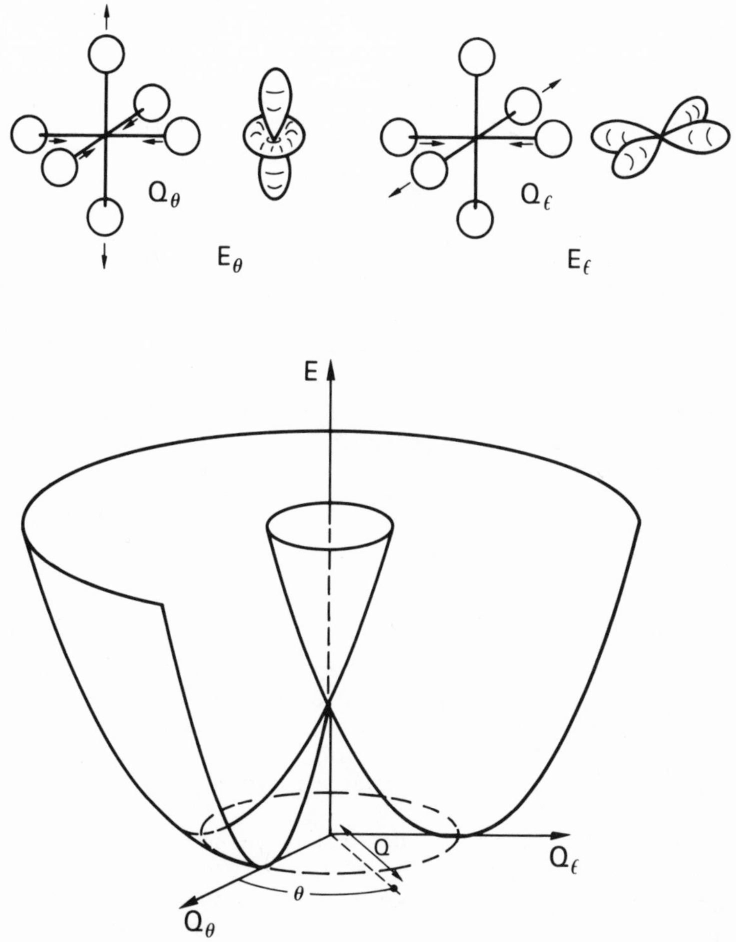

Figure 12 The distortions and d orbitals which transform as the E irreducible representation of a cubic point group. The resultant configuration-coordinate diagram for linear vibronic coupling is shown. A point in Q_θ-Q_ε space can be specified by its polar coordinates, Q and θ, as shown.

These features are also common to orbital doublets for trigonal and hexagonal symmetry. Orbital doublets in tetragonal symmetry behave as did our two-dimensional model of section III before we made the unphysical assumption of the equivalence of Q_3 and Q_4. Thus there are two minima in this case even for linear coupling.

Let us now consider the behavior of the doublet state in cubic, trigonal, or hexagonal symmetry when we include significant non-linear coupling in H_{JT} and, in addition, anharmonic terms in V_{VIB} are included. Then our simple configuration-coordinate diagram is altered somewhat. In particular the trough of lowest energy is no longer flat (i.e., independent of θ) but is warped in such a way as to have 120° periodicity. Thus the constant energy contours would appear as in Figure 13. The case of slight warping and a nearly flat trough leads to only slightly perturbed "rotational" behavior and corresponds to the dynamic JTE. A highly warped trough which localizes vibrations near the three minima leads to the static JTE. Note in the static case the 3 minima imply a three-fold degeneracy for very high barriers, whereas a two-fold degeneracy arising from the "rotational" motion characterizes the ground vibronic state for the dynamic JT effect. We will return to this point in section VII. It is, of course, necessary that behavior intermediate between these two extremes also occurs.

V.B. T Electronic States Coupled to E Modes

An orbital triplet (T) state can only occur in cubic symmetry. However it can couple, in a Jahn-Teller sense, to both E and T modes. For many systems one or another of these couplings appears to predominate and this is more frequently to an E mode. Thus we consider only coupling to one E mode at first.

For an electronic doublet, as described in section V.A., the configuration-coordinate diagram is double-valued; i.e., for any point in Q_θ - Q_ε space there are two values of energy or two sheets. For an electronic triplet there are three sheets and, for linear coupling to a single E mode and no vibrational anharmonicity, these are 3 identical disjoint,displaced paraboloids of revolution. This is illustrated in Figure 14. The paraboloids have the same shape as for zero coupling. The minima correspond to tetragonal distortions. The static Jahn-Teller effect does not require nonlinear coupling in this case, as it did for orbital doublets. Apart from the obvious differences in the shapes of the surfaces for orbital doublets and triplets, another very important difference

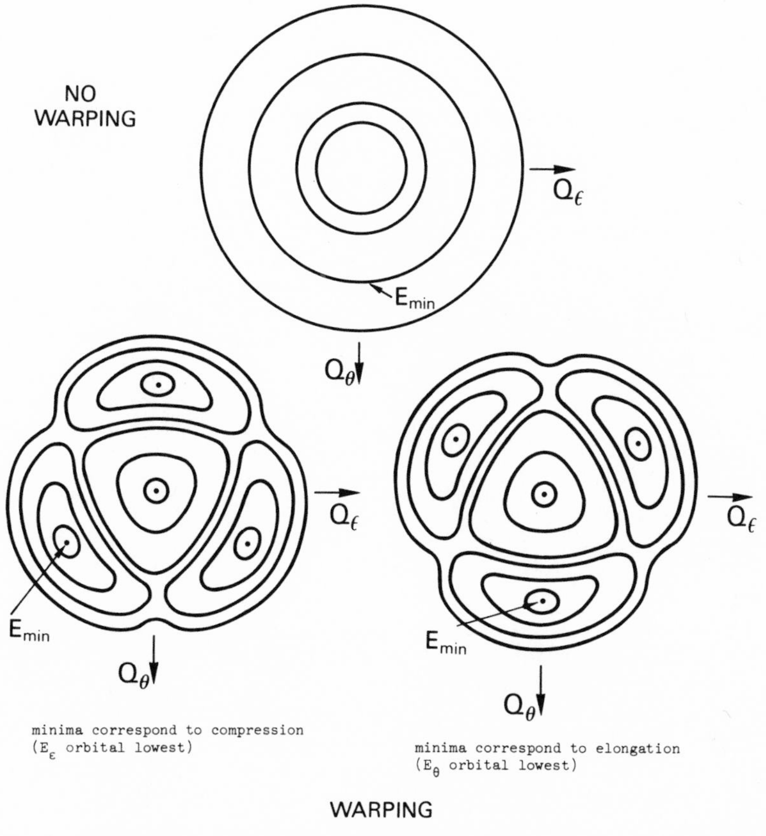

Figure 13 Constant energy contours for an E-state configuration-coordinate diagram without and with warping from non-linear coupling and anharmonicity.

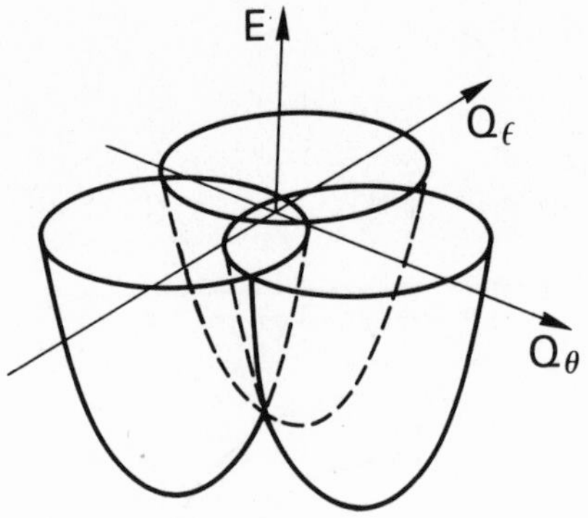

Figure 14 Configuration-coordinate diagram for a T electronic state coupled to an E normal mode for linear coupling and harmonic vibrations.

occurs. For any one of the three paraboloids the electronic wave function is independent of Q_θ and Q_ε and orthogonal to the electronic wave functions for the other two paraboloids.

V.C. T Electronic States Coupled to T Modes

When we attempt to consider the next simplest problem, coupling of the orbital triplet state to a T mode, we reach a substantial conceptual obstacle since we must now plot E against 3 coordinates and this requires a four-dimensional space in which to make this construction. Various 2- and 3-dimensional cuts through this 4-dimensional space are possible but it is no longer easy to visualize the results. We will therefore not attempt a discussion of the configuration-coordinate diagram but will make only two qualitative comments. First, the minima in energy correspond to trigonal or <111> distortions and without tunneling a 4-fold-degenerate vibronic ground state results. Second, the electronic wavefunction for any of the 3 sheets depends on the position in configuration-coordinate space (as was the case for an E state but not for a T state coupled to an E mode).

The problem of comparable coupling of an orbital triplet to one E and one T mode is considerably more complicated yet and no attempt to discuss this problem will be made. Both it and the coupling to 2 or more modes of the same symmetry would lead to n-dimensional configuration-coordinate diagrams with $n > 4$ and hence would be still more difficult to visualize.

VI. BROAD BAND OPTICAL SPECTRA

VI. A. Introduction

If no vibronic structure is resolved, but only broad bands are seen in absorption, emission, or Raman scattering, then one can employ the Franck-Condon principle and semiclassical arguments to infer many features of these optical spectra. The arguments proceed much as they do for the more familiar configuration-coordinate diagrams involving a single configuration coordinate. We may classify the types of spectral bands which may be obtained into three major types depending on the orbital degeneracies involved. The first and simplest are singlet-singlet transitions. These have been discussed extensively elsewhere and they do not involve the JTE. We will be primarily concerned with distinguishing transitions involving orbital multiplets from these. The primary feature which we will use for this purpose is that such transitions lead to a single broad band, although such characteristics as the shapes and Stokes' shifts are very useful also.

The second type of spectral band consists of singlet-multiplet transitions. The features of such transitions result by considerations of the configuration-coordinate diagram for the multiplet level and the paraboloid with its minimum at the origin of configuration-coordinate space for the singlet (the paraboloid cannot be displaced since the equilibrium corresponds to the high symmetry). We will make the reasonable assumption that the elastic energy is independent of the electronic level involved although most of our qualitative conclusions would be unchanged even if this were not the case.

The final type of spectral band involves transitions between two multiplet levels. In this case the spectra could become rather complex in principle. However in practice it appears that the coupling is much stronger to one of the two multiplets and thus we will treat the more weakly coupled multiplet as if it were a singlet. This case then becomes approximately the same as the previous one. Hence we will only discuss transitions involving a singlet and a doublet or a triplet.

VI. B. Transitions from a Doublet

To be specific let us take the doublet to be the lower state (usually the ground state) and discuss absorption. The configuration-coordinate diagram for the two levels must be obtained for a two-dimensional configuration-coordinate space, Q_θ and Q_ε. However it consists of two rotationally-symmetric diagrams, that for the doublet as shown in Figure 12 and a paraboloid of revolution for the singlet (A). We can plot this as a function of the magnitude of the distortion, Q, and obtain the entire diagram by rotating about the E axis. This is shown in Figure 15.

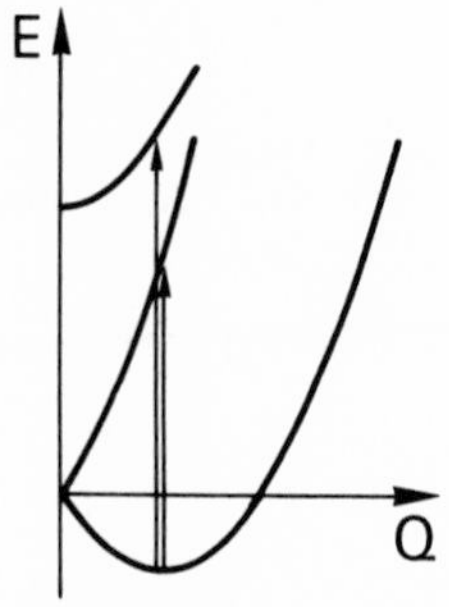

Figure 15 Configuration-coordinate diagram for E-A transitions. The actual diagram is three-dimensional and results from rotation of this figure about the E axis. The two possible absorption bands are shown schematically by the arrows.

Two possible absorption bands could occur as shown in Fig. 15. The higher energy one is the only actual transition between the E state and the A state (or a weakly coupled multiplet). The lower energy transition is between the two sheets of the E state diagram. Sturge (1) discusses the experimental evidence for the latter transition, however a definitive identification is difficult. Without it there is little difference between this case and transitions between two orbital singlets.

VI.C. Transitions to a Doublet

The configuration-coordinate diagram which we need is identical, except possibly for level ordering, to that in Fig. 15. In fact we may use that diagram as it is and consider emission from the upper, singlet level. In that case we might naively expect a single band as suggested by the arrow of Figure 16(a). However, we must realize that we are dealing here with a 2-dimensional

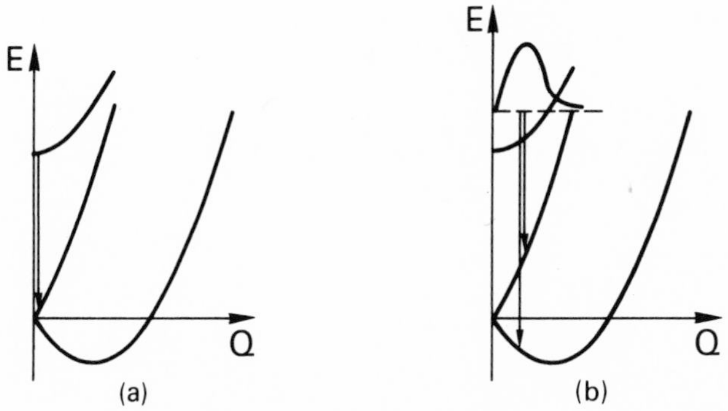

Figure 16 An illustration of why two bands may be observed for transitions from orbital singlets to orbital doublets. This results from the peak in the probability for non-zero Q as shown in (b) rather than the case shown in (a).

configuration-coordinate space and for the singlet the probability is not a maximum for $Q = \sqrt{Q_\theta^2 + Q_\varepsilon^2} = 0$ but rather for a non-zero Q as indicated by the dotted distribution curve for the zero-point vibrational state in Figure 16(b). Thus, as suggested by Fig. 16(b), there are two resolved broad bands. Experimental evidence up to about 1967 for this is also discussed by Sturge (1).

VI. D. Transitions Involving an Orbital Triplet Coupled to an E Mode

The actual A-T transitions are simply the superposition of three identical spectra corresponding to transitions to or from the three identical displaced paraboloids of Figure 14. No structure results because this differs only slightly from the case of A-A transitions analyzed for one configuration-coordinate. However it should be possible to see transitions between different paraboloids, i.e., transitions which do not involve the singlet. Nygren, Vallin, and Slack (6) have so interpreted the broad absorption band near 1000 cm^{-1} in ZnSe containing Cr^{2+}.

VI. E. Other Possibilities

It is not easy to discuss transitions involving orbital triplets if the coupling is to a T mode or to both E and T modes because of the absence of a simple configuration-coordinate diagram. Nevertheless, the simple situation of disjoint paraboloids which led to no structure for A-T transitions does not exist for these cases and broad-band structure can occur.

We should point out that we have ignored the spin-orbit interaction in our discussion of triplets so far. We can do so for doublets because no first-order effects occur, but we must be more careful for orbital triplets. The spin-orbit spitting may itself produce structure but additional structure may also result as can be seen by considering an orbital triplet coupled to an E mode. We indicated in section VI.D. that no structure should occur for A-T transitions if we ignore, as we did there,the spin-orbit interaction. If, on the contrary, the spin-orbit interaction is large, we may have the situation shown in Figure 17. In this case the

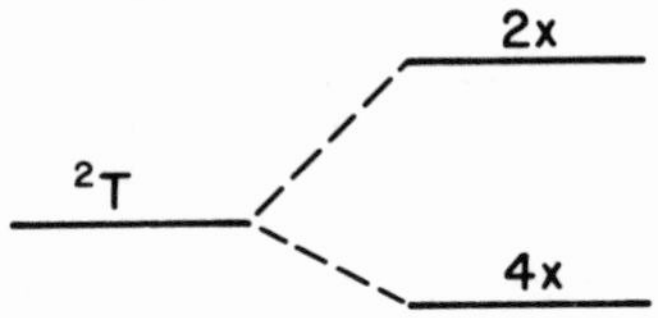

Figure 17 Splitting of a ^2T state by spin-orbit interaction. The degeneracies of the resultant states are shown.

resultant 4-fold degenerate state acts much like the orbital doublet (E) states we have considered earlier, hence we can have broad-band structure from that source as well as the spin-orbit splitting.

VII. VIBRONIC STRUCTURE IN ENERGY LEVELS AND OPTICAL SPECTRA

Considerably more information can be obtained from optical spectra in which the vibronic structure is resolved. The analysis of the theory, particularly for the cases of greatest interest, is much more difficult. We will approach this subject by using correlation diagrams to draw inferences about the behavior of systems of interest. The limits of these diagrams will be cases for which the theory is relatively simple. We will also mention experimental evidence for most of the interesting features. We will treat E states and T states separately.

VII.A. Vibronic Structure of E States

We will consider the case of an orbital doublet in cubic symmetry (the latter just for simplicity). The vibronic eigenvalue problem to be solved results from the Hamiltonian (see section IV)

$$H = [(H_o + V_c) + H_{JT} + V_{VIB}] + T_n$$

where V_c' is not present because of the cubic symmetry and H_{so} and H_p have been neglected because their effects are small (for H_{so} this arises because there is no first-order effect). The quantity in parentheses gives the orbital-doublet state of interest (a typical static crystal field calculation). Adding to it $H_{JT}+V_{VIB}$ gives the quantity in brackets and leads to the configuration-coordinate diagrams already discussed in section V.A. Thus we now wish to add the nuclear kinetic energy, T_n, to our configuration-coordinate-diagram description and determine the vibronic states. We will consider two simple limits depending on whether the effects of H_{JT} or T_n dominate. If H_{JT} is dominant and linear we have already suggested the results. In this case the Born-Oppenheimer approximation is valid and we have radial vibrations and rotational motion in the trough. For the opposite limit we merely ignore H_{JT} and then we have independent, i.e., uncoupled, vibrational and electronic states. Intermediate between these limits the behavior is more complicated (because, for example, the Born-Oppenheimer approximation does not hold). However we can get an impression of the behavior over the whole range of coupling strengths by examining the correlation diagram of Figure 18. The diagram consists of a succession of vibronic doublets with simple spacings only in the two limits. The ground state is always an E state, i.e., a doublet caused by symmetry rather than an accidental doublet consisting of an A_1

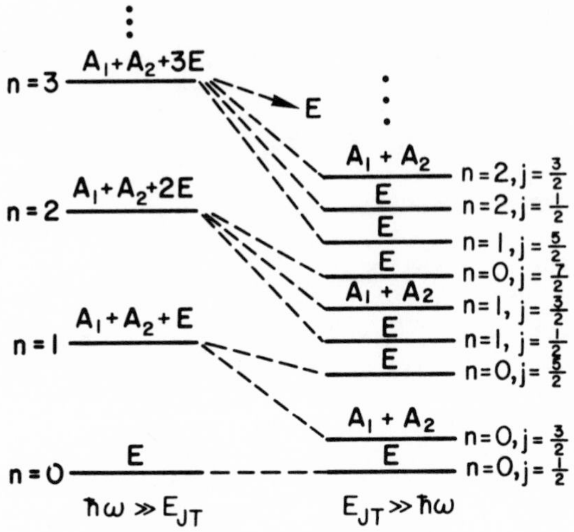

Figure 18 Correlation diagram for an orbital E state linearly coupled to an E vibrational mode. On the left the coupling is negligible and the states are the vibrational states of a two-dimensional simple harmonic oscillator times an electronic E state. On the right the levels are those for radial vibration with rotational structure because of the angular motion in Q_θ - Q_ε space. The rotational splitting is taken larger than the approximations allow in order to make the diagram easier to see. The resultant vibronic states are labeled by their transformation properties under the octahedral point group; A_1 and A_2 are singlets and E is a doublet. The symmetry assignments involve some group theory and/or physical arguments which are straightforward but lengthy and hence will not be discussed here.

and an A_2 state (the accident being the linear coupling).

If we now introduce the nonlinear coupling (and vibrational anharmonicity) but consider the strong-linear-coupling limit (the right side of Fig. 18), we obtain the correlation diagram of Figure 19. This is shown for only the ground radial vibrational state. The limit on the left of the figure is for negligible

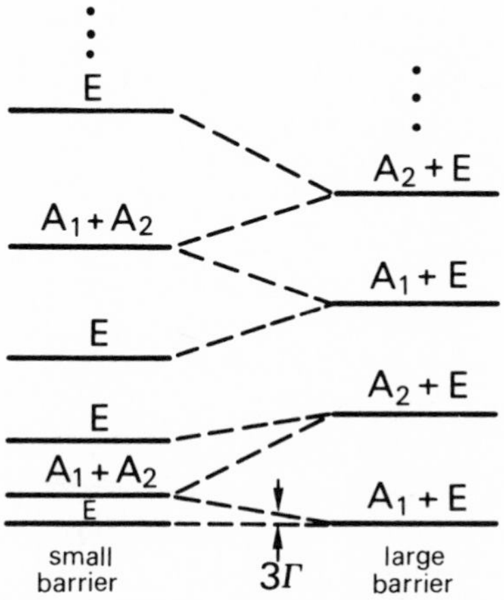

Figure 19 Correlation diagram for an orbital E state coupled to an E vibrational mode. A strong constant linear coupling is assumed and the barrier resulting from nonlinear coupling and vibrational anharmonicity is varied from very small to very large relative to the rotational level spacing. The levels on the left correspond to the rotational levels in the trough. Those on the right represent the angular vibrations about the three minima and would presumably have an even larger splitting than shown. The accidentally-degenerate triplet ground state on the right can be either an A_1+E or an A_2+E state depending on whether the energy minima represent tetragonal compression or elongation (see Figure 13). The spacing between the ground doublet and the first excited singlet is referred to as the tunneling splitting and is designated 3Γ.

nonlinearity or barrier whereas that on the right is for a barrier much larger than the rotational energy spacing. The small-barrier limit corresponds to the dynamic Jahn-Teller effect, i.e., unhindered rotation. The large barrier limit corresponds to the static

Jahn-Teller effect. In the intermediate region a singlet state approaches (as the barrier gets larger) the ground doublet. Their spacing, 3Γ, is the tunneling splitting. The accidentally-degenerate triplet ground (and excited) state in the static limit corresponds to the 3 possible static tetragonal distortions.

VII.B. Optical Evidence of E State Vibronic Structure

Two examples of this structure will be mentioned. The tunneling splitting between the ground vibronic doublet and the first excited singlet was observed in the emission spectra of Eu^{2+} in CaF_2 and other alkaline earth fluorides by Kaplyanskii and Przevuskii (7). The transition studied is from the $4f^6 5d^1$ configuration to the $4f^7$ ground configuration of the Eu^{2+} ion. The initial state is an orbital doublet because of the single 5d electron. The tunneling splitting is observed to be 15.3 cm^{-1} for CaF_2. Subsequently Chase (8) confirmed this by electron paramagnetic resonance studies of the excited E state.

The second example involves Raman scattering from an ion with an orbital E ground state, Cu^{2+} in CaO. In this case Guha and Chase (9) report observation of a number of excited vibronic states and they were able to fit their data quantitatively with an earlier calculation of O'Brien (10) for the intermediate region.

VII.C. Vibronic Structure of T States

To the Hamiltonian used to discuss orbital doublets in section VII.A., we must now add the spin-orbit interaction, H_{so}, in order to be able to analyze triplet states (since H_{so} produces a first order splitting of a triplet)

$$H = [(H_o + V_c) + H_{JT} + V_{VIB}] + T_n + H_{so} \quad .$$

We now have 3 parameters which characterize the spectrum of energy levels. They are the uncoupled normal-mode vibrational quantum, $\hbar\omega$, and the Jahn-Teller stabilization energy, E_{JT}, both of which occurred for the doublet problem as well (see Fig. 18), plus the spin-orbit coupling parameter λ. Several easily soluble limiting cases exist and we will correlate these to infer intermediate behavior.

We shall only consider the simplest case, an orbital triplet with a spin of 1/2. Since optical phonon energies are roughly a few hundred cm^{-1}, a value comparable to λ for iron group ions, we will consider the case for which λ and $\hbar\omega$ are comparable. The resultant correlation diagram is shown in Figure 20. No spin-orbit splitting occurs for the strong-coupling limit,

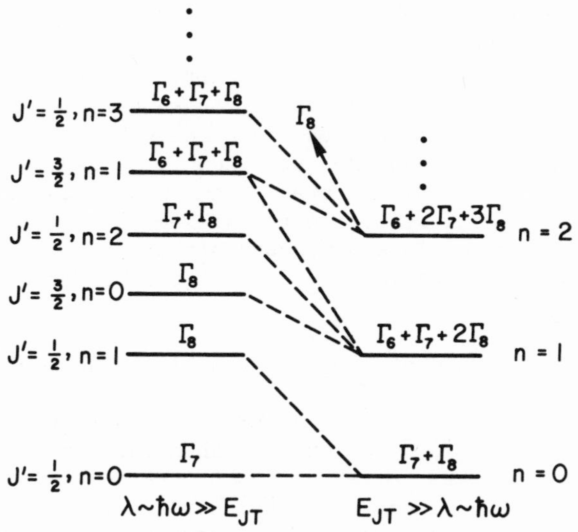

Figure 20 Correlation diagram for an orbital T state with s = 1/2 linearly coupled to an E mode. The states are labeled according to their symmetry including operations on the spin. The symbols Γ_6 and Γ_7 label Kramers' doublets and Γ_8 is a quartet. The right side is the strong-coupling limit and the states are simply the vibrational states associated with the three paraboloids times the two-fold-degenerate spin state times the appropriate orbital state. No spin-orbit splitting occurs, as explained in the text. The weak-coupling limit corresponds to independent electronic and vibrational behavior. Thus the energy levels correspond to vibrations in an undisplaced paraboloid with a spin-orbit splitting superimposed.

$E_{JT} \gg \lambda \sim \hbar\omega$, since the matrix elements of $\underline{L}\cdot\underline{S}$ are off-diagonal in the electronic states corresponding to the three disjoint paraboloids. As a consequence the matrix elements of $\underline{L}\cdot\underline{S}$ between the strong-coupling vibronic states are proportional to the vibrational overlap integrals for different paraboloids and this approaches zero as the coupling increases. This is an example of a reduction factor, quantities which arise because many electronic operators have smaller matrix elements when vibronic rather than electronic states are used. Similar reductions occur for trigonal strain effects and small trigonal crystal-field splittings for orbital triplets and for certain types of interactions for orbital doublets (2). As a consequence of the quenching of the spin-orbit interaction by the Jahn-Teller coupling, the strong-coupling energy levels are just those of a 2-dimensional harmonic oscillator, with proper state symmetries requiring allowance for the 3 paraboloids and spin. In the weak-coupling limit we ignore the vibronic coupling and obtain product states, harmonic oscillator states for vibration, spin-orbit states for the electronic part.

In the region intermediate between the limits we see that neither $(3/2)\lambda$ nor $\hbar\omega$ represent energy splittings. (Although this part of the diagram is not quantitative, it is clear from the required limits that such splittings would occur only by chance). In addition the character of the states is altered in the intermediate region. For example the first excited state (Γ_8) changes from an n = 0 to an n = 1 vibration for this choice of parameters ($(3/2)\lambda > \hbar\omega$). This results because the states having the same symmetry are mixed by H_{JT} in the intermediate region.

We mentioned at the end of section VI.E. the possibility that the J' = 3/2 level resulting from the spin-orbit interaction could undergo a Jahn-Teller distortion much as an orbital doublet does, but the J' = 1/2 level would not. This more complicated situation would refer to the relationships $\lambda >> E_{JT} >> \hbar\omega$. Thus the spin-orbit interaction can partially quench the Jahn-Teller effect but complete quenching requires that $\hbar\omega >> E_{JT}$.

Similar correlation diagrams can be constructed for trigonal crystal-field or strain effects.

VII.D. Optical Evidence of T State Vibronic Structure

Many experimental studies have demonstrated the features we inferred in the previous section, particularly the reduction of the spin-orbit splitting for strong coupling. For example Sturge (11) observed approximately a 3-fold reduction for the lowest $^4T_{2g}$ state of V^{2+} in $KMgF_3$. A second example is the roughly 6-fold reduction for the lowest $^4T_{1g}$ state of Mn^{2+} in $RbMnF_3$ observed by Chen, McClure, and Solomon (12). A third example is the roughly 2-fold reduction observed by Kaufman, Koidl, and Schirmer (13) for the lowest $^3T_2$ state of Ni^{2+} in ZnO and ZnS.

Of greater interest to us is the experimental evidence for the intermediate behavior, particularly the absence of well-defined progressions of harmonic oscillator levels. This shows up most clearly as a lack of similar "vibrational" structure in absorption and emission (these would usually be very similar because the elastic energy is nearly independent of the electronic state). The first such interpretation was made by Ham and Slack (14) to explain the difference between the absorption (15) and emission (16) of Fe^{2+} in ZnS. More recently intermediate behavior has been observed and discussed by several groups. Wittekoek, van Stapele, and Wijma (17) observed intermediate behavior for Fe^{2+} in $CdIn_2S_4$; Koidl, Schirmer, and Kaufmann (18) did for Co^{2+} in ZnS; and Ray and Regnard (19) did for Co^{2+} in MgO.

VIII. CONCLUDING REMARKS

There are many aspects of the Jahn-Teller effect and related physical phenomena which we mentioned only briefly or not at all. Also the very limited mention of actual experimental studies left out much of the literature. The choice of subjects and references was meant only to illustrate the development in the subject. No effort to review the Jahn-Teller effect was intended, but rather a simple-minded and primarily pedagogical discussion of a subject which this author has often found difficult to understand and explain.

REFERENCES

1. M. D. Sturge, in Solid State Physics, ed. by F. Seitz, D. Turnbull, and H. Ehrenreich (Academic Press, New York), Vol. 20, p. 91, 1967.

2. F. S. Ham, in Electron Paramagnetic Resonance, ed. by. S. Geschwind (Plenum Press, New York), p. 1, 1972.

3. R. Englman, The Jahn-Teller Effect in Molecules and Crystals (Wiley, New York), 1972.

4. H. A. Jahn and E. Teller, Proc. Roy. Soc. (London), A 161, 220 (1937).

5. H. A. Jahn, Proc. Roy. Soc. (London) A 164, 117 (1938).

6. B. Nygren, J. T. Vallin, and G. A. Slack, Solid State Communications 11, 35 (1972).

7. A. A. Kaplyanskii and A. K. Przevuskii, Opt. i. Spektroskopiya 19, 597 (1965) [Opt. Spectry. (USSR) 19, 331 (1965)].

8. L. L. Chase, Phys. Rev. Letters 23, 275 (1969); L. L. Chase Phys. Rev. B2, 2308 (1970).

9. S. Guha and L. L. Chase, Phys. Rev. Letters 32, 869 (1974).

10. M.C.M. O'Brien, Proc. Roy. Soc. (London) A281, 323 (1964).

11. M. D. Sturge, Phys. Rev. B1, 1005 (1970).

12. M. Y. Chen, D. S. McClure, and E. I. Solomon, Phys. Rev. B6, 1690 (1972).

13. U. Kaufmann, P. Koidl, and O. F. Schirmer, J. Phys. C6, 310 (1973).

14. F. S. Ham and G. A. Slack, Phys. Rev. B4, 777 (1971).

15. G. A. Slack, F. S. Ham, and R. M. Chrenko, Phys. Rev. 152, 376 (1966).

16. G. A. Slack and B. M. O'Meara, Phys. Rev. 163, 335 (1967).

17. S. Wittekoek, R. P. van Stapele, and A.W.J. Wijma, Phys. Rev. B7, 1667 (1973).

18. P. Koidl, O. F. Schirmer, and U. Kaufmann, Phys. Rev. B8, 4926 (1973).

19. T. Ray and J. R. Regnard, Phys. Rev. B9, 2110 (1974).

SPECTRA OF ASSOCIATED DONOR-ACCEPTOR PAIRS

D. CURIE

Luminescence Laboratory

University of Paris, France

ABSTRACT

This article deals with associated donor-acceptor spectra in semiconductors. Here the emphasis is put upon an elementary formalism. However it is found that some of the basic points of the theory (i.e., especially assuming that the matrix element of dipolar moment for the D-A transition is proportional to the overlap integral between the donor and the acceptor) surely need a more elaborate discussion.

I. INTRODUCTION

The interest of D-A centers comes from their high probability of electron-hole recombination. The capture cross-section of a conduction electron by a shallow donor is expected to be rather high; the same occurs for the capture of a free hole by a shallow acceptor. Once both carriers have been captured in that way, radiative recombination occurs necessarily (except at high temperatures producing thermal quenching). On the other hand, direct recombination probabilities of the trapped electron with a free hole are expected to be lower as a general rule.

Figure 1 employs the usual notation:

E_D is the depth of the Donor,

E_A the depth of the Acceptor, and

E_G the energy gap.

The energy of the emitted photons for a large D-A separation is:

$$h\nu_\infty = E_G - E_D - E_A. \tag{1}$$

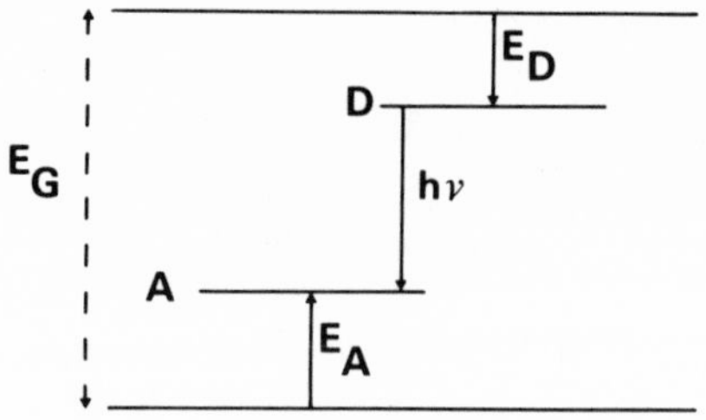

Figure 1 The D-A center (1).

For instance in GaP (E_G = 2.33 eV at 4 K) :
S(donor) substituted for P : E_D = 0.090 eV,
An(acceptor) substituted for Ga : E_A = 0.064 eV,
Si(acceptor) substituted for P : E_A = 0.050 eV;
whence for "type I" pairs Si-S : $h\nu_\infty$ = 2.19 eV
and for "type II" pairs Zn-S : $h\nu_\infty$ = 2.176 eV.

In addition, other pair spectra have been observed including the deep donor O (E_D = 0.896 eV).

Now for a pair made of a donor and an acceptor whose distance r_{DA} is finite :

$$h\nu\ (r_{DA}) = h\nu_\infty + e^2/\varepsilon\ r_{DA} + \text{other corrections.} \qquad (2)$$

As a result of the interactions contained in (2), the numerical values of photon energies computed from (1) are relevant to the low-energy tail of the zero-phonon line spectrum, not to the emission peak.

In equation (2), ε is the static dielectric constant. In this treatment, I shall not discuss the "other corrections" which appear in equation (2). It should be pointed out that it is not satisfactory to include only the Van der Waals term. Taking into account only the electrostatic term is correct for large r_{DA}; on the other hand the Van der Waals correction alone does not lead to a satisfactory experimental agreement for smaller values of r_{DA}. Good results are obtained by including configuration interaction (2), (3).

Some of the more important references dealing with experimental results are listed below:

line spectra in GaP : (4), (5), (6), (7) ;
edge emission in CdS : (8), (9), (10);
edge emission in ZnSe : (11);
large band spectra in ZnS : (12), (13).

Other interesting results have been obtained with germanium, silicon, SiC and diamond.

The D-A model was proposed originally by Prener and Williams (1) for green and red centers in ZnS, but as a matter of fact the most convincing experimental evidence is obtained for GaP and for "edge emission" of II-VI compounds. In these cases a series of emission lines is observed, occurring just below the exciton lines. These lines can be identified as D-A transitions by:

(a) their energy position: for instance in GaP the possible values of r_{DA} are

$$r_{DA} = a_o \sqrt{\frac{m}{2}} \quad \text{for "type I" pairs,} \qquad (a_o = 5.45\ \text{Å})$$

$$r_{DA} = a_o \sqrt{\frac{m}{2} - \frac{5}{16}} \quad \text{for "type II" pairs.}$$

The "missing lines" which occur in "type I" spectra for m = 14, 30, 46, 56, 62, ... give particularly convincing evidence.

(b) their intensities, assuming that for neighbouring r_{DA} separation pairs these intensities are proportional to the stastistical occupation probabilities of each site (7). Sometimes deviations from this "law" are observed, and we shall study them below.

In ZnSe strong $\hbar\omega_{LO}$ phonon replicas of the exciton lines I_1 and I_x are observed among the zero-phonon lines due to D-A centers.

For the green copper centers in ZnS:Cu, and the blue Ag centers of ZnS:Ag, the situation is quite different (12), (13). Phonon coupling is much stronger than in the above cases, and no line structure is observed. But changes in the peak position of the emission spectra occur :

(a) a shift towards higher energies for increasing excitation densities.

(b) a shift towards lower energies for increasing decay time following flash excitation (time-resolved spectroscopy); these shifts were first observed in GaP and behave the same way in ZnS:Cu and ZnS:Ag.

However we may reasonably assume that in these materials most of the observed bandwidth comes from electron-phonon coupling rather than from energy differences in the zero-phonon line emitted by the center according to (2).

II. THE MATRIX ELEMENT FOR D-A TRANSITION PROBABILITIES

In simple theories of the D-A spectra, it is generally assumed that D-A transition probabilities are proportional to the square of the overlap integral :

$$M_{DA} = \int F_D(\vec{r})\, F_A(\vec{r} - \vec{r}_{DA})\, d\vec{r} \tag{3}$$

where F_D and F_A are the envelope wave functions of the electron bound to the donor D and of the hole bound to the acceptor A.

Two derivations of this result have been given (2), (14). Both consider only the case of direct band gap semiconductors.

A) Williams uses the formalism in which the bound electron is described by a Bloch function of the bottom of the conduction band, modulated by $F_D(\vec{r})$:

$$\psi_{el} = F_D(\vec{r})\, u_{co}(\vec{r})$$

$$\psi_{hole} = F_A(\vec{r} - \vec{r}_{DA})\, u_{vo}(\vec{r} - \vec{r}_{DA})$$

Assuming that F_D and F_A extend over many lattice cells, and are approximately constant over one cell, we may write for donor and acceptor at equivalent sites :

$$\int \psi^*_{el}\, e\vec{r}\, \psi_{hole}\, d\vec{r} = \int F_D(\vec{r})\, F_A(\vec{r} - \vec{r}_{DA})\, d\vec{r} \cdot \int_{\text{unit cell}} u^*_{co}(\vec{r})\, e\vec{r}\, u_{vo}(\vec{r} - \vec{r}_{DA})\, d\vec{r}$$

An additional term is found for donor and acceptor at non-equivalent sites and presently we have no experimental evidence supporting such a difference between both types of pairs.

B) I used the developments of ψ_{el} and ψ_{hole} as series of Bloch functions, assuming that for shallow levels :

$$\psi_{el} = \sum_{\vec{k}} c_c(\vec{k})\, u_c(\vec{k},\vec{r}) \; ;$$

$$\psi_{hole} = \sum_{\vec{k}'} c_v(\vec{k}')\, u_v(\vec{k}',\vec{r}) \, .$$

The development of ψ_{el} contains only Bloch functions of conduction band states and ψ_{hole} only Bloch functions of valence band states. Non-zero contributions in the matrix element

$$M_{cv}(\vec{k},\vec{k}') = \int u_c(\vec{k},\vec{r})\ e\vec{r}\ u_v(\vec{k}',\vec{r})\ d\vec{r}$$

come from the terms $\vec{k}' \neq \vec{k}$. The result (3) is obtained if and only if we assume that $M_{cv}(\vec{k},\vec{k}'=\vec{k})$ does not depend appreciably on $\vec{k}$ near the extremum $\vec{k} = 0$. In this case we may write :

$$M_{DA} = M_{cv}(0,0) \sum_{\vec{k}} c_c(\vec{k})\ c_v(\vec{k})$$

and the relation

$$\sum_{\vec{k}} c_c(\vec{k})\ c_v(\vec{k}) = \sum_{\vec{R}_j} \vec{F}_D(\vec{R}_j)\ \vec{F}_A(\vec{R}_j)$$

comes from the properties of Fourier transforms.

Both derivations rely on different approximations, and no detailed discussion on the validity of these approximations has been performed until now.

III. CAPTURE CROSS-SECTIONS

The relative intensities of emission lines from different D-A pairs depend, not only on M_{DA} and on the number of available sites, but also on the cross-section of the first carrier to be trapped (15). Let us suppose that the electron is trapped first. If E_D means the depth of the donor when both A and D are neutral, then the binding energy for electron capture (i.e. the energy loss of the electron, which is either given to the lattice or appears as electromagnetic radiation) is:

$$E_{L1} = E_D - e^2/\varepsilon\ r_{DA} + \text{polarization terms} .$$

If now the electron has been trapped, then the binding energy for capture of the hole is $E_{L2} = E_A$ except for small polarization terms.

Dean and Patrick (15) have observed discontinuities in the relative intensities of neighbouring emission lines, and they suggest the following interpretation. If we consider that capture occurs by the Lax cascade mechanism via excited states of the donor, then an excited level which merges into the conduction band as a result of electrostatic and polarization corrections can no longer capture electrons. A sudden drop in the magnitude of the cross-section then occurs for smaller values of r_{DA}.

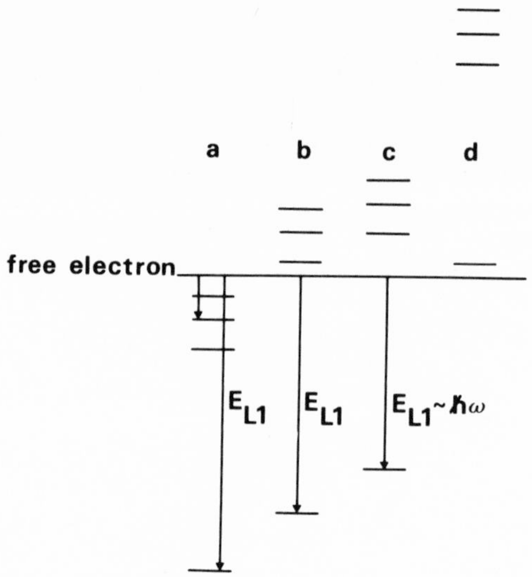

Figure 2 a) Large values of m : electron capture occurs, either directly in the ground state of the donor, or via an excited state.
b) Small values of m : capture can occur in the ground state only.
c) Resonant capture : E_{L1} turns out to be = $\hbar\omega_{LO}$ or $\hbar\omega_{TO}$ for a particular value of m.
d) Capture not possible.

IV. SHAPE OF THE SPECTRUM

A number of sophisticated theories of the spectral shift versus excitation intensity and decay time have been developed by several authors. One of the most recent papers on this topic is (16), which also contains references to older papers.

Apart from using additional parameters, these theories use the same formalism. We shall give a brief account of the simplest one.

We start from equations (1) and (2). Moreover it is generally assumed, following Thomas and Hopfield (6), that:

$$M_{DA} = \text{const.} \ \exp(- r_{DA}/a) \tag{4}$$

where a is the larger of the Bohr radii of D and A. For the above impurities in GaP, $a_A > a_D$; for Cu,Cl centers in ZnS, $a_D > a_A$.

If now we assume a random distribution of impurities, the number of available sites is given by :

$$N(r_{DA})\ dr_{DA} = N \ . \ 4\pi\ r_{DA}^2\ \ dr_{DA} \tag{5}$$

where N is the concentration of acceptors (in GaP) per unit volume.

More elaborate formalisms can be easily obtained by writing instead of (4) :

$$M_{DA} = \text{const.} \ \ r_{DA}^m \ \ \exp(-r_{DA}/a)$$

and instead of (5) :

$$N(r_{DA})\ dr_{DA} = \text{const.} \ \ \exp(- \frac{4}{3}\ \pi\ r_{DA}^3\ N) \ . \ \ 4\pi\ r_{DA}^2\ dr_{DA}$$

in the case of a random distribution, and :

$$N(r_{DA})\ dr_{DA} = \text{const.}\exp(\frac{\alpha}{r_{DA}})\ \exp\ (- \frac{4}{3}\ \pi\ r_{DA}^3\ N) \ . \ 4\pi\ r_{DA}^2\ dr_{DA}$$

in the case of preferential pairing.

In addition the capture cross-section of the pair increases with r_{DA} and one assumes :

$$\sigma(r_{DA}) = \text{const.}\ r_{DA}^z$$

(except for the above discontinuities).

On these bases, Thomas, Hopfield and Augustyniak (17) have performed accurate numerical calculations of the spectral shape after a decay time t (Fig.3).

Figure 3 is computed by assuming that a given pair emits only the zero-phonon line given by (2). Experimental agreement is not good, but the main points are well accounted for :

a) the spectral shape is asymmetric, and the half-width is larger for higher energies (this behaviour is opposite to that of Pekarian curves) ;

b) the high energy part of the spectrum , ascribed to short distance D-A pairs, decays first.

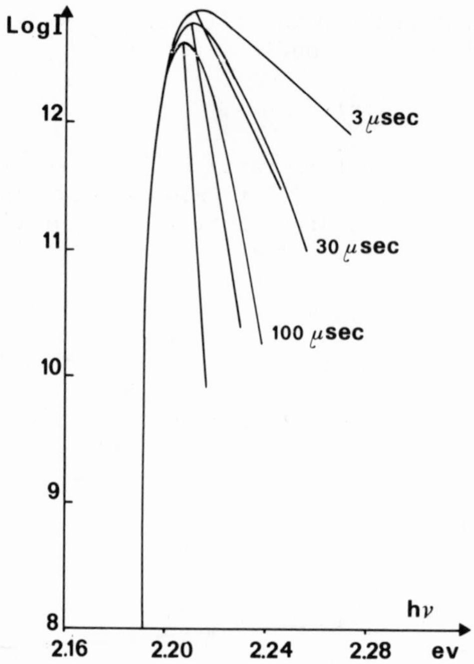

Figure 3 Theoretical spectral shape (17) for Si-S pairs in GaP, assuming reasonable values for N_D, N_A and the excitation intensity, also : $m = 0$, $\alpha = 0$, $z = 2$. Note the rectilinear diameter.

To obtain a good agreement with experiments, one must take into account the additional enlargement of emission lines due to electron-phonon coupling, and in the present case the shape of the emission band from a given pair is rather intricate as a result of the distribution of involved phonons. Of course in such a multi-peaked distribution one can not expect to observe a "rectilinear diameter".

REFERENCES

1. J.S. Prener and F.E. Williams. J. Phys., 17, 667 (1956).
2. F. Williams, Phys.Stat.Sol., 25, 493 (1968).
3. L. Mehrkam and F. Williams, Comm. Leningrad International Conference on Luminescence, 1972.
4. J.J. Hopfield, D.G. Thomas, M. Gershenzon, Phys.Rev.Letters, 10, 1621 (1963).
5. D.G. Thomas, M. Gershenzon, F.A. Trumboke, Phys.Rev., 133, A 269 (1964).
6. D.G. Thomas, J.J. Hopfield, K. Colbow, 7th International Conference on the Physics of Semiconductors, Radiative Recombination in Semiconductors (M. Hulin, ed.), Dunod, Paris 1964, p.67.
7. P.J. Dean, C.C. Henry, C.J. Frosch, Phys.Rev., 168, 812 (1968).
8. L.S. Pedrotti and D.C. Reynolds, Phys.Rev., 120, 1664 (1960).
9. R.E. Halsted, "Edge Emission" in Physics and Chemistry of II-VI Compounds (Aven and Prener, eds.), North Holland 1967, p.385.
10. E. Gutsche and O. Goede, J. of Luminescence, 1-2, 200 (1970).
11. P.J. Dean and J.L. Merz, Phys.Rev., 178, 78 (1969).
12. S. Shionoya, Proceedings of the Intern. Conf. on II-VI compounds, Providence (D.G. Thomas, ed.), Benjamin Inc, New York 1968, p.1.
13. S. Shionoya, K. Era and Y. Washizawa. J. Phys.Chem.Sol., 29, 1827 (1968)
14. D. Curie, J. Phys., 28, suppl. C3, 105 (1967).
15. P.J. Dean and L. Patrick, Phys.Rev. B, 2, 1888 (1970).
16. W.E. Hagston, J. of Luminescence, 5, 285 (1972).
17. D.G. Thomas, J.J. Hopfield, and W.M. Augustyniak, Phys.Rev. 140, A 202 (1965).

ZERO-PHONON AND PHONON-ASSISTED RADIATIVE TRANSITIONS OF DONOR-ACCEPTOR PAIRS IN LUMINESCENT SEMICONDUCTORS*

Lester Mehrkam and Ferd Williams

Department of Physics, University of Delaware

Newark, Delaware 19711

ABSTRACT

The theoretical basis for the zero-phonon line and phonon-assisted broad band luminescence of donor-acceptor pairs is considered. For the known cases of pair spectra the electronic orbital time is shown to be short compared to the period of polar lattice modes. We derive detailed justification for the use of the static dielectric constant in the Coulomb interaction term for the line spectra. The effective Hamiltonian relevant to the zero-phonon transitions is formulated. The theory including configuration interaction and effects of lattice polarization is formulated for symmetrical pairs which have equal electron and positive hole orbital radii, and applied to the spectrum of gallium phosphide doped with tin and zinc.

I. INTRODUCTION

Radiative transitions of donor-acceptor pairs (DA) in phosphors and semiconductors have been extensively investigated (1,2). The well-resolved line spectra of distant DA pairs have been usually interpreted by the following equation

$$h\nu(R) = E_g - (E_D + E_A) + e^2/\kappa_s R \qquad (1)$$

where $h\nu(R)$ is the radiative recombination energy for pairs with DA distances R, E_g is the band gap of the semiconductor, E_A and E_D are the electronic binding energies for the isolated acceptor and donor, respectively, and κ_s is the static dielectric constant. Despite recent advances in the theory of donor-acceptor pair spectra including the effects of configuration interaction and correlation (3), the widespread use of the static dielectric constant in the Coulomb interaction term for the line spectra of distant pairs has not been justified rigorously. Hopfield et al. (4) refer briefly to a calculation based on a Born cycle for electrons and holes relevant to the use of κ_s; Thomas et al. (5) claim that κ_s is justified for distant pairs with R large compared to orbital electronic radii; and Dean (2) correctly emphasized that eq. (1) refers to zero-phonon transitions but gave no detailed justification for κ_s in the Coulomb term. Earlier, Williams (6) in a detailed analysis of the electrostatic interactions of DA pairs showed that κ_s is the correct dielectric constant to use for both optical absorption and luminescent emission if the electronic orbital times are long compared to the period of polar lattice modes (these modes follow the motion of the electronic particles), but that the proper dielectric constant to use is different for absorption and emission and that for emission the use of the optical dielectric constant κ_o is a good approximation for the phonon-assisted transition if the electronic orbital time is short compared to the period of polar lattice modes (these modes adjust to the stationary distributions of the electronic particles).

In the following we shall analyze data on the electronic binding energies of donors and acceptors in III-V and II-VI luminescent semiconductors, and then consider both zero-phonon and phonon-assisted radiative transitions using a configuration coordinate model. The line spectra of pairs will be shown to arise with the lattice in an improbable state of polarization. An effective Hamiltonian is formulated to obtain zero-phonon transitions. Characteristics of charged dopants, which in DA pairs yield line spectra, are discussed. The theory, including configuration interaction, is then formulated for symmetrical DA pairs, in which the electron and positive hole have equal radii and then applied to the spectrum of GaP:Sn,Zn.

II. ANALYSIS OF EXPERIMENTAL SPECTRA

We shall consider spectra for both gallium phosphide and zinc selenide, representative of III-V and II-VI luminescent semiconductors. In Table I, we show the experimental electronic binding energies for various donors and acceptors in these materials. The electronic orbital times τ_o for each dopant are also shown. The τ_o are obtained from the electronic binding energies: $\tau_o = h/2E$, derived from the orbital radius, $a = \hbar/\sqrt{2m^*E}$ and orbital velocity, $v = \sqrt{2E/m^*}$.

TABLE I

material	dopant	type	E(mev)	τ_o(sec)x10^{14}	τ_ℓ(sec)x10^{14}
GaP	O	D	896	0.23	9
GaP	S	D	104	2.0	9
GaP	Sn	D	69	3.0	9
GaP	C	A	46	4.5	9
GaP	Zn	A	62	3.3	9
ZnSe	Aℓ	D	26	7.9	17
ZnSe	Ga	D	28	7.4	17
ZnSe	In	D	29	7.2	17
ZnSe	Li	A	114	1.8	17

The periods of longitudinal optical phonons for these two materials are also shown on Table I. In general, for these systems $\tau_o < \tau_\ell$ and, therefore, the most probable luminescent emission for pairs of these dopants involves κ_o in eq. (1). For both these semiconductors, there is indeed a broad emission band on the long wavelength side of the line spectrum, which is obviously phonon-assisted. The line spectrum involves improbable zero-phonon transitions. In Fig. (1) we show a configuration coordinate model for DA pair transitions for pairs in which $\tau_o < \tau_\ell$ and, therefore, the adiabatic approximation discussed in a previous chapter applies. In the ground electronic state of the DA pair there is no electron nor positive hole bound to the donor or acceptor core, respectively, and, therefore, the medium is polarized in the region of these charged dopants in accordance with κ_s. The minimum for the ground state, therefore, corresponds to a condition of lattice polarization involving κ_s. For the excited electronic state an electron is bound to the donor core and a positive hole is bound to the acceptor core, thereby screening the core charges; the lattice, therefore, is to a good approximation unpolarized and the minimum for the excited state corresponds to a condition of polarization corresponding to κ_o. The most probable optical excitation $h\nu_a$, the most probable radiative de-excitation $h\nu_e$ and the zero-phonon transition $h\nu_o$ are also shown on Fig. (1). These quantities are related by the following inequalities:

$$h\nu_a > h\nu_o > h\nu_e \tag{2}$$

We shall now consider the zero-phonon transition in detail. This transition obviously has the same transition energy in absorption or emission. As evident from Fig. (1), the condition of polarization is different from the initial and final state. The electronic part of the effective mass Hamiltonian for the excited state is, therefore, the following:

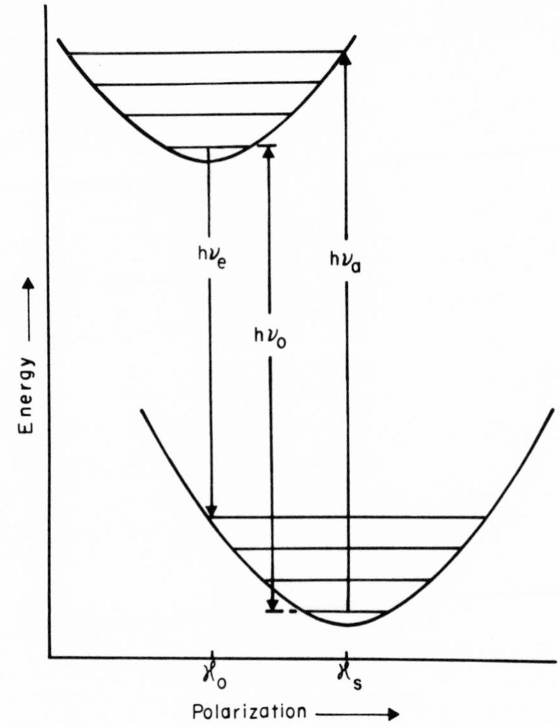

Fig. 1 Configuration coordinate model for donor-acceptor pair transitions including zero-phonon and phonon-assisted transitions.

$$H = -\frac{\hbar^2}{2m_e^*}\Delta_r - \frac{\hbar^2}{2m_h^*}\Delta_{r'} - \frac{e^2}{\kappa_o}\left[\frac{1}{r} + \frac{1}{|r+R-r'|} - \frac{1}{|R+r|}\right]$$
$$- \frac{e^2}{\kappa_s}\left[\frac{1}{r'} - \frac{1}{|R+r'|}\right] , \qquad (3)$$

where m_e^* and m_h^* are the effective mass for electron and positive hole, respectively, and the interparticle distances are defined in Fig. (2). The electron interacts with the ion cores in accordance with κ_o; the positive hole interacts with the cores in accordance with κ_s because the polarization corresponds to no screening of cores by the hole; and the electron and hole interact with each other in accordance with κ_o. The corresponding two-particle effective mass function is:

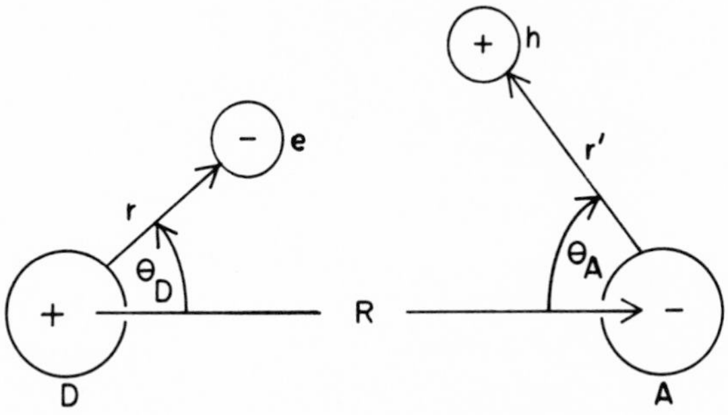

Fig. 2 Effective mass model for donor-acceptor pair with interparticle distances and angles defined.

$$F(r,r') = F_D(r\ )F_A(r'), \tag{4}$$

and the effective mass function for the electron is of the following form:

$$F_D(r) = (1+\sigma_D^2)^{-\frac{1}{2}} a_D^{-3/2} e^{-r/a_D}[1 + \sigma_D(r/a_D)\cos\theta_D] \tag{5}$$

where again the distances and angle are defined in Fig. (2), and σ and Z are variational parameters. The $F_A(r)$ for the hole has a similar form. The Bohr radius for the electron is the following:

$$a_D^o = \frac{Z_D^2 e^2}{2E_D^o} \left[\frac{1}{\kappa_o} + \frac{1}{16\kappa_s}\right] \tag{6}$$

in accordance with the work of Lehovec (7) and Curie (8). The Bohr radius of the positive hole is, however, the following:

$$a_A^s = Z_A^2 e^2/2E_A^s\kappa_s \tag{7}$$

since the lattice is polarized by the acceptor core in accordance with κ_s, without screening by the stationary hole distribution.

We substitute eq. (6) and (7) into equations for $F_A(r')$ and $F_D(r)$ and then substitute into eq. (4). If the effective mass Hamiltonian H given by eq. (3) is applied to F(r,r') and for pairs with $a_A = a_D$ the variational principle is applied with respect to σ and Z, the following equation is obtained:

$$h\nu_o(R) = E_g - Z^2(E_D^o + E_A^s)$$
$$+ \frac{e^2}{\kappa_o R} \frac{1}{1+2\sigma^2+\sigma^4} \{A+B\sigma+C\sigma^2+D\sigma^3+E\sigma^4\}$$
$$+ \frac{e^2}{\kappa_s R} \frac{1}{1+\sigma^2} \{A' + B'\sigma + C'\sigma^2\} \quad (8)$$

where

$$A = (Z-1)\rho + e^{-2\rho}\left(\frac{\rho^3}{6} + \frac{3}{4}\rho^2 + \frac{3}{8}\rho\right)$$

$$B = \frac{2}{\rho} - e^{-2\rho}\left(\frac{\rho^4}{3} + \frac{11\rho^3}{6} + \frac{37\rho^2}{12} + 4\rho + 4 + \frac{2}{\rho}\right)$$

$$C = -\frac{11}{\rho^2} + \frac{2Z}{\rho} - \frac{3}{2}\rho + e^{-2\rho}\left(\frac{7\rho^5}{30} + \frac{89\rho^4}{60} + \frac{489\rho^3}{120} + \frac{55\rho^2}{6} + \frac{347\rho}{24} + 22 + \frac{23}{\rho} + \frac{11}{\rho^2}\right)$$

$$D = \frac{2}{\rho} + \frac{36}{\rho^3} - e^{-2\rho}\left(\frac{\rho^6}{12} + \frac{7\rho^5}{12} + \frac{113\rho^4}{60} + \frac{697\rho^3}{120} + \frac{3097\rho^2}{240} + 28\rho + 52 + \frac{74}{\rho} + \frac{27}{\rho^2} + \frac{36}{\rho^3}\right)$$

$$E = (2Z-1)\frac{\rho}{2} - \frac{3}{\rho^2} - \frac{54}{\rho^4} + e^{-2\rho}\left(\frac{\rho^7}{140} + \frac{141\rho^6}{2520} + \frac{19\rho^5}{70} + \frac{893\rho^4}{840} + \frac{17089}{67220}\rho^3 + \frac{4491\rho^2}{640} + \frac{23691}{1280}\rho + 42 + \frac{78}{\rho} + \frac{111}{\rho^2} + \frac{108}{\rho^3} + \frac{54}{\rho^4}\right)$$

$$A' = (Z-1)\rho + 1 - e^{-2\rho}(1+\rho)$$

$$B' = -\frac{2}{\rho} + 2e^{-2\rho}(\rho^2 + 2\rho + 2 + \frac{1}{\rho})$$

$$C' = Z\rho + 1 + \frac{3}{\rho^2} - e^{-2\rho}(\rho^3 + 3\rho^2 + \frac{11\rho}{2} + 7 + \frac{6}{\rho} + \frac{3}{\rho^2})$$

and

$$\rho = \frac{ZR}{a} .$$

For $a_A \neq a_D$, four variational parameters are required.

Eq. (8) reduces to the following in the limit of large R where there is no electronic overlap, and $\sigma \to 0$ and $Z \to 1$:

$$h\nu_o(R) = E_g - (E_D^o + E_A^s) + \frac{e^2}{\kappa_s R} \quad (9)$$

where E_D^o and E_A^s are the following:

$$E_D^o = \frac{m_e^* e^4}{h^2 \kappa_o}\left[\frac{11}{16\kappa_s} + \frac{5}{16\kappa_o}\right] ; \; E_A^s = m_h^* e^4/h^2\kappa_s^2, \quad (10)$$

in accordance with Lehovec (7) and Curie (7).

Our eq. (9) corresponds to eq. (1) with a significant difference. The proper donor and acceptor energies, E_D and E_A, for interpreting pair spectra correspond to different states of polarization for the two charged defects, that is, E_D^o and E_A^s.

As noted earlier, the most probable radiative transitions are phonon-assisted. Their transition energies are given by the following for absorption and for emission:

$$h\nu_a(R) = E_g - [E_A^s + E_D^s] + e^2/\kappa_s R \quad (11)$$

$$h\nu_e(R) = E_g - [E_D^o + E_A^o] + e^2/\kappa_o R.$$

Comparison of equations (9), (11), and (12) show that they are inter-related in accordance with eq. (2).

III. APPLICATION OF THE THEORY

The most symmetrical DA pair occurs with GaP:Sn,Zn. We have applied the theory with the results shown in Fig. (3). The data are from Dean et al.(9). Substantial improvement results from including configuration interaction and the detailed effects of lattice polarization. The effect of correlation is apparently small.

We have applied eq. (8) to other pairs including those for GaP:C,S and ZnSe:Ga,Li. The agreement with experiment is less perfect than for GaP:Sn,Zn. We believe this results from $a_D \neq a_A$ for these pairs so that eq. (8) involving only two variational parameters is inadequate. A more complete analysis for asymmetrical pairs will be reported elsewhere.

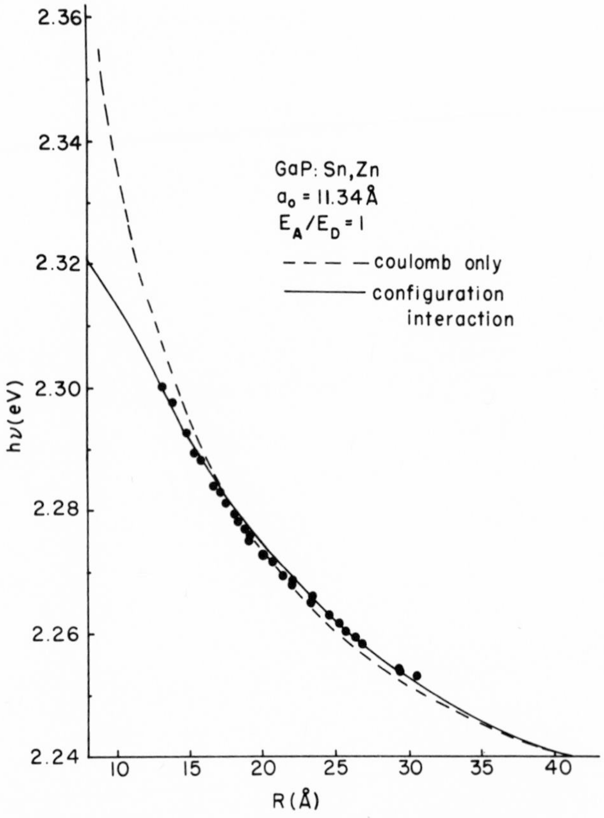

Fig. 3 Zero-phonon transition energy versus separation of donor and acceptor in DA pairs, experimental points and theoretical curves.

REFERENCES

*Supported by a grant from the U.S. Army Research Office-Durham.

1. F. Williams, Phys. Stat. Sol. 25, 493 (1968).

2. Recent review: P. J. Dean, Progress in Solid-State Chemistry 8, 1 (1973).

3. L. Mehrkam and F. Williams, Phys. Rev. 6B, 3753 (1972); "Luminescence of Crystals, Molecules, and Solutions," Edited by F. Williams (Plenum, N. Y., 1973),p. 628-637.

4. J. J. Hopfield, D. G. Thomas and M. Gershenzon, Phys. Rev. Lett. 10, 162 (1963).

5. D. G. Thomas, M. Gershenzon and F. A. Trumbore, Phys. Rev. 133A, 269 (1964).

6. F. Williams, J. Phys. and Chem. Solids 12, 265 (1960).

7. K. Lehovec, Phys. Rev. 92, 253 (1953).

8. D. Curie, Compt. rend. Acad. Sci. Paris 238, 579 (1954).

9. P. J. Dean, R. A. Faulkner and S. Kimura, Phys. Rev. 2B, 4062 (1970).

PRESENT TRENDS IN LUMINESCENCE RESEARCH

Ferd Williams

Department of Physics, University of Delaware

Newark, Delaware 19711, USA

ABSTRACT

The present trends in research on luminescence are discussed with regard to unusual materials, recent phenomena and applications. The principal emphasis appears to be on cooperative effects at high intensities of excitation. Complex associated defects are increasingly investigated. Luminescent properties of magnetic materials and of amorphous substances are being studied. In general, the research is becoming more sophisticated, both in concepts and techniques, but is also becoming more applied.

I. INTRODUCTION

During the past 30 years, the character of research on luminescence has changed several times quite drastically. Prior to the Second World War, the research was largely empirical. The invention of the transistor in 1948 had a direct impact on luminescence research in bringing the concepts of semiconductor physics into certain areas of luminescence research. For example, activator ions were identified as acceptors and co-activator ions were identified as donors in semiconducting phosphors. Electroluminescence created enthusiastic interest in both fundamental and applied research and focused attention on the relevance of electronic transport and dielectric breakdown to some areas of luminescence. The invention of the laser in 1961 broadened the discipline to include both coherent emission as well as spontaneous emission, and made available tools for excitation at high radiation densities. Other important developments in instrumentation include vacuum ultraviolet spectrometers, spectroradiometers, equipment for time-resolved spectroscopy even to picoseconds, and X-ray photo-electron spectrometers for measuring the binding energies of electrons. In general, luminescence research has become broader and more sophisticated, involving a greater diversity of materials and more quantitative rigorous interpretations.

II. UNUSUAL MATERIALS

The materials which have interesting, and in many cases useful, luminescent characteristics are becoming increasingly complicated. A wide diversity of rare earth doped materials, both trivalent and divalent rare earths,are being developed. Some of these have complex crystal structures; others have complex associated defects. The latter is also the case with luminescent semiconductors.These semiconductors were initially primarily binary compounds, but increasingly ternary and quaternary compounds or alloys are being investigated. Three- and even four-center defects are of interest,in addition to donor-acceptor pairs and other two-center associates. Magnetic materials are of interest in part because of coupling of the electronic transitions with quantized spin waves. The luminescence of amorphous materials can provide information on inhomogeneous spectral broadening,as well as contributing to the development of high power lasers. Studies of luminescence of organic materials and of inorganic materials are increasingly unified. Organic ions in solutions and in solids are currently being investigated.

III. RECENT PHENOMENA

Cooperative luminescent phenomena involving the interaction of excitations with each other are observed with high excitation intensities. These include multiphoton excitations via virtual states and radiative transitions arising from the simultaneous annihilation of excitations on interacting excited ions.Sequential transitions may also occur at high excitation intensities. For example, one can find sensitizer ions which sequentially transfer energy to a fluorescer which radiates a single anti-Stokes photon, after being multiply excited.

Quantum efficiencies less than but approaching unity, have been traditional in luminescence. There has been for almost a decade controversial evidence for quantum efficiency greater than unity with excitation in the vacuum ultraviolet. Recently, well-defined materials with well-understood energy levels have been found capable of cascade de-excitation via real intermediate states following ultraviolet excitation, and quantum efficiencies exceeding unity were measured.

Research on injection electroluminescence is involving unusual electronic states of dopants. In addition to the radiative recombination at donor-acceptor pairs, recombination at isoelectronic dopants is investigated. These dopants are in the same charge state as the ions they replace, and alternately capture electronic particles of opposite sign. Enhancement of the luminescence from isoelectronic dopants, whose states are related to continuum states in certain regions of k-space,has been reported for mixed crystals of a composition such that states of the dopant have strong resonant interaction with continuum states from a different region of k-space.

IV. APPLIED RESEARCH ON LUMINESCENCE

The phenomena described above allow new applications of luminescence. Cooperative phenomena and multiphoton transitions can be applied to infrared-to-visible converters and possibly to the more efficient conversion of ultraviolet radiation into visible radiation. The development of complex rare earth doped materials has led to the application of these narrow band fluorescers to color television. Related materials have been proposed for high color quality, high efficiency fluorescent lamps. Injection electroluminescence is now well-established in light emitting diodes for information display and may yet be applied to solid state lamps for illumination. Improved fluorescopic screens continue to be developed with new materials.

The present concern with new energy sources and with energy conservation has added to the incentives for the development of new and more efficient luminescent materials and systems for lighting and for information display.

LONG SEMINARS (ABSTRACTS)

APPLICATION OF VIBRONIC SPECTROSCOPY -- STRONTIUM TITANATE AS AN EXAMPLE

Richard C. Powell

Department of Physics, Oklahoma State University

Stillwater, Oklahoma 74074, U.S.A.

ABSTRACT

Investigations of the vibronic spectra of ions in crystals can be used to study the lattice dynamics of the host crystal and the effects caused by the interaction between the lattice phonons and the transition electron on the impurity ion. In this seminar we review the results of such studies of ions in strontium titanate crystals. This is an interesting host to study because of the presence of "soft" phonon modes which are responsible for the interesting dielectric properties and structural phase transitions of the material. Our recent work on $SrTiO_4:Cr^{3+}$ will be presented in detail as an example of a research project on vibronics. Most of the structure in the low energy vibronic sideband at low temperature can be identified using selection rules derived from group theory and comparing the results with those obtained from infrared absorption, Raman and neutron scattering. An iteration process was used to obtain a computer fit to the data from which one-phonon, two-phonon, and multi-phonon contributions to the sideband were derived. It was found that quadratic coupling between an impurity-induced local mode and the lattice phonon modes had to be included in order to obtain a fit. A very simple model is used to obtain a phonon density of states from the one-phonon vibronic sideband and this is found to agree fairly well with the density of states obtained from neutron scattering. The high energy vibronic sideband is shown to be useful in observing the low frequency phonons including the soft modes. The temperature dependences of the widths and positions of both the zero-phonon lines and the impurity-induced local mode are discussed in terms of the theoretical fittings predicted using a Debye phonon distribution, the effective phonon distribution obtained from the vibronic sideband, and an Einstein phonon distribution describing coupling to only a soft phonon mode.

RADIATIONLESS DECAY OF IMPURITY IONS IN SOLIDS

R. Englman

Soreq Nuclear Research Centre

Yavne, Israel

ABSTRACT

The excitation energy of an ion in a solid may be passed over either to the photon continuum superimposed on the ground state, in which case an emission is said to occur, or to the phonon quasi-continuum, which case is that of a radiationless decay process. In the latter instance the natural separation of the degrees of motion is into electronic ones and to those of all the nuclei in the solid.

A further separation of the nuclear motion is into active and passive modes, such that the latter modes are responsible for the different relaxed configuration of the excited state from that of the ground state, while the former modes admix the two electronic states. When we have in mind high excitation energies (so that the energy gap is many times the characteristic vibrational energy of the lattice) then the natural description of the relaxation process is in terms of the involvement of many passive phonons and of one active phonon.

The resulting expressions for the decay rate take different forms in case of very strong coupling between the electronic states and the passive modes and in case of weak-moderate coupling. Strong coupling leads to an Arrhenius-type activation law, which becomes increasingly more valid as the temperature rises. Weak-moderate coupling gives the energy-gap law which also characterizes intra-molecular non-radiative decay. This law contains as factors the attenuation factor for transitions between the vibrational states of different relaxed configurations (the Debye-Waller factor at zero temperature), then the N factors for the probability of emission of N passive phonons, which are needed for the energy balance, some other statistical factors and finally the factor involving the active phonons.

This last factor represents a first order perturbation in the active phonons. The application of perturbation theory to the first order in active phonons represents, from a more fundamental viewpoint, the replacement of the t-matrix by its lowest order term, namely, the perturbational potential. The validity of this approximation depends on the magnitude of the ratio: the rate of relaxation over the width of the (passive) phonon-coupling between the initial and final states, as a function of the energy. For the large active phonon coupling or fast conversion this ratio may not be small at all and a different approach is called for.

We introduce a procedure which separates out from the normal modes a small set of modes which are strongly coupled to the electronic states, whereas the remainder of the modes is coupled to the electronic states and to the interacting modes only by a residual coupling. This procedure is physically equivalent to the alternative description of radiationless transitions, wherein one has a discrete set of vibronic states, involving the electronic states and some interaction coordinates, and the lattice modes induce transitions between the vibronic states.

In the procedure adopted, the strongly coupled modes are derived by a variational procedure and it is shown that the residual coupling (essentially due to the fact that the strongly coupled modes are not normal modes) is between the transformed lattice modes and the strongly coupled modes as well as the electronic states. An important conclusion arising from our procedure is that the residual coupling is subject to an upper limit, proportional to the <u>effective</u> dispersion of the normal mode frequencies, i.e., to the root-mean-square deviation of the frequencies weighted by the (passive) coupling strength to the appropriate modes. Because of this upper limit the residual coupling belongs to the weak-moderate regime, whereas before the introduction of the strongly coupled modes the coupling appeared as strong.

LUMINESCENCE FROM YAG:Cr^{3+}, MgO:V^{2+}, and MgO:Cr^{3+}

G. F. Imbusch*

Department of Physics, University College

Galway, Ireland

ABSTRACT

At 77°K the luminescence from YAG:Cr^{3+} consists of two sharp no-phonon lines due to transitions from ^2E to $^4A_2$ states of the chromium ion in the trigonally distorted octahedral site. In addition one-phonon vibrational sidebands are seen at longer wavelength. The no-phonon lines are weak magnetic dipole transitions(1) while the sidebands are presumably vibronically induced electric dipole transitions - since at low temperatures they have an intensity four times as large as the no-phonon lines. Since the difference in the Cr^{3+} - lattice coupling strength in the ^2E and $^4T_2$ states is small, this $^2E \rightarrow {}^4A_2$ transition can be considered as occurring in the weak-coupling limit (Huang-Rhys factor, S, less than unity). Consequently we expect (i) that the oscillator strength for the no-phonon lines is approximately a constant independent of temperature, and (ii) that only one-phonon sidebands appear. As the temperature is raised anti-Stokes sidebands also appear and the no-phonon lines and their sidebands appear to sit on top of a broad band continuum. Above 300°K most of the luminescence occurs in this broad band whose oscillator strength is found to increase strongly with temperature.

One can, with some degree of error, separate out the no-phonon lines (whose intensity we label I_R), the true sidebands of the no-phonon lines (I_S) and the broad band (I_T), and we observe how I_S/I_R and I_T/I_R increase with temperature. We observe that I_S/I_R varies approximately as Σ_i coth $(\hbar\omega_i/2kT)$ where the ω_i's are the phonon frequencies and this is theoretically expected. We attribute the broad band to luminescence from the $^4T_2$ state which becomes populated at higher temperatures. The intensity data are in agreement with this model since I_T/I_R appears to vary as $\exp(-\Delta/kT)$, where Δ is the separation between the zero vibrational levels of ^2E and $^4T_2$. (The $^4A_2 \rightarrow {}^4T_2$ no-phonon line cannot be clearly resolved in this material but the energy separation is around 1000 cm^{-1}(2)).

At any one temperature all parts of the luminescence spectrum have the same decay time, τ, and τ is found to vary as

$$\tau^{-1} = \tau_R^{-1} + \tau_S^{-1} + \tau_T^{-1} \exp(-\Delta/kT)$$

where τ_R, τ_S and τ_T are decay times for the no-phonon transitions alone, sideband processes alone, and τ = 20 μs which is appropriate to a $^4T_2 \rightarrow {}^4A_2$ radiative process, respectively. Thus the lifetime data are consistent with a model where the broad band part of the luminescence is due to Stokes-shifted luminescence from the $^4T_2$ level a distance Δ = 1000 cm^{-1} above ^2E and where this $^4T_2$ level becomes weakly populated from room temperature upwards (2).

An examination of the luminescence data on MgO:V^{2+} (3,4) suggests that appreciable $^4T_2 \rightarrow {}^4A_2$ emission could also be occurring at higher temperatures in this material even though Δ = 1840 cm^{-1}. The intensity and lifetime data are in substantial agreement with this interpretation. However, further analyses are being made on this material (5).

In MgO:Cr^{3+} one finds in addition to $^2E \rightarrow {}^4A_2$ transitions from different chromium sites a broad band transition which has been attributed to $^4T_2 \rightarrow {}^4A_2$ transitions (6). What is unusual in this material is the fact that the ^2E luminescence occurs at millisecond lifetimes while the $^4T_2$ luminescence occurs at a much faster lifetime (≈50μs). Further work is being done on the detailed analysis of this luminescence.

*Work supported by the Irish National Science Council.

REFERENCES

(1) M.O. Henry (to be published).
(2) M.O. Henry, et al., J. Phys. C. (1974), in press.
(3) B. Di Bartolo and R. Peccei, Phys. Rev. 137A, 1770, (1965).
(4) B. Di Bartolo and R.C. Powell, Nuovo Cimento 66B, 21, (1970).
(5) B. Di Bartolo and G.F. Imbusch (to be published).
(6) J.H. Parker, Jr., in Optical Properties of Ions in Crystals, edited by H.M. Crosswhite and H.W. Moos (Wiley, 1967).

SPECTROSCOPY OF 5f - SYSTEMS: ACTINIDES VERSUS LANTHANIDES

R. G. Pappalardo

GTE Laboratories Inc.

Waltham, Massachusetts 02154, U.S.A.

ABSTRACT

Electrodeless microwave discharges usually provide the first-order and second-order spectra from actinide halides (1). The classification of the corresponding electronic energy levels by means of their J-quantum number, their LS-content and their electronic configuration relies heavily on the study of the Zeeman effect of the magnetic hyperfine splitting and of the isotope shift. In contrast to the case of the lanthanides, the normal configuration of the neutral actinide elements contains a d-electron up to Np, suggesting a similarity in binding energy of the 5f and 6d electrons at the beginning of the series (1). This similarity in binding energy and the decreased ionization energies provide the explanation for the wide range of stable oxidation states (from trivalent to hexavalent, or even hepta-valent) observed in the first half of the $5f^n$ series. This is to be contrasted with the common oxidation state of three in the lanthanide series. A convergence in the chemical behavior of lanthanides and actinides is only observed toward the middle of the two series.

The available free-ion spectra of actinides are not accurately described by the model of a single electronic configuration. The effect of configuration mixing is conveniently introduced as a correction factor on the diagonal matrix elements of the electro-static interaction for the dominant electronic configuration (2). Other effects (orbit-orbit and spin-other-orbit interactions) arising from a relativistic solution of a many-electron Hamiltonian involve similar correction factors on the diagonal matrix elements of the dominant electrostatic and spin-orbit interactions(2). These modifications of the single-electronic-configuration model are of relevance to the corresponding interpretation of condensed-phase spectra.

The basic model used to explain the spectra of trivalent rare earths in the condensed phase considers the electrostatic and spin-orbit interactions as dominant, while the inter-ionic effects are viewed as much weaker perturbations. The trivalent-actinide spectra are interpreted using basically the same model. The problem of energy-level identification in trivalent actinides is commonly carried out using two different procedures.

In one approach the fine structure of the condensed-phase spectra is averaged out to provide the baricenter of the corresponding line group. This baricenter is assumed to give the energy position of the corresponding spherical-symmetry J-level in the limiting case of vanishing interatomic interactions. The positions of these "free-ion" J-levels are the basic input for a fitting procedure involving as freely-varying quantities the parameters describing radial averages (over the 5f-electronic wave-function) of the dominant interactions (namely, F_2 and ζ; or F_2, F_4, F_6, and ζ; or F_2, F_4, F_6, ζ and additional parameters describing configuration interaction and other effects discussed above). Solution spectra were originally used as the experimental data for this procedure of J-label identification (3), while now thin-film absorption spectra at various temperatures provide a wider range of more accurate spectroscopic information (4,5,6).

The internal consistency of the model can be checked using the intensity theory developed by Judd and Ofelt for lanthanides (2). The intensity of forced electric-dipole transitions can be expressed via three parameters implicitly containing the radial dependence and the scale factors of the mechanisms producing the radiative coupling of $5f^n$ levels. The basic input for these intensity predictions is the intermediate-coupling LS-composition of the $5f^n$ levels. A proper matching of predicted and observed relative intensities of transitions between J-levels can be viewed as a good internal check on the overall level assignment (4). An additional check for internal consistency of the level assignment is the correlation of large overall spread of a line-group with a corresponding high value of the J-label.

A more rigorous derivation of the $5f^n$ energy-level scheme relies on single-crystal spectra. Here the key feature in the level assignment procedure is the polarized nature of absorption and emission spectra in hosts like $LaCl_3$ and $LaBr_3$, with high point-symmetry at the $5f^n$-ion site. The polarized spectra provide the identification of crystal-levels of definite symmetry properties , which on the one hand should reliably lead to the J-label of the spherical —symmetry parent term, and in addition give the crystal-field parameters by a fitting procedure of the observed fine structure (7,8). Krupke's and Gruber's work (8) on $LaBr_3:Np^{3+}$ is an excellent example of this rigorous treatment of the two-fold question of deriving a J-level energy distribution and a set of values for the crystal-field parameters. The observation of the Zeeman effect and of emission from some specific levels can provide tests for the internal consistency of the assignments.

This overall elaboration of the spectral data on trivalent actinides results in deriving F_K parameters of electrostatic interaction in the 5f-series that are roughly 2/3 of the corresponding lanthanide parameters, while spin-orbit effects in the 5f-series are twice as strong as the corresponding effects in lanthanides. The stronger spin-orbit coupling causes pronounced LS coupling even in the ground levels of the various trivalent ions.

Considerable controversy surrounds the physical interpretation of the crystal-field parameters in trivalent lanthanides (and actinides, by extension). An alternative model (9) of the crystal-field parameters, seen as a measure of σ-antibonding effects was proposed many years ago. While this model predicates a pronounced increase in the spread of the crystal fine-structure in covalent systems, this effect was not observed in volatile compounds like the lanthanide cyclopentadienides (10). On the other hand, the rather marked changes observed in trivalent actinide spectra in going from halides to tricyclopentadienides suggest that covalency effects may be important for $5f^n$ systems (11,12).

While the spectroscopy of trivalent actinides can be satisfactorily interpreted by a model similar to that used for the lanthanides, the presence of stronger crystal-fields in tetravalent actinides requires a different analytical procedure.

A test case for the existence of a strong crystal-field is provided by U^{4+} in octahedral halide coordination, where the two lowest crystal levels are very widely separated, by ~1000 cm^{-1}, as indicated by magnetic and spectroscopic measurements.

The absorption spectra of U^{4+} in octahedral coordination provide an excellent example of vibronic spectra (13,14). The identification of the corresponding cubic-symmetry electronic levels relies on well-obeyed selection rules for vibronic transitions. A reasonable discription of the observed spectra can be derived from a simultaneous diagonalization of the electrostatic, spin-orbit and crystal-field matrices and from a fitting procedure, with six parameters (F_2, F_4, F_6, ζ_{5f}, $A\langle r^4\rangle$ and $B\langle r^6\rangle$) (15). The fourth-order crystal-field potential is roughly one order of magnitude larger than that deduced for trivalent lanthanides in lanthanum chlorides. For a more meaningful comparison one should really compare two isoelectronic ions, a trivalent lanthanide and a tetravalent actinide in the same octahedral coordination. A detailed study of U^{4+} in $ZrSiO_4$ only gave a fair matching of observed with predicted electronic energy levels (16) using a fitting procedure similar to that used for UX_6^{2-}. More work on octahedrally coordinated Pu^{4+} and Np^{4+} is needed to confirm the results obtained for U^{4+}.

REFERENCES

1. M. Fred, Advances in Chemistry Series 71, 180 (1967).
2. B.G. Wybourne, Spectroscopic Properties of Rare Earths, John Wiley & Sons, New York, 1965.
3. W.T. Carnall and P.R. Fields, Advances in Chemistry Series 71, 86 (1967).
4. R.G. Pappalardo, W.T. Carnall and P.R. Fields, J. Chem. Phys. 51, 1182 (1969).
5. W.T. Carnall, P.R. Fields and R.G. Pappalardo, J. Chem. Phys. 53, 2922 (1970).
6. W.T. Carnall, S. Fried and F. Wagner, Jr., J. Chem. Phys. 58, 3614 (1973).
7. J.B. Gruber, J. Chem. Phys. 35, 2186 (1961).
8. W.F. Krupke and J.B. Gruber, J. Chem. Phys. 46, 542 (1967).
9. C.K. Jörgensen, R.G. Pappalardo and H.H. Schmidtke, J. Chem. Phys. 39, 1422 (1963).
10. R.G. Pappalardo, J. Mol. Spectros. 29, 13 (1969).
11. W.T. Carnall, P.R. Fields and R.G. Pappalardo, Proc. Intern. Conf. Coord. Chem. 11th, Haifa, Israel, Sept. 1968.
12. R. Pappalardo, W.T. Carnall and P.R. Fields, J. Chem. Phys. 51, 842 (1969).
13. R.A. Satten, D.J. Young and D.M. Gruen, J. Chem. Phys. 33, 1140 (1960).
14. R.G. Pappalardo and C.K. Jörgensen, Helv. Phys. Acta 37, 79 (1964).
15. R.A. Satten, C.L. Schreiber and E.Y. Wong, J. Chem. Phys. 42, 162 (1965).
16. I. Richman, P.Kisliuk and E.Y. Wong, Phys. Rev. 155, 262 (1967).

ELECTRONIC AND VIBRATIONAL TRANSITIONS OF LEAD AZIDE AND EFFECTS THEREON OF PHOTODECOMPOSITION*

Ferd Williams

Department of Physics, University of Delaware

Newark, Delaware 19711

ABSTRACT

Lead azide is a well-known explosive; azides are also used in air bags for auto collisions. The structure of PbN_6 is orthorhombic with 12 Pb^{2+} and 24 N_3^- per unit cell, divided among two types of Pb^{2+} sites and four types of N_3^- sites. We shall review very recent research on the electronic states and on the lattice modes and on the intra-azide vibrations of PbN_6, and the effects thereon of photodecomposition; and the discovery of low temperature photoluminescence. All measurements were made on thin films prepared by evaporation of the metal followed by conversion to the azide in the vapor of HN_3(1).

The near ultraviolet absorption and photodecomposition involve creation of Frenkel excitons. The main transition is broad and maximal at 330 nm, having some characteristics of charge transfer and some of intra-cation excitons. A less-intense peak at 375 nm is sensitive to thallium doping and to photodecomposition(2). The band gap of PbN_6 is $\sim$ 4 e.v.; and the valence band is multiple, consisting of states of the four types of N_3^- and the two types of $Pb^{2+}(\ldots 6s^2\ {}^1S_0)$ with configuration interaction.

The principal far infrared absorption peaks at 62, 72 and 83 cm^{-1} are the fundamental restrahlen bands of PbN_6 involving the motions of the Pb^{2+} sub-lattice against the N_3^- sub-lattice, which can be roughly described as vibrations along the three perpendicular axes of the crystal. With 330 nm photodecomposition the peaks become less intense but remain well-resolved. With 15% N_2 evolved, the far i.r. transmission at these peaks increased from 44 to 55% for a 0.54 μm film, which is equivalent to $\sim$ 30% of the film no longer contributing to lattice absorption. These results and the time-dependence of the N_2 evolution are explained by the decomposition occurring where the u.v. is absorbed and with regions of the azide film which are 10% or more decomposed contributing nothing to these far i.r. peaks. This product is therefore amorphous. Thus, mechanisms involving preferential decomposition at the free surface or homogeneous decomposition resulting from energy transfer throughout thick films are eliminated. Decomposition is not diffusion-limited(3).

The near i.r. absorption has maxima at 2020, 2030 and 2055 cm^{-1} with a shoulder indicating a weaker band at 2080 cm^{-1}. We attribute these to the antisymmetric stretching mode of N_3^- at the four types of azide sites. The coupling between N_3^- is weak so that these crystal modes are well-approximated by those of single N_3^-. The type I site has N_3^- with equal N-N distances R at equilibrium; types II, III and IV have unequal N-N distances, R_1 and R_2; all are linear. We calculate $\delta\nu/\nu$ as a function of $\delta R_1/R$ and $\delta R_2/R$, assuming $E(R) = aR^{-n} - bR^{-m}$, $n > m$, and small vibrations, and confirm that the 2080 cm^{-1} shoulder arises from the symmetrical type I N_3^- and that the 2055, 2030 and 2020 cm^{-1} peaks arise from the asymmetrical N_3^- at type II, III and IV sites, respectively. With 330 nm photodecomposition the resolved maxima are replaced by a broad structureless absorption band peaking at 2020 cm^{-1}, indicating destruction of the distinct sites characteristic of the PbN_6 crystal structure and its replacement by an amorphous structure(4). Doping with Tℓ and/or Bi affects the near i.r. but not the far i.r. absorption (5).

The near ultraviolet absorption also becomes less structured with photodecomposition, with an Urbach exponential tail characteristic of a disordered structure developing.

From a kinetic analysis of the time-dependence of photodecomposition, we conclude that N_3^o is an intermediate of long life in the overall mechanism(4).

Photoluminescence has been found for lead azide films at 15°K(6). The emission extends from 370 to 400 nm, with five or six maxima separated by ∿ 0.03 ev. Photodecomposition occurs concurrent with the photoluminescent excitation, resulting in loss of the luminescence. This also indicates substantial structural changes during decomposition. Photodecomposition at 15°K is itself interesting because the decomposition product, N_2, is, of course, solid at this temperature. The separation of contiguous peaks in the emission is too great to correspond to the restrahlen modes but may correspond to modes involving the motion of the different azide sub-lattices against each other, local modes, and/or crystal field splitting.

In conclusion, we have measured and interpreted the u.v., near and far i.r. spectra of PbN_6 films, found low temperature luminescence, and used these spectra to probe the structural changes occurring during photodecomposition.

REFERENCES

*Supported by the Solid State Branch, Feltman Research Lab through ARO-Durham.

1. H. Fair and A. Forsyth, J. Phys. and Chem. Solids 30, 2559 (1969).

2. R. B. Hall and F. Williams, J. Chem. Phys. 58, 1036 (1973).

3. S. P. Varma, F. Williams and K. D. Möller, J. Chem. Phys. 60, 4950 (1974).

4. S. P. Varma and F. Williams, J. Chem. Phys. 60, 4955 (1974).

5. S. P. Varma and F. Williams, J. Chem. Phys. 59, 912 (1973).

6. J. Schanda, B. Baron and F. Williams, J. Luminescence (in press).

SHORT SEMINARS (TITLES ONLY)

JAHN-TELLER QUENCHING FACTORS IN A SOLID

R. Englman
Israel Atomic Energy Commission
Soreq Nuclear Research Centre
Yavne, Israel

LIFETIME MEASUREMENTS OF DONOR-ACCEPTOR PAIRS AND IONS IN SOLIDS

K. Luchner
Physics Department
Technical University
Munich, Germany

SPECTROSCOPY ON ATOMS AND IONS ISOLATED IN RARE GAS SOLIDS

H. Micklitz
Physics Department
Technical University
Munich, Germany

ENERGY TRANSFER IN YbAlG:Er^{3+}

D. Pacheco
Physics Department
Boston College
Chestnut Hill, Massachusetts, U.S.A.

LUMINESCENCE, ABSORPTION, AND EMISSION LIFETIME MEASUREMENTS ON Cr^{3+} HEXA-UREA SINGLE CRYSTALS

H. Yersin
Chemistry Department
University of Regensburg
Regensburg, Germany

LIST OF CONTRIBUTORS

D. Curie, Faculté des Sciences, Paris, France
B. Di Bartolo, Boston College, Chestnut Hill, Massachusetts, U.S.A.
R. Englman, Soreq Nuclear Research Center, Yavne, Israel
T. L. Estle, Rice University, Houston, Texas, U.S.A.
G. F. Imbusch, University College, Galway, Ireland
D. S. McClure, Princeton University, Princeton. N.J., U.S.A.
L. Mehrkam, University of Delaware, Newark, Delaware, U.S.A.
R. L. Orbach, Tel Aviv University, Tel Aviv, Israel(*)
R. G. Pappalardo, GTE Laboratories Inc., Waltham, Massachusetts, U.S.A.
R. C. Powell, Oklahoma State University, Stillwater, Oklahoma, U.S.A.
K. K. Rebane, Institute of Physics, Tartu, U.S.S.R.
L. A. Rebane, Institute of Physics, Tartu, U.S.S.R.
W. A. Wall, U.S. Army LWL, Aberdeen, Maryland, U.S.A.
R. K. Watts, Texas Instruments Inc., Dallas, Texas, U.S.A.
F. Williams, University of Delaware, Newark, Delaware, U.S.A.

(*) Permanent Address:
University of California, Los Angeles, California, U.S.A.

SUBJECT INDEX